T0215474

Biological Systematics

Species and Systematics

The *Species and Systematics* series will investigate the theory and practice of systematics, phylogenetics, and taxonomy and explore their importance to biology in a series of comprehensive volumes aimed at students and researchers in biology and in the history and philosophy of biology. The book series will examine the role of biological diversity studies at all levels of organization and focus on the philosophical and theoretical underpinnings of research in biodiversity dynamics. The philosophical consequences of classification, integrative taxonomy, and future implications of rapidly expanding data and technologies will be among the themes explored by this series. Approaches to topics in *Species and Systematics* may include detailed studies of systematic methods, empirical studies of exemplar taxonomic groups, and historical treatises on central concepts in systematics.

For more information visit:
www.crcpress.com/Species-and-Systematics/book-series/CRCSPEANDSYS

Biological Systematics

History and Theory

Igor Ya. Pavlinov

CRC Press
Taylor & Francis Group
Boca Raton London New York

CRC Press is an imprint of the
Taylor & Francis Group, an **informa** business

First edition published 2021
by CRC Press
6000 Broken Sound Parkway NW, Suite 300, Boca Raton, FL 33487-2742

and by CRC Press
2 Park Square, Milton Park, Abingdon, Oxon OX14 4RN

Routledge is an imprint of the Taylor & Francis Group, an informa business

© 2021 Taylor & Francis Group, LLC

Library of Congress Cataloging-in-Publication Data
A catalog record has been requested for this book

ISBN: 9780367654450 (hbk)
ISBN: 9780367671945 (pbk)
ISBN: 9781003130178 (ebk)

Typeset in Times
by Newgen Publishing UK

If you don't know what to search for, how will you find it?

If you don't know how to search for, what will you find?

Igor Ya. Pavlinov [paraphrase on a Chuang-tzu theme]

Contents

Contents

Preface

Living is easy with eyes closed.

John Lennon

The well-known statement that "In the beginning was the Word," with which the Gospel of John begins, is not fully correct: of course, in the beginning was the classification. For a word is but one of the forms of being of knowledge. With this, any knowledge begins with a distinction between "one" and "another": good and evil, living and inert matter, plants and animals—and so forth to infinity. And all such distinctions constitute an essential part of the classification.

So, classification stands at the beginning of both knowledge and the "words" shaping it. The very ability to generalize and classify is one of the most important parts of any cognitive activity—and scientific knowledge provides no exception in this respect. All scientific disciplines classify in one way or another, some to a lesser and others to a greater extent; in some of them, classifications play an auxiliary role, while in others they constitute the very goal of the cognitive activity.

Biology is one of the most "classifying" natural sciences; as a matter of fact, it includes a special division devoted exclusively to the development of classifications, namely, *biological systematics*. This is a peculiar specificity of biology: all sciences classify, but it is only biology that developed and termed such a special classifying division.

Systematics takes a fundamental position in biology, for it shapes the subject area of this natural science branch by recognizing and fixing in living matter, in a certain way, specific structural units called *taxa*. Only reference to the latter makes all biological knowledge objective and concrete—and thereby scientific. Meanwhile, an attitude towards systematics among many biologists (and not only among them) is quite disrespectful. The prevailing belief among non-specialists is that systematics is engaged in something futile like "stamp collecting"; it just divides organisms by species and genera and then assigns names to them. Such an impression of systematics as a "defective science" seems to be one of the principal causes of the current "taxonomic impediment" that many taxonomists worry about. Unfortunately, such an attitude is supported indirectly by those taxonomists who present systematics in just such a scanty, purely empirical manner.

However, such a viewpoint is undoubtedly wrong. In fact, biological systematics is a fairly developed scientific discipline with its specific theoretical background that was being elaborated for centuries in the form of particular research programs. It is this background that defines, by and large, what "taxonomic reality" is as an essential part of the subject area of biology, how it is explored by specific taxonomic means, and how both theoretical and empirical taxonomic knowledge is substantiated and shaped. With this, theoretical foundations of systematics arise not by themselves but within certain philosophical contexts, so it is hardly possible, without the latter's consideration, to comprehend what, how, and why systematics investigates. Further, it is of importance to recognize that theoretical knowledge of systematics is a developing cognitive system, so it cannot be understood outside the historical contexts of its development. Thus, the general framework of comprehending the science of biological systematics can be presented in the form of the fundamental triad *"theory + philosophy + conceptual history."*

This book represents the author's ideas about historical development and essential content of the theoretical foundations of biological systematics considered within the context of this fundamental triad. It provides a kind of summary of my more extensive books published in 2012 and 2018 in Russian, so I am happy with an opportunity to present my ideas in English as part of the Species and Systematics Series of CRC Press, thanks to the positive attitude of Kip Will (Editor in Chief) and Chuck Crumly (Science Publisher).

The approach to consideration of the theoretical foundations of systematics adopted here is caused by the author's commitment to the ideas of non-classical philosophy of science, including evolutionary epistemology that emphasizes both philosophical and historical conditioning and pluralistic treatment of theoretical and, consequently, empirical knowledge [Popper 1975; Hübner 1988; Radnitzky and Bartley 1993; Stepin 2005]. According to this "non-classical" viewpoint, the theoretical foundations of systematics are shaped by changing philosophical and historical contexts. So the main task of this book is not just to review the diversity of taxonomic concepts but to analyze their basic premises: on what philosophical and scientific theoretical grounds particular taxonomic theories and concepts are founded and why and how they arise and function.

This historical-philosophical standpoint assumes the realization that particular taxonomic theories seem to appear rather regularly in certain philosophical and historical contexts in due time and die out with them in the respective time, although not completely, leaving some "conceptual traces" in the overall contents of systematics. Therefore, we should not blame our predecessors for doing or thinking "wrong" but give them our respect for their having laid the foundations of biological systematics with their ideas as they understood them in those times.

With this, such a position has one particular feature: it presumes that, in the theory of systematics, there is hardly anything that might be thought of as established once and forever. Adherents of every particular conception, believing it is "most true" and therefore "final" while others are "wrong" or "obsolete," seem to look at the entire subject of biological systematics "with eyes wide shut." And do contemporary theoreticians, thus believing, differ greatly in this respect from those who believed in their own conceptions in the 18th and the 19th centuries? And isn't it wiser to remember King Solomon's "everything passes—and this will also pass" and to realize that any current taxonomic knowledge is but a transition between the past and future developments?

So, readers are welcome to plunge into the variety of theoretical foundations of biological systematics, considered from broad historical and philosophical perspectives.

Igor Ya. Pavlinov
Zoological Museum at Lomonosov Moscow State University
Moscow, Russia

Introduction

The structure of the book is determined by its main task—to represent the historical roots, philosophical foundations, and theoretical content of biological systematics, considered in its full capacity.

Chapter 1 is of an introductory kind: the structure of systematics is characterized, its main sections (theoretical, practical, and applied) and tasks are briefly outlined. The general notion of the Natural System is characterized as a fundamentum for the entire systematics, defining the latter's theoretical content and its historical development. The theoretical branch of systematics (*taxonomy*) deals basically with the developing cognitive situation and the theoretical foundations of this discipline; one of its principal tasks is to connect taxonomic theory with the philosophical scientific context of natural science. Practical systematics deals with the elaboration of particular taxonomic systems based on the respective taxonomic theories; among its important tasks is the development of systematic collections and herbaria. Applied systematics provides various users with reliable working instruments for solving their particular research and applied tasks related to the diversity of organisms.

Chapter 2 considers prehistory and the conceptual history of systematics from the point of view of evolutionary epistemology. Folk systematics is characterized as an initial stage of the cognitive activity dealing with the elaboration of primordial classifications based on primarily pragmatic motivation. The proto-systematic stage (Antique to Renaissance) includes the initial development of the rational cognitive program and analytical methods, including the basic logical categories and the genus–species classification scheme. The scientific systematics emerged in the 16th century when these categories and scheme were mastered. Subsequent scientific revolutions and principal stages are recognized in the conceptual history of systematics: post-scholastic, evolutionary, positivist, and post-positivist. The main research programs are briefly characterized as a result of these revolutions, viz. scholastic systematics, natural systematics, early typology, taxonomic "esotericism," the beginning of evolutionary interpreted systematics; development of phenetic, numerical, and biosystematics as responses to the positivist challenge; and the rebirth of phylogenetic systematics.

Chapter 3 briefly considers key issues of the philosophical background of systematics. The three-partitioned structure of the cognitive situation is characterized, including its interrelated ontic, epistemic, and subject components, with the metaphor of cognitive triangle being its adequate representation. Any cognitive activity is selective by addressing the particular conceptually construed manifestations (*Umwelts*) of Nature (*Umgebung*). Particular *Umwelts* constitute an ontic component of the respective cognitive situations in which research programs develop and function. The principal cognitive regulators of the taxonomic research are briefly considered.

Chapter 4 outlines the author's idea concerning a possible approach to elaborating taxonomic theory (TT), in its most general sense, as a specific quasi-axiomatic. It is represented as a hierarchically arranged conceptual pyramid, with general TT at its top and particular TTs at its lower levels. General TT provides a conceptual framework for the entire cognitive situation of systematics, while particular TTs specify its statements for particular *Umwelts*. The following principal categories of particular TTs are recognized: aspect-based, relational, object-based, and episteme-based. Most fundamental aspect-based and episteme-based partial TTs shape respective research programs (typological, phylogenetic, numerical, etc.), while object-based TTs deal with the most fundamental notions (hierarchy, taxon, character, homology, species, etc.). The key quasi-axioms/presumptions and inference rules (principles) of systematics are briefly outlined. The fundamental concepts of the taxonomic reality and the classification system are represented in an original manner.

Chapter 5 considers the contemporary research programs in systematics that develop particular TTs. These programs are characterized based on a standard scheme, including its historical roots, basic ontic and epistemic principles, key concepts and notions, pros and cons. The following research programs are considered: classification phenetics, rational systematics (with its onto-rational and episto-rational versions), numerical program (with numerical phenetics and numerical phyletics as its principal versions), classification typology (its principal contemporary developments are characterized), biomorphics (elaborating classification of life forms), biosystematics (considering taxonomic diversity at lower levels), phylogenetics (with evolutionary taxonomy and cladistics as its basic contemporary versions), and an evolutionary ontogenetic program (based on the evo-devo concept).

Chapter 6 deals with the most fundamental problematic issues in systematics considered as specific "taxonomic puzzles." It is emphasized that distinguishing between natural and artificial classifications (taxonomic systems) depends on the contents of the particular TTs and thus cannot be universal. The elaboration of particular taxonomic systems is based on an iterative procedure including altering precedences between judgments about taxa and their characters. The general arrangement of taxonomic systems can be either hierarchical or parametric, and their hierarchies can be either rankless or ranked. A complex interrelation between similarity and kinship is considered, emphasizing the non-objective status of the former. The concept of (arche)type is characterized as supported by contemporary essentialism. The diversity of the contemporary ideas about homology reflects the impossibility of a unified system of partonomic structure of organisms and presumes the need for the elaboration of a hierarchically arranged generalized concept of multifaceted correspondences. The contemporary species problem is caused by a controversy between its monistic *vs.* pluralistic treatments; it can be resolved based on acknowledging the "specieshood" as an integral part of the natural history of organisms, with its particular manifestations corresponding to the latter's specifics.

1 A Brief Introduction to Systematics

> Any science is ordering, and if systematics is equivalent to ordering, then systematics is synonymous with science.
>
> George Simpson

If we try to characterize Nature, briefly but at the same time sufficiently profoundly, as a sphere of application of cognitive activity, then perhaps the most appropriate "formula" may be as follows: Nature is an ordered diversity of its phenomena (objects and processes). From a scientific viewpoint, acknowledging the orderliness of Nature as its fundamental feature is of prime importance: cognizable can be only what is ordered. Thus, all scientific disciplines are essentially engaged in the same enterprise: they investigate various manifestations of the diversity of Nature and look for a certain order in this diversity.

Two main analytical approaches are usually distinguished in how the ordered diversity of Nature can be comprehended: *parametrizing* and *classifying* [Hempel 1965; Rozov 1995; Subbotin 2001]. In the first approach, the main emphasis is on the order as such: certain parameters (variables, etc.) are fixed and their gradients are linked by a single formula. Illustrative examples are the ratio of mass and energy in physics, the relationship between reaction rate and concentration of substances in chemistry, the relationship between body size and age of a multicellular organisms in biology. All diversity lying outside each such "formula" is ignored as irrelevant to the revealed law-like "nature of things." The second approach focuses on the diversity as such: the task is to present it as comprehensively and irreducibly as possible in some generalized form. This is usually accomplished by developing classifications that give a certain idea of the ordered structure of the diversity itself. Examples are also well known: classifications of elementary particles in physics, of cosmic bodies in astronomy, of substances and their compounds in chemistry, of organisms in biology.

These two fundamental ways of understanding and describing the ordered diversity of Nature allowed Francis Bacon, 17th-century philosopher, to distinguish between two basic domains of classical natural science, *natural philosophy* and *natural history*. The former represents its knowledge in the form of *parametric systems* (formulas), while the latter does it in the form of *classificatory systems* (classifications). Accordingly, within the framework of natural philosophy, predominantly parametrizing disciplines emerged (physics, chemistry, astronomy, etc.), and within the framework of natural history, classifying ones began to dominate (biology, geography, geology, etc.). As can be assumed, such a division is not accidental; it reflects a fundamental idea that different aspects of the ordered diversity of Nature can be most adequately represented by different descriptive systems—some by parametric and others by classificatory [Whitehead 1925].

It is of importance to note that there is no contradiction between these two ways of studying and describing the ordered diversity of Nature: in many cases, they complement each other in describing the same phenomenon. A simple illustrative example is the color scale, which can be represented qualitatively as a classification by enumerating traditionally distinguished colors (red, blue, green, etc.), or quantitatively by reference to a continuous scale of wavelength values. Another, not so trivial, example is a thermodynamic system with transitions between quasi-discrete phase spaces under continuous variation of some key parameter: its structure can be described both by an equation expressing that parameter and by a qualitative description of quasi-discrete states. In biology, an example similar to the second one is provided by species diversity: according to the Darwinian model of evolution, a continuous process of speciation yields quasi-discrete diversity of biological species. So, the speciation process is described by a "formula" of continuous transformations of some characters of organisms, while resulted species diversity is described by respective classification.

Biology is one of the most "classifying" natural sciences. And it is probably not accidental that a special discipline was formed in it, *biological systematics*, dealing exclusively with the study and description of the ordered "qualitative" diversity of organisms. This specificity of biological science is a really striking fact of its "biography"—as was just stated, all sciences classify their objects in one way or another, but it seems to be just in biology that systematics appeared as a separate area of research. As a matter of fact, almost all of biology, in the early period of its history, emerged as a "classifications creator," i.e., as systematics, and was, by and large, subsumed by it in many respects. At present, the importance of systematics is not so overwhelming, as modern biology is very diverse, with molecular biology, physiology, and ecology, each exploiting respective parametric systems, being in the first place. But all of them and other biological disciplines cannot do without appealing to classificatory systems provided by systematics, which describe respective diversity in a qualitative manner: such is one of many manifestations of the *complementarity principle*.

This introductory chapter describes, in the shortest form, what systematics is: what and it studies and how, and why the results of its research are in demand.

1.1 WHAT IS THE NATURAL SYSTEM?

Everything is System, and systems are everywhere: this is what most prominent systemologists say [Bertalanffy 1968; Urmantsev 1988]. This statement is true with respect to both Nature and scientific knowledge about it—to the extent that every system is essentially an ordered diversity.

So, Nature is a system—and if so, then this statement can be tweaked: Nature, in a sense, is the *System of Nature*. Such a natural-philosophical understanding of Nature has a special reason: it focuses on the integrity and orderly nature of Nature, encourages one to reveal a certain general order in it, and, on this basis, to try to uncover its causes. Since the 17th century (although probably even earlier), this idea has dominated the minds of European thinkers who understood Nature as a "diversity in unity and unity in diversity." The author of this aphorism, the philosopher and mathematician Gottfried Leibniz, published a short essay in 1695 under an iconic title "A new system of nature..." [Leibniz 1900]. Half a century later, the naturalist Carl Linnaeus published the first version of his "System of Nature..." in 1735, which underwent a dozen increasingly complete reprints (the last within his lifetime was in 1768). Finally, the philosopher Paul-Henri Holbach, an active participant in the Enlightenment, published his work "The System of Nature..." in 1770 (under the pseudonym of M. Mirabeau), in which he summed up the "laws of both the physical world and the spiritual world" under a common systemic denominator [Holbach 1770].

So now, what is the System of Nature? The answer to this question in its most general form can be twofold. On the one hand, it is an ordered Nature as such, a global systemic object (in the sense of Urmantsev). On the other hand, this is a certain law that orders Nature, makes it this very systemic object. Natural philosophers from various times, trying to get to the fundamentals of the Universe and to find out what this "law" is, asked the same question of the most general order: what is the "law" that causes the orderliness of Nature? For religiously minded thinkers (like Leibniz, Linnaeus), the Divine Plan of Creation appears in this capacity as the causal basis of all that exists. For materialistic natural philosophers (like Holbach), Nature is self-sufficient, its cause is in itself (*causa sui*). Holbach himself, who imagined Nature as a well-construed "mechanism," looked for the causes of its systemic essence in universal natural laws like those of Newtonian mechanics. Since the end of the 18th century and especially in the 19th century, one of the fundamental causes of the ordered diversity of the System of Nature was thought to be global evolution; this view had a particular impact on understanding the diversity of living nature as a consequence of biological evolution.

By the time that systemic natural philosophy was clearly shaped, systematics had already been developing for almost two centuries. After the release of the first works of Linnaeus, the concept of the System of Nature (*Systema Naturae*) became a key for comprehension of wildlife. As a result, the very designation of our discipline as biological systematics acquired a special, deep meaning: it is "biological" and it is "systematics" just because it studies the System of Living Nature. This natural-philosophical concept turned into a more operational concept of the Natural System (*Systema naturalis*)—just like that, with a capital letter to emphasize its special significance—as a kind of integral mental image of the System of Nature. The natural classification (*Classificatio naturalis*) elaborated by systematicians—in this case with a lowercase letter—was recognized as the best way to represent it.

It should be noted, for the sake of fairness, that the concept of the System of Nature served at that time only as one of two basic forms of representation of the ordered diversity of Nature. Another one was the equally fundamental natural-philosophical idea of the Ladder of Nature (*Scala Naturae*), which had a strong influence on the minds of natural scientists. The difference between them is that the System of Nature implies a hierarchical ordering of diversity, while the Ladder of Nature is predominantly linear. Nevertheless, in the "ladderists" community, the general conception of the Natural System was as fundamental as it was among the "systemists."

Thus, since the middle of the 18th century, systematics was aimed to uncover the Natural System of living beings. In different natural philosophical doctrines, this general concept is filled with different content—"systemic" (Linnaeus), "ladder" (Buffon), organismic (Oken), typological (Cuvier, Baer), genealogical (Darwin, Haeckel)—but these peculiarities do not change the main point.

This main point is as follows. Wildlife is part of the general System of Nature and can be thought of as an ordered diversity of organisms. This orderliness is manifested in law-like interrelationships, either hierarchical or linear, between organisms and their attributes. A fundamental property of these interrelationships is that their hierarchy is quite evident while some linear "parameterized" ordering is but weakly manifested. All this taken together—both diversity, and its orderliness, and the latter's non-linearity—is what systematics should uncover and present in a form of the Natural System of organisms—or, less pretentiously, in a form of the natural classification.

The Natural System, being a product of cognitive activity, is not identical to the System of Nature as such, but should be isomorphic to it, i.e., reflect it as precisely as possible. In this capacity, the Natural System may be thought of as a kind of "absolute truth," an ideal towards which systematics should strive. However, it is obvious that this ideal, like any other one, is in principle unattainable. Therefore, the concept of the Natural System serves for systematics as a kind of "lighthouse" that directs taxonomic research in a certain direction. This direction does not represent a single main path; as mentioned above, both the System of Nature itself and the Natural System representing it can be understood in different ways, so that the "lighthouse" is not a monolithic illuminator, but rather something composite with its pieces highlighting different aspects of the System of Nature. In the language of modern systematics, these "pieces" are treated as different *research programs*: they develop classifications filled with different content. Each such classification can be considered natural to an extent that it approximates a particular aspect of the Natural System.

1.2 WHAT IS BIOLOGICAL SYSTEMATICS?

It is customary to define scientific disciplines through their subject areas by indicating what exactly they study: physics, chemistry, biology, geography, linguistics, etc. Within each of these domains of knowledge, sub-areas are distinguished on a similar basis: microphysics and cosmology, economical and physical geography, histology and embryology, etc. It can be reasonably assumed that this general principle is true for biological systematics.

As can be seen from the previous section, systematics, if considered natural-philosophically, is associated with exploration of the System of Living Nature, with its ultimate end being elaboration of the Natural System or, more correctly, natural classification representing it in some way. From a quite empirical standpoint, systematics is defined as a branch of biology studying diversity of organisms (biological diversity, *aka* biodiversity) in all and any of its manifestations, representing the results of the study in the form of some "omnispective" classifications as a kind of reference system [Blackwelder 1967]. Both treatments, despite a significant difference in their backgrounds, are too broad: it turns out that systematics is the science of biological diversity "in general" and, accordingly, should be rightfully called *biosystematics* (in fact it is sometimes defined and called this way). In this case, what about ecology or biogeography also dealing with the same biodiversity, though in their own manner? On the other hand, there are significantly narrower definitions of systematics, based on their particular backgrounds: for example, it is defined sometimes as a "science of species" [Mayr 1969; Wheeler 2009]. In this case, a question arises: what about the supraspecific taxa? Does not and should not systematics study them?

From this, it becomes clear that the definition of systematics as a scientific discipline through the definition of its subject area is not an easy task. To say that it studies the System of Living Nature "in general" or biodiversity "in general" is too vague. But it is too narrow to associate it with some particular task (as in the case of the study of species). Systematics is actually a branch of biology studying the diversity of organisms and elaborating classifications to represent it in one way or another—but what kind of biodiversity does it actually deal with?

Taking all the above considerations into account and not going more deeply into this issue, it seems correct enough, from a theoretical perspective, to treat systematics as a branch of biology that develops:

- specific ways of understanding the System of Living Nature (typological, phylogenetic, organismic, etc.) as particular manifestations (aspects) of the ordered diversity of organisms
- the methods and criteria by which specific classifications can be elaborated within the contexts of those particular understandings
- the classifications that might be treated as natural with respect to those particular understandings.

1.2.1 THE STRUCTURE OF SYSTEMATICS

A well-thought-out "stratification" of systematics as a scientific discipline should, first of all, provide for the delineation of its main divisions. What exactly and how exactly should be reflected in classifications, how they should be elaborated, by which criteria they could be considered natural—these fundamental questions are posed and answers sought for by the theoretical division of systematics. Various kinds of theoretical propositions are implemented in specific classifications by its practical division. Finally, methods of bringing such classifications into the forms suitable for use by other disciplines are being developed by the applied division of systematics.

Theoretical systematics was denoted as *taxonomy* by the botanist Augustin Pyramus de Candolle at the beginning of the 18th century [de Candolle 1819]; another recent designation is *taxonology* [Zuev 2015]. This understanding of taxonomy is most widely accepted in contemporary literature, but there are other interpretations of it. Some authors identify taxonomy with the entire systematics [Mayr 1942; Rogers 1958; Griffiths 1974a,b; Zuev 2015]. Others call taxonomy the practical issues dealing with the identification of particular taxa [Blackwelder and Boyden 1952; Blackwelder 1967; Wheeler 2001].

Considering understanding of possible meanings of taxonomy as dealing basically with theoretical knowledge, it makes sense to distinguish between its two principal levels, universal and biological [Wilkins 1998; Zuev 2015]. *Universal taxonomy* develops general principles of classificatory activity and can be considered a part of logic; it is nearly the same as "philosophical" taxonomy [Humberstone 1996], or classiology [Kozhara 2006; Pokrovsky 2014], or "the doctrine of any classifications" [Meyen and Schreider 1976]. *Biological taxonomy* is a special subject branch of universal taxonomy that shapes the theoretical section of biological systematics.

The most important and most general task of biological taxonomy is the development of the philosophical and theoretical foundations of biological systematics. The whole of this book is devoted to these foundations from which one can conclude that this objective is very extensive and multifaceted. Specifying this objective the following most significant points are to be indicated.

The first point, in both order and importance, in that general task is the development of the *cognitive situation* in which systematics operates, and within which framework the basic onto-epistemic foundations of this discipline are developed. As a matter of fact, the construction of the entire edifice of systematics as a scientific discipline begins with setting and solving theoretical problems within the framework of this principal point. Obviously, they are based on the general principles of the philosophy of science, so establishing certain "contacts" with the latter is also an important task of the taxonomy. According to the structure of the cognitive situation, the following two main groups of more particular tasks should be distinguished here:

- the *ontology* of systematics deals with the correct definition of the subject area of taxonomy, that is, what it studies; herewith, it is defined rather informally as *taxonomic diversity* (TD: see below)
- the *epistemology* of systematics deals with developing principles of systematic research, that is, how to investigate the subject area thus defined; one of the aims of this is a validation of the scientific status of taxonomic knowledge developed by systematics.

The second point is the development of *taxonomic theory*, which constitutes the main content of theoretical knowledge in systematics. It has a rather complex structure: general and particular theories are distinguished in it; the latter are divided according to the principles of both delineating their particular subject areas and elaborating methods of their exploration.

A correct delineation of the research programs in systematics is closely related to the previous task. They (at least some of them) are sometimes called "schools," "theories," or even "philosophies" of systematics. In the understanding accepted here, these programs serve as a means of implementing particular taxonomic theories. Each of them actualizes and implements general theoretical concepts in its own way and brings them to an operational state suitable for use in the practice of taxonomic research.

Another key point of taxonomy, which no scientific discipline can do without, is the development of its conceptual apparatus (thesaurus). The matter is that the System of Nature (in any meaning) is given not in personal perceptions but in general concepts and notions, including definitions of the subject area, classification, taxon, character, homology, etc. Therefore, this thesaurus shapes the entire cognitive situation of systematics, in which particular taxonomic research is conducted.

Finally, as the modern philosophy of science stresses a close connection between history and theory of science, one of the specific tasks of taxonomy is the reconstruction of the conceptual history of systematics [Pavlinov 2018]. This allows one to find out how changes in the philosophical, general scientific, and even socio-cultural contexts yield corresponding changes in theoretical foundations of systematics in the course of its historical development.

Practical systematics implements the ideas elaborated by taxonomic theories. These theories form the context of empirical systematic studies, serving as a prerequisite for setting their tasks, methods, choice of characters, ways of representing the structure of TD, etc. Empiricists are unlikely to agree with such an assessment of the relations between theoretical and empirical aspects of systematics: for them, practical systematics is self-sufficient and shapes the basis of this discipline. And nevertheless, it is quite proper to emphasize here that, according to the modern philosophy of science, empirical taxonomic knowledge is meaningless outside a certain biologically sound theoretical knowledge.

According to this, practical taxonomic research starts with bringing theoretical ideas to the point where they can be applied in the procedures of elaborating particular classifications. For example, in phylogenetic systematics, this includes specific definitions and methods of assessing similarity and kinship, choosing characters as indicators of the latter, methods of reconstructing phylogenetic history, and, finally, ways of presenting the results of this history in the form of phylogenetic classifications.

This is not a practice yet, but rather (if it may so called) a "semi-theory." The main practical task is to develop specific classifications and present them in a format that makes them available for further use. This general task is deconstructed into the following components:

- conducting particular taxonomic research, including revision of existing and development of new classifications, with the descriptions of new taxa and "closure" of those not confirmed by new research
- elaboration of identification keys that allow particular organisms to be allocated to particular taxa
- publication of the results of this research (classifications and keys) in articles or monographs.

One of important tasks of practical systematics is the elaboration and application of taxonomic nomenclature dealing with the regulations and manipulations of scientific names of taxa. It is governed by a number of codes in botany, zoology, etc. [Jeffrey 1992; Pavlinov 2015].

Another important task of practical systematics the development of its empirical basis in the form of scientific systematic collections. Systematics was and remains a "museum" science; this is its fundamental specificity. The reason is that collections, in their epistemic status, are analogous to standard experiments in physics and chemistry: they provide both repeatability and, thereby, testability of taxonomic knowledge, making it scientific. Therefore systematics cannot do without its collections, just like physics or chemistry cannot do without their experiments [Whewell 1847; Mayr 1982; Pavlinov 2016]. Long-existing scientific collections kept under standard conditions represent one of the key information resources for research on the structure of bio-diversity [Miller 1985, 1993; Graham et al. 2004; Berendsohn 2007; Ariño 2010; Drew 2011]. Therefore, systematics must develop systematic collections, as physics and chemistry develop their experimental base. The strategy of collections development is determined by the fact that the general structure of the global collection pool should be adequate to the diversity of organisms studied by systematics (Cotterill 2002 2016; Pavlinov 2016).

One of the very important tasks of practical systematics is specific pedagogical activity: it is called on to ensure the reproduction of the taxonomic community to conduct taxonomic research in different groups of organisms. No science can do without it; in systematics, ensuring a continuous "relay" of transfer of practical knowledge about the diversity of organisms between generations is of crucial importance. The reason is that such knowledge is largely descriptive; it contains a large portion of accumulated experience regarding particular groups of organisms and methods of their research. And no one theory, however perfect and detailed, can replace such experience: its loss means, in fact, that a certain group of organisms "falls out" of the sphere of taxonomic research.

Applied systematics provides a junction between systematics proper and all those spheres of human activity that involve contact with biological diversity. It is based on the results of practical systematics, and its main task is information support of that activity.

1.2.2 WHAT DOES SYSTEMATICS STUDY?

This question is asked and answered by the ontology of a scientific discipline, in which cognitive activity is aimed at comprehending the "nature of things." It can be reduced to finding out what diversity of these "things" is and why they are different. Why are there different cosmic bodies—galaxies, stars, planets? Why are there different microparticles—say, hadrons and leptons? Why are there different chemicals—for example, alkalis and acids? Why are there different ethnic groups and different languages? In biology in general and in systematics in particular, cognitive activity is associated with description and explanation of the diversity of organisms: it begins with questions about how, to what extent, and why they are different or similar, and ends with answers.

Thus, consideration of systematics as a scientific discipline begins from this point—what exactly it studies. Obviously, the understanding of the tasks solved by systematics depends on this. Therefore this point was designated in the previous section as one of the first among the principal issues in developing the theoretical foundations of systematics. Currently, there is no single satisfactory understanding of its subject area: depending on the preferences of respective authors, its definitions either cover the entire diversity of organisms without differentiating its manifestations or are too narrow and reduce the scope of systematics to one or another particular manifestation of this diversity (phylogenetic, typological, species, etc.).

The key question for systematics about its object (subject area) is considered in Section 4.2.1, so here a few important points should be noted. Its definition should be sought at a theoretical level, and not be limited to indicating certain manifestations of biodiversity. In general, the object of systematics is denoted as TD; it is not identical to biological diversity as a whole, but is one of its manifestations. Its other manifestations, which systematics is not directly involved in, are ecological diversity (studied by ecology, biocenology), biochorological (studied by biogeography), partonomic diversity within organisms themselves (studied by anatomy, developmental biology, physiology, histology, etc.), biosocial (studied by social biology, ethology), etc. Of course, all of them, in one way or another, fall within the sphere of interests of systematics as "suppliers" of characters for elaborating classifications, but they do not form its own subject area.

The TD itself is structured, so its own different manifestations (aspects, elements, etc.) can be distinguished as subject areas of respective research programs. Of the aspects of TD, the following are most clearly outlined:

- The *typological aspect* includes diversity of structural features of organisms (structural plans, ontogenetic patterns, etc.), studied by typology.
- The *phylogenetic aspect* includes the hierarchy of monophyletic groups, studied by phylogenetics.
- The *biomorphological aspect* includes diversity of biomorphs (life forms), studied by biomorphics.
- The *population aspect* includes diversity of populations and species, studied by biosystematics (in its narrow sense).

One of the important issues related to a general understanding of TD is to find out what the different elements of its structure are. In this case, we are talking primarily about taxa (for example, what is a species?), about ranks in taxonomic hierarchy, and about partons (what is homology?).

1.2.3 How Does Systematics Study?

The questions of how systematic research is carried out are considered by epistemology at a theoretical level. At more practical levels, they are summarized as "principles of systematics" in many manuals, with varying detail according to the theoretical positions of their authors. After reading them, one is convinced that there are too many of these principles and their list is hardly exhaustive. The reason is that, if a taxonomic research procedure is to be worked out rigorously and with sufficient detail, certain working principles need to be formulated for each particular research task.

An overview of the most important principles that form the backbone of the epistemic foundations of taxonomy is presented in the theoretical chapter of this book (see Section 4.1.2). Here several important issues are pointed out.

First is the *choice of a certain theoretical background*. Contrary to what empiricists usually assert, such a choice is always present initially, even though in an implicit form; in fact, rejection of theorizing is also a kind of "theorizing," albeit very reduced. Thus, any study begins with a choice, whether consciously or not, of some classification approach that best suits the researcher's preference—empirical, typological, phylogenetic, etc. This choice affects preferences as regards characters and methods, which in turn affect the results of taxonomic research.

The *choice of characters* is an important part of composing a research sample, to which some exploratory method is to be applied. This choice is largely dictated by the requirements of the respective theoretical background chosen initially: in phylogenetic research characters should allow monophyletic groups to be recognized; in typological research they should allow body plans to be uncovered, etc.

The *choice of method* is a very important issue in taxonomic research because the latter's result largely depends on the method applied. This choice depends, in turn, on many attendant circumstances, including the theoretical context. The latter defines, in particular, which method can be considered scientific and if it is compatible with the particular theoretical background of the research. At a more "technical" level, of prime importance is the correct application of the method chosen. In sum, if a method is incompatible or incorrectly applied, the desired result will not be achieved: although the resulting classification may look quite respectable, it will present a highly biased estimate of a certain aspect of TD studied.

1.2.4   WHY DOES SYSTEMATICS STUDY?

Any classification in any area of human activity performs two kinds of function. Within the taxonomic domain, classification is the goal of cognitive activity: it epitomizes the knowledge for which all research activity is carried out. Another function is to serve as a means of accomplishing other forms of activity in whatever scientific and applied fields.

In cognitive activity, taxonomic classifications developed by systematics appear as a primary form of ordering knowledge about certain manifestations (aspects, fragments, etc.) of biodiversity. Their task is to represent the structure of respective diversity in a form that is suitable for research in other biological disciplines (biogeography, biocenology, etc.). These latter use taxonomic classifications as initial data for their own research and build upon them their own classificatory or parametric systems to represent other manifestations of biodiversity. From this standpoint, classifications elaborated by systematics can be considered *primary*, while other biological disciplines use *secondary* classifications or parametrizations.

> This is also true for those natural sciences in which parametrizing is considered basic. In them, detailed qualitative classifications are developed to serve as a basis for the subsequent application of quantitative parametric methods. Separation of different states of matter (solid, liquid, gas, plasma) in classical physics, separation of groups of elementary particles (leptons, hadrons, photons, etc.) in quantum mechanics, separation of groups of cosmic bodies (galaxies, stars, planets, etc.) in astrophysics—these are all well-known examples of this kind of primary classification.

One very important conclusion can be drawn from the previous consideration. Although in various disciplines, including the "parameterizing" ones, primary classifications do perform some kind of service function, no other function seems to be possible without them. The reason is quite obvious: any extrapolations based on secondary classifications and/or parametric descriptions are most reliable within certain classes that unite objects with certain properties. Therefore, such classes should not be distinguished arbitrarily, but to reflect the "nature of things" of whatever understanding. To emphasize this fact, such classes, following the philosopher John Mill, are usually called *natural kinds* [Magnus 2014]. Thus, respective primary classifications should be "natural" in the very classical sense that was incorporated in this concept in the 17th–18th centuries and traditionally adopted in systematics. This is what makes it possible to use "primary" classifications as so-called *reference systems*, i.e., those reference to whom justifies correctly solving various kinds of "secondary" research and applied tasks.

As for the taxonomic classifications developed by systematics, their reference function is determined by the fact that different users, speaking figuratively, look at the diversity of organisms through the eyes of practicing taxonomists. What is distinguished in the latter's classifications is accepted by all others as a kind of particular "taxonomic reality" (see Section 4.2.1), which is chiefly evident in the case of species. Indeed, it is only after a certain systematist identified and named a certain species, that others began to study it, recognize its biological properties, and, if necessary, protect it. But if another, more authoritative taxonomist decides that this is not a distinct species, but just an intraspecific form of another species, it drops from the user's attention—it ceases to be a part of the above "taxonomic reality."

It is clear from this that taxonomists should be interested in expanding the spheres of competent applications of taxonomic knowledge. The reason is that the demonstration of the practical significance of results of their research in the eyes of various kinds of users and especially decision makers serves as a "justification" for the very existence and development of systematics as a scientific discipline.

Below are indicated the areas of activity in which the results of taxonomic research are used most actively. This includes both scientific disciplines that use these results in their studies of biodiversity, and forms of strictly applied activity.

In *biogeography* dealing with biogeographic regionalization, the initial material is provided by faunistic and floristic lists, as well as general lists by taxonomic groups, if they include distribution data. In regionalization, strictly speaking, it is not taxa themselves that are analyzed and compared, but their ranges, although it is obvious that distribution data largely depend on how the taxa were initially recognized [Morrone 2018]. Therefore, biogeography is interested in having full-scale natural taxonomic classifications as reference systems to warrant recognition of natural areas of different ranks in biogeographic classifications. With this, particular research programs in biogeography evidently rely upon taxonomic classifications developed within taxonomic theories of respective content [Nelson and Platnick 1981; Santos and Amorim 2007].

In *biostratigraphy* dealing with the correlation of geological structures and events by the composition of fossils in them, taxonomic classifications of extinct organisms linked to the time scale are of great importance. In stratigraphic analysis, the main units are species and genera, whereas in paleogeographic reconstructions, taxa of higher ranks are usually taken into account. It is evident that taxonomic classifications of extinct organisms are to be as natural as possible to make historical reconstructions as reliable as possible [Meyen 1988; Forey et al. 2004].

In *biocenology* studying composition of ecosystems from structural and functional viewpoints, lists of taxa within them are used to develop various kinds of biocenological classifications. For instance, syntaxa and guilds are distinguished in such classifications; the former are diagnosed by dominant species, while the latter combine species with similar ecological functions [Mirkin 1985]. In some sections of autecology, the emphasis is placed not on species as such, but on biomorphs, which may correspond to several ecologically and morphologically indistinguishable species [Chernov 1991; Krivolutskiy 1999]; biomorphics is engaged in their study [Pavlinov 2010a].

In *nature conservation*, according to the current strategy, one of the main emphases is placed on biodiversity conservation [Wilson 1988]. The latter is most often understood in a simplified form as species diversity [Claridge et al. 1997; Wheeler et al. 2012; Sigwart 2018; Costello 2020]. Respectively, species classifications tied to specific biotic complexes and geographic regions are of key importance [Vogel et al. 2017]. Based on the lists of species presented by systematics, an assessment of the degree of their endemism (overlap with biogeography), abundance, and vulnerability (overlap with ecology), recommendations are made to identify protected natural areas and local biomes.

Various kinds of *quarantine services* (customs, agricultural, sanitary, etc.) monitor the movements of organisms and their derivatives, which are subject to certain restrictions (pests, vectors of dangerous diseases, protected ones, etc.), through control points. For them, as in the previous case, detailed classifications of species and subspecies with high-quality identification keys developed for respective groups of organisms, are of particular importance.

The main link between systematics and various users of the results of taxonomic research are three main types of publications: (a) reviews (checklists) of particular taxa on a global scale; (b) faunistic and floristic reviews for particular regions; and (c) identification keys or tables. In the first two, lists of taxa are provided, usually supplemented with their characters, distribution, etc.; the latter serve for allocation organisms to certain taxa, and many (the best) publications combine reviews with keys.

The quality of such publications prepared by taxonomists determines how useful the results of their research look from users' viewpoint. And this, in turn, determines how much society is ready to support biological systematics as a science. Therefore, taxonomists should be interested in regularly publishing the results of their research in a form that is available for use in various research and applied activities.

2 Conceptual History of Systematics

> Only then can you understand the essence of things when you know their origin and development.
>
> Heraclitus of Ephesus

Science is a non-equilibrium information macrosystem. According to a theoretical model developed by synergetics, one of the important inherent properties of such a system is its development, providing its dynamic stability: without development, a non-equilibrium system simply cannot exist [Prigogine and Stengers 1984]. The trajectory along which the developing system moves in the space of its possible states is branching and thus inevitably leads to a gradual complication of the system due to diversification of its realized states. In each transition from one state to another ("bifurcation point"), a certain mixture of continuity and novelty occurs: the first provides a certain conservatism in the development of the system, while the second is responsible for the latter's moving off from the previous states. Continuity means that the past of the developing system influences its future to a certain degree. Novelty means that there is some uncertainty in the choice of the next trajectory among several possibilities; this makes the development of a non-equilibrium macrosystem as a whole not entirely predictable.

The development of science is its history. The above consideration means, on the one hand, that scientific knowledge achieved at a certain stage of its history depends to some extent on its previous theories and concepts. On the other hand, one of the inevitable consequences of such historical development of science is its diversification—multiplication of its theories and concepts, with their emergence, generally speaking, being hardly predictable, although usually explainable *a posteriori*.

According to the ideas of evolutionary epistemology supported by contemporary non-classical philosophy of science [Popper 1975; Hübner 1988; Radnitzky and Bartley 1993; Merkulov 1996; Abachiev 2004; Stepin 2005], the above synergetic interpretation of the historical development of science as a non-equilibrium macrosystem is strengthened by its likening to biological evolution [Hull 1988]. One of the key provisions arising from this evolutionary metaphor is an appreciation that the history of science is not something "outside" in relation to science proper, but is an integral part of it—just like the evolutionary development of biota is not something "external" to it. So, without considering the evolutionary history of science, it is impossible to understand why scientific knowledge at certain stages of its development is just so and not something other.

Obviously, this conclusion is true for systematics as a part of natural science. Therefore, in order to understand its theoretical foundations, it is necessary to consider them not statically, but in their historical dynamics [Hübner 1988]. Such historical consideration of theoretical knowledge is very significant for an understanding of how systematics functions, and how and why its key ideas emerged and changed at different stages of its historical development. Therefore, recent interest in the theory of systematics inevitably generates interest in its conceptual history [Hull 1988].

This chapter presents the author's concept of the history of biological systematics shaped within the framework of evolutionary epistemology [Pavlinov 2018, 2019, 2020]. First, a general idea of what the conceptual history of systematics is and how it may look from different theoretical positions is presented; then the major stages and directions of the development of systematics are considered. The main focus is on the development of theoretical concepts, and the most important events in the history of taxonomy are characterized as scientific revolutions.

2.1   SOME PRELIMINARY CONSIDERATIONS

The realized history is unique—but it can be narrated in different ways. With this, the content of its narration is different; it reflects a particular narrator's interests: the focus is on what seems worthy of attention, while everything else is overlooked [Rozov 2002]. Thus, using the figurative language of the historian philosopher Robin Collingwood, it can be said that taxonomists use to write their "own histories" of systematics following their views and inclinations [Collingwood 1994]. So, there is no "history in general"—or, more precisely, historiography, but rather there are its particular interpretations and narrations according to particular concepts of both the theory and history of systematics. Therefore, it is hardly surprising that, in different manuals and textbooks on systematics, its history is presented in different ways.

Slightly coarsening the situation, it is reasonable to identify two main ways of considering the historical development of systematics, empirical and theoretical ones. They correspond to two ways of understanding its principal contents.

Empiricism sees in systematics mainly generalized facts, namely particular classifications. Accordingly, the *empirical history* of systematics is reconstructed as a chronicle, the main events of which are successive updates of these classifications due to widening of empirical knowledge. Obviously, such an approach gives little understanding of why and how these events happened conceptually and why and how they influenced the subsequent development of taxonomic theory. And even more so, it does not provide an understanding of why different classifications even arise that do not fit a single trend of development of more and more "complete" and "perfect" empirical taxonomic knowledge.

In contrast, a predominating theoretical view of systematics presupposes a similarly theoretical view of its historical development, which in this case appears as the *conceptual history* associated not so much with changes of classifications, but with the development of taxonomic theories, concepts, etc. As Ernst Mayr rightly noted, "the most important aspect of the history of systematics is that it is, like the history of evolutionary biology, a history of concepts, not facts" [Mayr 1982: 144]. These concepts, in turn, do not arise spontaneously, but in a certain general scientific—more broadly, in a scientific philosophical—context, which itself is also subject to change. Therefore, it seems reasonable to suppose that one of the main driving forces in the development of theoretical contents of any scientific discipline, including systematics, is the historical dynamics of the general scientific philosophical context [Putnam 1981; Rozov 2002; Ilyin 2003].

These two ways of understanding and describing the history of systematics can be presented as an emphasis on *extensive* and *intensive* paths of its development, respectively. They can be considered as the implementation of two main cognitive programs: the first embodies the "*collectionist*" (empirical) program, whereas the second embodies the "*methodical*" (theoretical) program [Zuev 2002, 2015; Pavlinov 2018, 2019]. Extensive development expands a descriptive picture of the diversity of life, providing an empirical basis for an application of the concepts and stimulating their development in its own way. Intensive development is problematic: transitions from one stage of history to another are always caused by solutions of old and formulations of new problems. This intensive path is connected, on the one hand, with the deepening and expansion of ideas about the structure and general causes of the diversity of living beings, and, on the other hand, with the deepening and expansion of ideas about how to study this structure with the help of certain theoretical "tools." These two modes of historical development and the narrations corresponding to them complement each other: one cannot exist without the other.

2.1.1 HOW A CONCEPTUAL HISTORY OF SYSTEMATICS CAN BE WRITTEN

The content of conceptual history of systematics is made up, generally speaking, of significant changes in its cognitive situation (see Section 3.2), which are marked by changes in its key research programs [Pavlinov 2018, 2019]. As was noted in the previous section, these changes occur not by themselves, but according to one or another "dictate of the time," provided by the development of the philosophical scientific context. In the 16th century, this was the emergence of a rational way of describing the diversity of organisms stimulated by the emergence of modern European science. At the turn of the 18th–19th centuries, it was caused by ideas of philosophical empiricism. In the second half of the 19th century, the fully fledged assimilation of the evolutionary idea became this "dictate." In the first half of the 20th century, systematics develops mainly under the influence of positivist ideas, while in its second half, it was under the influence of post-positivist philosophy of science; accordingly, the metaphysical component of taxonomic knowledge is expelled from it or again becomes legalized in it.

Unavoidable placement of certain accents in the reconstruction of the conceptual history of systematics causes a serious dilemma of *presentism vs. antiquarism*, each obliging an evaluation in different ways of theoretical constructs that appear at different times [Foucault 1970; Rozov 2002; Kuznetsova 2009]. In the first case, a certain theory that arose in the past is considered in the context of a modern understanding of the content of the scientific discipline and its philosophical background. In the second case, such a theory is considered in the scientific and philosophical contexts that existed at the time of its emergence and supposedly gave birth to it. For example, presentism obliges contemporary taxonomists, devoted to the evolutionary idea, to evaluate taxonomic theories of the 17th–18th centuries according to their *post factum* evaluation of their contribution to the development of this idea, as it is now understood. From the point of view of antiquarianism, such consideration is hardly correct: the taxonomists who formulated theoretical foundations of systematics at that time were not thinking about evolution but rather about creation, and accordingly developed concepts as they then understood them—and only later did their concepts receive an evolutionary interpretation. For example, the alchemical affinity of "everything with everything" was transformed into the kinship relations of organisms.

2.1.2 HISTORY OF SYSTEMATICS AS AN EVOLUTIONARY PROCESS

The metaphor of the conceptual evolution of science implies the following. If in biological evolution, speciation is considered its key evolutionary act, then in conceptual evolution, it is a *"conceptiation,"* which means the emergence of new concepts and theories. With this consideration, theoretical concepts are an analog of different genotypes, while specific classifications arising on this basis are analogous to the phenotypes; and a selecting environment for them is a scientific community guided by both philosophical and pragmatic contexts [Hull 1988]. This point of view sets specific emphasis on the understanding of some mechanisms of the formation of research programs in systematics [Mishler 1990, 2009; Pavlinov 2018].

With such consideration of the conceptual history of systematics, two main components may quite naturally be recognized in it, viz. *anagenetic* and *cladogenetic*. The first means a sequential progression of taxonomic theories from less to more developed, whereas the second means their fragmentation and multiplication. Together they presume an upgrowing branching tree-like metaphor of the historical development of taxonomic theories and concepts. The anagenetic component orders the development of theoretical systematics within the framework of a general *unification trend*, which is determined by a fundamental idea of cognition of the System of Nature in its most general understanding. The cladogenetic component corresponds to a *diversification trend* shaped by different interpretations of this general idea. As systematics develops as an open information system, a *network* component is naturally superimposed on this basic two-component evolutionary model: it means exchanges and combinations of different ideas, as an analogy of horizontal gene transfer in the biological evolution.

Two general concepts of the philosophy of science, classical and non-classical (see Section 3.1), differently assess the contribution of each of these components to the conceptual history of systematics. Accordingly, assessments of both the whole process and the significance of taxonomic theories are different. The "classics" emphasizes the anagenetic component: development of systematics is conceivable as a steady progress of taxonomic knowledge from less to more complete, guided by successively elaborated more perfect and consistent theories within the framework of a single dominating idea [Shatalkin 2012]. With this, those theories that do not fit this progress are thought of as "delusions" doomed to elimination and oblivion. In contrast, the "non-classics" considers both the ana- and cladogenetic components of the conceptual history of systematics of equal significance: this means that diversification of theories is just as inherent in its evolution as its consistent progression. As noted above, the development of systematics is associated, among other things, with a growing understanding of taxonomic diversity being complexly structured and multi-faceted, generated by various complexly interacting causes—and such understanding is also included in the overall assessment of what progress and completeness of the taxonomic knowledge are. Accordingly, with a growing understanding of the complex structure of taxonomic diversity, the conceptual space of systematics also becomes more complex and structured due to the emergence of particular taxonomic theories following new visions of multiple manifestations of that diversity.

The evolutionary interpretation of the historical development of systematics allows the inclusion of the *concept of adaptive zones* focusing on certain important regulating factors of biological evolution [Pavlinov 2018]. It is presumed that, in the conceptual history, various kinds of general scientific and philosophical ideas act as "adaptive zones" stimulating the development of systematics in one direction or another. So, in the second half of the 19th century, a new, very extensive "adaptive zone" was opened by an evolutionary idea that gave rise to the now dominant evolutionary-interpreted systematics. At the beginning of the 20th century, such a "zone" was formed by the positivist philosophy of science, which directed the main trend in the development of systematics towards the phenetic, numerical, and partly experimental pathways. Accordingly, if an idea loses its relevance, its respective "adaptive zone" narrows and the taxonomic theory developed in it loses its significance. In the 19th century, this happened to the organismic natural philosophy, while an example from the 20th century is classification phenetics. And yet both of them, figuratively speaking, did not "sink into oblivion": the first became one of the foundations of classical phylogenetics, while the second became a source of certain important ideas for numerical phyletics.

2.1.3 MAJOR STAGES OF THE CONCEPTUAL HISTORY OF SYSTEMATICS

Different emphases on the conceptual history of systematics inevitably lead to a significantly different understanding of its main stages that are acceptably considered scientific revolutions. Thus, botanists usually distinguish two main stages in this history, viz. the epochs of artificial and natural systems [Sachs 1906]. Proponents of the evolutionary idea divide the entire systematics into pre-evolutionary and evolutionary [Mayr 1942, 1969, 1982]. A variant combining these two versions is a partition of the history of systematics into artificial, natural, and phylogenetic stages [Starostin 1970]. Within the framework of empirical history, it is popular to distinguish three main stages, denoted as *alpha-*, *beta-*, and *omega*-systematics: the first (initial) corresponds to descriptions of local fauna and flora, the second to the construction of general classifications, and the third (final) to the development of a comprehensive overall classification [Mayr 1942, 1969; Davis and Heywood 1963; Stace 1989]. The ideologists of biosystematics distinguish between descriptive, systematic, and biosystematic stages in the history of systematics [Valentine and Löve 1958]. Theoreticians of the numerical program believe that its development in the second half of the 20th century became the most significant achievement almost since the time of Linnaeus [Sneath 1995; Vernon 2001], while the devoted cladists argue that the most significant is the cladistic revolution that also occurred in the second half of the 20th century [de Queiroz 1988].

Based on the above general ideas about what the history of systematics is and how it can be presented, four main stages can be quite naturally distinguished: pre-systematics, proto-systematics, scholastic systematics, and post-scholastic systematics [Pavlinov and Lyubarsky 2011; Pavlinov 2018, 2019]. The first two are united in the "prehistory" of the scientific systematics, whereas the last two correspond to the development of the scientific systematics proper. The principal feature of the latter is that research programs dealing with rational exploration of the Natural System of living nature begin to develop. As noted, changes in such programs, dominating at one stage or another in the development of theoretical systematics, have the character of scientific revolutions, signifying transitions from one historical stage to another.

Pre-systematics begins with the history of all human classification activities; it is known as *folk systematics*, characteristic of communities of indigenous people. It is developed on various local ethnic bases and, in its foundations and aspirations, is not abstractly cognitive but mostly empirical and pragmatic, without explicitly expressed onto-epistemic premises.

Proto-systematics is immersed in a broader context of ideas about both the ordered Cosmos and ordered ways of its knowing; an idea of Cosmos means that proto-systematics, in contrast to pre-systematics, is by its scope global rather than local. Within the framework of this context, two key tasks are solved, laying down two already-mentioned basic cognitive programs: (a) "methodical" is associated with the rational development of the first onto-epistemic principles of any cognitive activity; and (b) "collectionist" is associated with the primary systematization and generalization of information about living nature, partly in a "cosmic" and partly in a utilitarian sense. In proto-systematics, several stages can be quite naturally distinguished, viz. primary (Antique), scholastic (Medieval), and herbal (Renaissance). In the first, certain general principles of logical classification and related basic notions are developed, viz. genus, species, essence, difference, etc. In the second, the main rational categories of onto-epistemology, basic argumentation schemes, and classification methods become more advanced; all of them will be inherited by theories of early scientific systematics. Renaissance herbalistics mainly develops the "collectionist" program, thus laying the foundations of descriptive systematics.

Scientific systematics begins with the shift from natural theology to natural philosophy (in its broad sense) in the 16th century, with the development of a new cognitive situation and cognitive paradigm of exploration of Nature. The main goal becomes the latter's rational cognition aimed at uncovering a certain universal natural law that organizes the entire material world into the System of Nature. In systematics, the key task becomes the recognition of the natural groups of organisms in which this System is manifested. According to the way this general cognitive intention is filled, the conceptual history of systematics can be divided into several stages, each initiated by the respective scientific revolution.

The first stage covers the 16th–18th centuries and is largely associated with the development of classification methodology borrowed from Medieval scholasticism. It begins with a specific initial rationalization of the branch of natural philosophy exploring plants and animals, which leads to the *scholastic revolution* and to the emergence of systematics as a scientific discipline. As a result, *scholastic systematics* appears to deal with a global rational systematization of living nature instead of just listing its facts and uses. The respective classifications of plants and animals are elaborated based on the logical genus–species scheme, with the essences of organisms being used to recognize their natural groups. An adaptation of this scheme to the exploration of living matter becomes the characteristic trend of the conceptual history of scholastic systematics, with the main debates being about the principal essences of organisms. As one of the important results of this trend, the hierarchical system of fixed ranks is developed to replace the rankless hierarchy of the genus–species scheme.

Post-scholastic systematics begins to emerge in the second half of the 18th century as a result of the *post-scholastic revolution* that becomes a kind of protest against the previously dominant scholastic approach to the comprehension of the Natural System. It is flagged by a transition from the main emphasis on the scholastic method of cognition of Nature, with organisms being characterized by a few essences, to Nature itself in all its manifestations, with organisms being characterized by all available characters. This new cognitive situation yields a certain harmony of "methodical" and "collectonist" programs and means "biologization" of systematics, thus making it *biological* in a full sense. This change is largely due to the empirization of taxonomic research: the predominantly deductive ("top-down") classification method is replaced by the predominantly inductive ("bottom-up") one. This general trend gives rise to a significant variety of particular taxonomic theories, differing in their natural philosophies (from "systemic" to "ladder," from typological to early transformist, from organismic to numerological, etc.) and methodologies (the ratio of inductive and deductive elements, different approaches to character weight). The most noticeable and influential become natural systematics (in its narrow, mostly botanical content) and classification typology, each claiming to be most "natural" in its classification theory and method.

In the second half of the 19th century, an important step in the development of post-scholastic systematics brings it from an early to a more mature stage. It is marked by an active assimilation of the transformist natural philosophy in its historical understanding and, on this basis, by the adoption of the genealogical meaning of the Natural System. This step marks the *evolutionary revolution* in the conceptual history of systematics giving birth to *evolutionary interpreted systematics*. Darwin's model of evolution attracts taxonomists' interest in intraspecific categories, while Haeckel's model focuses mainly on high-rank groups.

Throughout the 20th century, the development of systematics is largely associated with the search for answers to the challenges of new versions of scientific rationality. In the first half of that century, ideas of the positivist philosophy of science become dominant in systematics: this philosophical shift causes a *positivist revolution* in systematics and the emergence of the positivistically oriented taxonomic thought. The latter's development first leads to the formation of biosystematics based on the Darwinian evolutionary model. Somewhat later, classification phenetics and numerical systematics are formed in conjunction, and to a certain extent lead to the debiologization (formalization) of the positivist systematics.

In the second half of the 20th century, there is an active rethinking of the phylogenetic concept, which marks the next significant step in the conceptual history of contemporary systematics deserving being called a *post-positivist revolution*. The main tone is set by the cladistic version of phylogenetics, which absorbs significant elements of positivist reductionism. Towards the end of the 20th century, its development is governed mostly by its active "molecularization," i.e., the assimilation of molecular data together with new numerical methods [Lee 2004]. This results in the absolute dominance of genosystematics (better known as molecular phylogenetics); because of ignorance of all other biologically significant features of organisms (such as morphological, physiological, developmental, etc.), this appears to be another manifestation of the debiologization of our discipline.

Along with these "mainstream" research programs of post-scholastic systematics, each pretending to be "revolutionary," taxonomic theories of more particular significance emerge over the two centuries of its conceptual history. In the 19th century, these are "esoteric" theories like numerology and organismic natural philosophy. In the 20th century, several variants of rational systematics (in its broad sense) are formed, implementing in different ways the general idea of nomotetization of this discipline. Along with them, biomorphics emerges, paying most attention to the life forms (ecomorphs). Finally, in recent years, a new taxonomic theory seems to begin to assert itself, based on the ideas of evolutionary developmental biology (evo-devo).

2.2 PREHISTORY OF SYSTEMATICS

As emphasized above, the history of systematics as a natural science branch begins with a purposeful exploration of the diversity of living nature using a rationally organized method. Viewed from this perspective, the previous stage in the development of classification activity, in which wildlife is somehow involved, can hardly be considered part of the conceptual history of systematics in its scientific meaning. Rather, this is its *prehistory*, in the course of which certain prerequisites for the formation of proper scientific systematics appear.

2.2.1 AN INITIAL STEP: FOLK SYSTEMATICS

A capacity to classify, that is, to recognize similar and different objects and to group and divide them on this basis, is inherent in animals as something like a "classification instinct" [Atran 1990; Ellen 2008]. It constitutes an important part of biological activity associated with the need to adapt to an environment by recognizing in it "ours" and "others," viz. edible and inedible, companions and enemies, sexual partners and competitors, etc., and to react to them accordingly.

Members of the biological species of *Homo sapiens* inherit this form of primordial cognitive activity, with its inherent "classification instinct," from their ancestors. Conscious classifying as an initial form of intellectual comprehension of the surrounding world grows out of it [Lévi-Strauss 1966; Foucault 1970]. With becoming able to think and speak, people begin to actively "invent" notions to denote not individual concrete things, but their generalized mental images. For example, the word "tree" does not mean this particular tree in front of one's eyes, but a "tree in general," i.e., any tree differing from various "non-trees." Such generalizing notions, while multiplying more and more, together constitute primary classifications.

Until recently, it was generally acknowledged that reasonable cognition, as a prerequisite of modern science, originated in Antiquity, so there is nothing particularly interesting in the earlier forms of cognitive activity of archaic people. However, in the middle of the 20th century, this attitude changed under an influence of cognitology, which studies the primary forms of cognition [Klix 1983; Velichkovsky 2006]. One of its branches became ethnobiology, dealing with the relations of primitive humans with plants and animals, including their categorization [Maddalon 2003; Newmaster et al. 2006].

The development of such categorization gives rise to so-called *folk systematics*, which is defined above as a pre-systematics, as it develops no explicit theoretical and methodological constructions and thus remains basically irrational. Its background is intuitive and pragmatic, dictated primarily by the need to survive "here and now" and, for this, to comprehend the diversity of surrounding wildlife. The main product of folk systematics is *folk classification*—a primitive mental image of the structure of biota existing in a form of networked notions and names [Raven et al. 1971; Berlin 1973, 1992; Atran 1990; Pavlinov and Lyubarsky 2011; Lyubarsky 2018; Pavlinov 2018].

In interpretations of folk systematics, the above-mentioned dilemma of "presentism *vs.* antiquarism" clearly occurs. In this case, presentism means that, in describing folk classifications, researchers borrow the proper analytical apparatus from contemporary systematics. In this way, they impose their cognitive schemes on the knowledge of indigenous people [Hays 1983; Atran 1990, 1998; Taylor 1990; Ellen 1993, 2008; Atran and Medin 2008; Pavlinov 2018]. Therefore, in studying folk systematics, presentism inevitably takes the form of a specific concept-centrism: researchers consider folk classifications from their own much more advanced theoretical positions. However, it does not seem right to consider from the standpoint of modern conceptual systems the folk classifications that arose in the context of local pre-scientific cultures.

There are two main points of view among researchers of folk systematics on the basic motivation for developing folk classifications: *utilitarians* believe they emerge due to certain practical needs, whereas *intellectualists* think they are mainly motivated cognitively. From the first perspective, the structure of folk classifications reflects certain human consumer attitudes to local environments. In particular, the organisms most important for the survival of an indigenous community are classified first of all; therefore, such classifications are strongly determined by certain local needs and therefore are subject-centered, highly selective, and locally specific [Hays 1982; Morris 1984]. This subject-centrism manifests itself in ethnocentrism, therefore folk classifications are reasonably called *ethnobiological* [Brown 1986; Atran 1990, 1998; Berlin 1992; Newmaster et al. 2006], and such local folk classifications are not relevant outside the specificity of the local folk cultures producing them [Steward 1955; Hunn 1977; Hays 1982; Ellen 1993, 2008; Maddalon 2003; Dwyer 2005]. The second interpretation presumes that there is a certain basic cognitive model inherent to the species of *Homo sapiens* operating with universal categories, including universal classification schemes minimally depending on local specifics of the cognizing subjects [Atran 1981, 1990, 1998; Berlin 1992; Mithen 2006]. From this perspective, cognitively motivated folk classifications reflect some fundamental structure of biota regardless of the local peculiarities of its perception, so they have some fundamental features in common [Berlin 1973, 1992; Berlin et al. 1973; Atran 1990, 1998; Lopez et al. 1997; Medin and Atran 2004; Atran and Medin 2008; Lyubarsky 2018].

The "factual" grounds for developing folk classifications significantly differ, thus providing quite diverse categorical schemes of division of local biotas. It is supposed that groupings of organisms by characters, which can hardly be explained by reference to any utilitarian criteria, produce some "general" folk classifications [Berlin 1973, 1992; Berlin et al. 1973]. Their high-rank groups are distinguished based on two kinds of characteristics of organisms: some are intrinsic (morphology, etc.), whereas others are extrinsic (ecology, etc.). In the first case, for instance, woody *vs.* herbaceous plants or hairy *vs.* bare-skinned animals or winged *vs.* legged *vs.* legless animals are recognized; in the second case, aquatic *vs.* terrestrial or arboreal *vs.* fossorial organisms are distinguished. At lower levels, particular anatomical features (color details, shapes of body parts, etc.) are usually taken into account. This general classification principle will be inherited from folk systematics by nearly all proto-systematics.

Folk classifications are predominantly hierarchical: they include groups (*folk taxa*) of different levels of generality. In the simplest case, their hierarchy is treated as rankless: "primary" (higher), "secondary," etc. groups are distinguished [Berlin et al. 1973; Bulmer 1974]. Besides, other variants of orderings in the form of series, networks, and block schemes are recorded; however, hierarchical classifications are considered more "advanced" in comparison with others [Berlin 1992; Hunn and French 2000; Lyubarsky 2018]. With this, it is assumed that folk categories are cognitive universals that can be correlated with certain basic ontic categories of cognition [Atran 1990, 1998; Lyubarsky 2018]. Therefore, following "Linnean" terminology, hierarchical folk classifications lacking pragmatism are sometimes referred to as "natural," whilst all others are "artificial" [Berlin 2004].

In more elaborated schemes based on the same "Linnean" tradition, ranked classifications with certain universal *folk categories* are considered characteristic of different local folk classifications [Berlin 1973, 1992; Atran 1990, 1998, 2008; etc.]. From four to six such categories (ranks) are usually distinguished; they are given the same names as in contemporary classifications: life forms, genera, species, etc. Such interpretation of folk classifications is obviously concept-centric and therefore may be erroneous in reflecting not so much the characters of perception of nature by primitive systematicians, as rather the point of view of their advanced scholarly interpreters [Ellen 1993, 2008; Coley et al. 1997; Ghiselin 1998; Pavlinov 2018].

Nevertheless, great importance is attached to an analysis of the rank hierarchy of folk classifications; with this, special attention is paid to identifying those categories that are recognized first of all by indigenous people as the most distinguished for one reason or another [Brown et al. 1976; Rosch et al. 1976; Brown 1986; Atran 1990; Ellen 2008]. "Intellectuallists" substantiate the special importance of folk life forms by supposing that, for an archaic perception, Nature is structured into the typologically most clearly recognizable large blocks, which are trees and grasses, animals and birds, reptiles and fish, etc. [Rosch et al. 1976]. A special significance of genera and species (or "generic species") is substantiated from a utilitarian position [Atran 1999]: it is believed that archaic people first single out groups of lower ranks as the most significant for their survival "here and now" [Berlin 1992].

One of the issues occupying the attention of the researchers of folk classifications is the latter's correspondence to the scientific ones; the main task is to reveal certain common features in them. The basic motivation for this is an assumption that, if the same groups are distinguished in the respective classifications, this can be taken as indirect evidence of their "reality." In this way, contemporary scientific classification concepts serve as a general standard for such comparisons: for example, the "natural-ness" of folk taxa is sometimes assessed from the point of view of the "evolutionary relationships" of their representatives [Berlin 1992]. Such criteria are obviously quite alien to folk systematics, so this approach means nothing more than a clear concept-centrism, making these comparisons hardly correct [Hunn 1982; Brown 1986, 2004; Pavlinov 2018]. From this point of view, a method of "objectification" of biological species recognized by scientists by referring to folk taxa [Mayr 1969, 1988; Ludwig 2017] is hardly consistent, especially taking into consideration that scientists them-selves treat species quite differently (see Section 6.7).

It should be noted that folk systematics is not only an initial stage of the prehistory of the scientific systematics but also a certain rather stable trad-ition. It is continued by many empiricists who rely on their personal experi-ence and reject theorizing and formalization (possibly, because of their lack of understanding). This general attitude was clearly outlined by the zoolo-gist Philip Darlington, who, protesting against the dominance of "numerists," advocates "coming back [...] to a taxonomy that is correlated with reality" [Darlington 1971: 363]. And this can be achieved if taxonomists realize their "aspiration for truth without engaging any theoretical reasoning" [Stekolnikov 2003: 367].

2.2.2 Becoming Aware of the Method

The scientific style of cognitive activity is characterized primarily by rationality, i.e., by its subordination to a certain explicitly construed formalized method. Its natural-philosophical background is initially shaped by presuming that a true method leads to true knowledge; positivists will later add to this that a properly organized method makes the results of research subject-free and testable. This does not mean certain technical algorithms: rather, *the Method* (Aristotle's Organon) is meant as a rigorous logical argumentation scheme common to all branches of knowledge about Nature.

This rational scheme lays the foundation for proto-science in Antiquity; it will be refined by scholastics in the Middle Ages, from which contemporary science will begin to grow with its unified understanding of both research tasks and techniques [Gaydenko 1980; Lindberg 2007].

An original rational idea of the Method (Organon) is loaded with a profound natural philosophy, which makes it very productive. Its key inspiration is that the main goal of cognition is the understanding of what Nature itself is and what makes it what it is; from this naturally follows the question about what the Method of comprehension of Nature should be. An answer to this complex question seems to be quite obvious to antique thinkers; it is substantiated as follows [Akhutin 1988]. Nature is uniformly and rationally ("reasonably") organized in its bases. Therefore, a uniformly and rationally organized Method of exploring Nature must correspond to Nature itself: this means that, for the Method to lead to comprehending Nature, its logic must correspond to the logic (dialectics) of Nature. This metaphysical unity of Nature and Method allows one to expect that, through a properly organized and applied true Method, the true "nature of things" is certainly uncovered; it is formalized by the contemporary *principle of onto-epistemic correspondence* (see Section 3.2.1).

The key constituent of the Method thus understood is a qualitative categorization of all that is known about Nature, with a general classifying algorithm being thought of as its most adequate implementation. Its main task is to distinguish natural (true, existing in Nature) and artificial (false, "invented" by man) groups of objects employing true (logically consistent) judgments about them. Its rationality presumes that these judgments are inferred based on a syllogistic scheme construed as a set of uniformly fixed rules of logic. On this basis, a *classification* version of the "methodical" cognitive program is formed. It is first developed by antique thinkers followed by neo-Platonists and scholastics; in the 16th century it will be embodied in the fundamental ideas of the natural classification and the natural method leading to it.

The conceptual history of *proto-science* as an initial form of a rationally organized cognitive activity covers a long period from Antiquity to the Renaissance. Its principal advancement is in developing the "methodical" cognitive program as an onto-epistemic foundation of the entire cognitive activity, including the essentialist vision of Nature and deductive scheme of its description. At the same time, a "collectionist" program also develops as a continuation of the folk systematic tradition, but in a more advanced version of the systematization of plants and animals of the entire Oecumene. These two advancements mean the gradual maturation of *proto-systematics* associated with the development of some important ideas and general methods of classification activity.

Early Antiquity is customary to consider as the beginning of the development of the Method in its general understanding, outlined above. Two great philosophers of the 4th century BC, Plato and Aristotle, a teacher and his disciple, are most influential in this respect. They are antipodes in most of their natural-philosophical views, and nevertheless their ideas, somehow combined in the later works of neo-Platonists and scholasticists, will enter conjointly the body of theoretical knowledge of the early scientific systematics.

Plato's natural philosophy is based on a conception of *eidos* (εἴδοσ, idea, plural *eide*) as permanent and eternal supra-material basic units of the Universe (Cosmos). A totality of *eide* arises as a result of the successive emanation of the One as a creative principle and driving force of all that exists: the One gives rise to the most general *eide*; they emanate lower-order *eide*, until the latter are embodied in concrete material things. In some contexts, Plato mentions *eidos* as a form or model or template (τυπος), based on which concrete material things are made: this is how the notion of "type" first appears in descriptions of the world of ideas and things [Hammen 1981]. Cosmos thus understood is ordered by a descending cascade of causal relations among *eide* of different levels of generality (steps of emanation): more general phenomena cause more particular ones. According to this, the Method must reproduce the sequence of this emanation and causation: it begins with the identification of most general categories and divides them sequentially into more particular ones to reach the lowest levels of the Universe. This is the general natural-philosophical background for what will be subsequently denoted as a *deductive* (from general to particular) argumentation scheme.

For example, keeping in mind the subject of our book, an idea of "animality" first arises at some stage of emanation of the One, giving rise to an idea of "four-footedness"; the latter emanates an idea of "horseness," and the latter is embodied in concrete horses. Therefore, to understand what a horse is and what its place is in the Universe, it is to be comprehended what is "horseness" and what is its position in the hierarchy of *eide*, from "animality" through "four-footedness," etc., downward. From this general ontological scheme of the hierarchy of Cosmos, an epistemic classification algorithm is inferred to describe its structure [Makovelsky 2004]. This algorithm allows the recognition of certain classes of objects that are manifestations of particular *eide* of certain levels of generality. Of prime importance is that, according to the underlying natural philosophy, these classes, i.e., *universalia* ("animals in general," "quadrupeds in general," "horses in general") are no less real than observed individual horses, i.e. *naturalia*. In modern natural science, eidically given universals correspond, with some reservations, to natural kinds (in the sense of [Quine 1996]), and in systematics they are represented by taxa of different ranks.

Aristotle's natural philosophy is fundamentally different. It is based on an idea of *ousia* (ουσία, plural *ousiai*): like Plato's *eidos*, it is unchanging and eternal, but unlike the latter, it is embodied in the things (*primary ousia*) and in their natural grouping (*secondary ousia*), and does not lay outside and above them. Primary *ousiai* are the "bricks" of the Universe; they are hidden in the things and determine their "whatnesses"; the latter make the material (observable) things what they are in the Universe [Gotthelf 2012]. One of the interpretations of the "whatness" is the "soul" of a thing in its Aristotelian understanding [French 1994; Tipton 2014]. Variations on his "theme of the soul" will form the basis for understanding and searching for organismic "whatnesses" in scholastic systematics.

One detail should be noted in the later interpretations of Aristotelian natural philosophy, which is significant for understanding the origins of the classification method of early systematics. In scholastics, an original concept of *ousia* will be denoted by its Latin equivalent *essentia*, which corresponds to secondary *ousia*, i.e., "horses in general," "horsesness." Later, as the scholastic understanding of the essence develops, an important difference between Plato's *eidos* and Aristotelian secondary *ousia* will disappear; they will be referred to uniformly as *essence* [Balme 1962, 1987a; Winsor 2003; Berti 2016]. Be that as it may, such an interpretation will predetermine the focus of earlier systematicians on the search for essences; their comprehension will be thought of as a necessary condition for comprehending the true "nature of things."

According to Aristotle, Nature is organized hierarchically, but in an order directly opposite to Plato's Cosmos: it is arranged by an ascending cascade of causal relations, in which more particular *ousiai* are the causes of more general ones. According to this, primary *ousiai* give rise to secondary, tertiary, etc. higher-order *ousiai*, so all of Nature represents a hierarchy of *ousiai* of different orders [Gaydenko 1980; Sokolov 2001; Tipton 2014]. Turning back to the above "horse" example, this means that there is a primary *ousia* of concrete horses, based on which a secondary *ousia* of "horseness" arises, then an *ousia* of "four-footedness," etc. To this scheme of the Universe corresponds such a Method that is aimed at identifying, first, the primary *ousiai* and, following the ascending cascade, allows an understanding of the *ousiai* of higher orders. Although Aristotle himself prefers deductive argumentation, for which all his categories are designed [Pellegrin 1982; Falcon 1997], his ideas concerning ascending causalities of *ousiai* will later lay the foundations of the *inductive* (from particular to general) argumentation scheme.

It is important to keep in mind that Aristotle understands *ousia* functionally and teleologically, i.e., as a goal, a purpose of existence of a certain part of an animal or plant: for him, these parts exist for the sake of a work for which they are designed by their respective *ousiai* [Aristotle 2001]. Such a functional understanding of *ousiai* and how they are embodied in certain parts, organs, etc. of particular organisms will become one of the foundations of the natural method of scholastic systematics and some later taxonomic theories, up to the modern ones (such as biomorphics; see Section 5.5).

In all of Aristotle's natural philosophy, the *principles of perfection* and *plentitude* are very important [Lovejoy 1936; Balme 1987; French 1994; Gotthelf 2012]. They are manifested jointly in an idea of the *Ladder of Perfection*, in which each *ousia* occupies a certain "place," they all relate to each other in a consistent and strict degree of affinity, and there is no "gap" between them. According to the ascending cascade of causal relations among *ousiai*, the orderliness of the Ladder is also ascending. To comprehend how the Ladder is ordered and hence how all of Nature is organized, it is necessary to understand how specific *ousiai* relate to each other within the global Ladder; this yields a comprehension of natural interrelations between animals and plants embodying the respective *ousiai*. In the Middle Ages, this general idea will turn into an idea of the Great Chain of Being, or the Ladder of Nature (*Scala Naturae*) [Lovejoy 1936]; in natural science of the 17th–18th centuries, it will give birth to a conception of the Natural Order, which has a significant impact on the ideas of early post-scholastic systematics.

One may wonder what all this ancient natural philosophy has in common with modern systematics, in which the dominant idea is the construction of phylogenetic schemes based on calculating similarity by molecules. Oddly enough, it actually does have a very direct link. Indeed, why do some taxonomists believe that it is necessary to develop phylogenetic classifications, while others think of typological ones? Obviously, because the former are committed to evolutionary and the latter to typological natural philosophies. And how do phylogeneticists know that by building tree-like schemes, they actually get a certain representation of the real process of phylogenesis? Obviously, because their algorithms for tree inference reproduce in a way certain important aspects of the real processes of phylogenesis; this is how some modern methods of phylogenetic reconstructions are justified (see Section 5.7.1). Otherwise, what makes a phylogeneticist believe that a phylogenetic tree really reflects the phylogeny and is not just accidentally related to it? And if this is not hidden following the ancient natural philosophy with its natural-philosophical unity of Nature and Method, then what is it?

In the writings of Aristotle, the categories that are fundamental for the entire classification method of cognitive activity are developed in detail: *genus* (γένοσ), *species* (έίδοσ), *common* (συναγωγή), and *difference* (διαίρεσις) [Grene 1974; Pellegrin 1982, 1987; Wilkins 2009]. Thanks to their subsequent developments by neo-Platonism and scholastics, they will firmly enter the generic–species scheme for describing the world of living beings—and with it, the method of systematics.

When reconstructing the conceptual history of systematics, it is necessary to take into account that the meanings of categories of genus and species in Aristotle's writings are very polysemous and context-dependent. On the one hand, in his work "On the Parts of Animals," these categories relate basically to the essential properties of the objects [Lennox 1980], that is, they are partonomic, not taxonomic. According to this, for example, "species" is not the species in its current biological understanding as a specific group of organisms, but rather a particular quality (essence) of organisms of the respective natural group. Accordingly, the primary exploratory task is to identify the essences of organisms (a partonomic issue), while the delineation of their groups endowed with a common essence (a taxonomic issue) is secondary to it. So, it is more a method of definition rather than a method of classification, though it works well in both cases [Falcon 1997]. In such an understanding, the Aristotelian method will form the basis of the generic–species scheme of scholastics, and then it will be incorporated in the natural method of typological systematics. On the other hand, the same two categories correspond to the groups of organisms of different levels of generalities, just as in contemporary taxonomic classifications. As a result, it is important to realize that Aristotle, in his "Categories," separates terminologically two understandings of such groups: he uses the terms γένοσ and έίδοσ to designate them as the "formal" classification units, while the terms ομογενείσ and ομοφιλείσ designate them as the "natural" units [Wilkins 2009]. For him, the latter units are not a result of logical division, but rather real groups, each with its own secondary *ousia* and with its own "genesis" or "phylesis" (γενείσ, φιλείσ) [Grene 1974; Pratt

1984; French 1994]. It is curious enough that, because of the scholastic legacy, contemporary taxonomists will denote the respective units uniformly as "genera" and
"species" regardless of their naturalness, while Aristotle's notions of "homogenes"
and "homophyles" will be forgotten.

In both interpretations, Aristotelian genera and species, even in their "taxonomic"
connotation, do not have fixed ranks and precise biological meaning. For example,
in his "On The Parts of Animals," Aristotle recognizes the "genus" of animals with
blood, divides it into "species" of viviparous and oviparous beasts, and recognizes
among the latter the "lower species" of crocodiles and snakes [Aristotle 2001]. At
the same time, he designates the "genera" of oviparous and viviparous reptiles in his
"The History of Animals" [Aristotle 1910]. Thus, Aristotle can hardly be considered a
forerunner of something like the contemporary biologically sound concept of species,
which role he is often assigned in many textbooks on the history of biology and
systematics.

Aristotle's position regarding the classification procedure is also highly ambiguous;
in contrast to widespread opinion, the logical *principle of single basis of division* does
not look as strict to him as it will be later developed by scholastics [Falcon 1997].
In his works on natural history, he emphasizes that the main task is distinguishing
between natural groups (genera), rather than formally separating them by dichotomous division by a single feature; and since genera are defined by several essential
features, it is more correct to divide them according to several characters [Aristotle
2001]. So, from the contemporary (presentist) viewpoint, Aristotle's method at this
point looks more natural than that of scholastic systematics of the 16th century.

The closest followers of Aristotle develop the emerging natural history mainly
within the framework of the "collectionist" program. Of these, his direct disciple Theophrastus is to be mentioned for his famous complete botanical encyclopedia, "Enquiry into Plants." It contains a clear understanding that Nature acts
following its own design, and not in order to be useful to humans. Accordingly,
Theophrastus classifies plants based on the totality of their features (in their essential understanding), which together provide an integral and clear appearance of the
entire plant [Theophrastus 1916]. This natural "basis of division" will be forgotten
by herbalists and scholars of the 14th–18th centuries and will later be recalled by
theoreticians of natural systematics by the end of the 18th century. Following his
main idea, Theophrastus divides all plants into four "main species," which are trees,
shrubs, perennial shrubs, and grasses. This division is quite consistent with the main
life forms of plants already distinguished by folk systematics; it will be substantiated
by A. von Humboldt at the beginning of the 19th century as one of the important (in
long-term assessment) developments of natural systematics [Humboldt 1806]. One
peculiarity of Theophrastus' understanding of "species" arising from Aristotelian
metaphysics is noteworthy: he believes it is possible for plants to shift from one
"species" to another if they are modified after their living conditions change. This is
because the "matter" becomes clothed in different "forms," i.e., it becomes different
"species" under different circumstances [Zirkle 1959]. A similar idea of transmutation of organisms from one species to another will be rather common in Medieval
natural philosophy [Amundson 2005]; something similar can be found in the 19th
century in Edward Cope's classification theory [Cope 1887].

Neo-Platonism is the next important stage in the formation of the Method, which will form the basis of future systematics. Late Antique neo-Platonist philosophers are engaged in developing and detailing ideas of Plato and Aristotle, bringing them to a state of the "method" in its narrow sense, i.e., as a specific algorithm. They borrow deductive method and fundamental classification categories from their early antique predecessors [Makovelsky 2004]. The neo-Platonists' lengthy comments on the works of Aristotle become the main link between his teachings and early (scholastic) systematics [Sorabji 1990; Wilkins 2009; Richards 2016]. One of the central figures here is Porphyry, best known for his "Introduction to Aristotle's Categories," which summarizes Aristotle's method; another key person is Boethius, with his "Commentaries on Porphyry" [Boethius 1906].

The main contribution of neo-Platonism to the foundations of early systematics is a detailed elaboration of the deductive *generic–species scheme* as a specific classification algorithm based on a sequential division of logical "genera" into logical "species" of different levels of generality. The basis of this scheme is a universal Aristotle's formula of the definition of any notion (except for the most general) by indicating its "generic common and species particular" (*per genus proximum et differentiam specificam*) [Pellegrin 1982]. This formula makes the logical notions of "genus" and "species" fundamental for any deductive argumentation schemes, with the latter's famous illustration being a classification tree-like graphical scheme, the so-called "*tree of Porphyry.*" This scheme will become the basic method of scholastic systematics, so its key notions will become basic elements of the taxonomic thesaurus, and the "tree of Porphyry" will appear in most of the works of the 16th–18th centuries.

For neo-Platonists, who are mostly concerned about the logical consistency of the genus–species scheme, the notions of genus and species lose their Aristotelian multifaceted content. They appear to be reduced to logical predicates (judgments) as elements of the classification scheme, so their "natural" understanding as primarily the elements of Nature appears almost washed out [Wilkins 2009; Richards 2016]. Their logical interpretation will then be strengthened by scholastics and from them will pass into scholastic systematics. Accordingly, the future anti-scholastic revolution will be largely associated with the rejection of the logical interpretation of these notions in favor of the biological one.

And nevertheless, neo-Platonic natural philosophy obliges its adherents to discuss seriously the problem of the ontological status of genera and species. In this regard, one of the most important questions becomes as follows: when using the generic–species scheme and distinguishing "genera" and "species," do we mean something real and existing in Nature outside of us and beside us, or are they only conditional and therefore arbitrary categories, behind which there is no natural background? In particular, neo-Platonists and after them scholasticists worry about species, which is expressed by Boëthius' aphorism "if we do not know what species is, nothing would secure us from delusion" (translated after [Boëthius 1906: 1.04.10]). In this respect, the contemporary discussants of the species problem are all "Boëthians."

Medieval scholasticism covers a very long period of the development of European cognitive culture: it follows the neo-Platonic phase of Antiquity and lasts more than a thousand years. It will be usually mentioned in manuals on the history of natural science as a "dark era"—in the sense that it generates a minimum of new knowledge about material things because of being engaged basically in the issues of their supra-material causes. The reason is that the transformation of Christianity into an official state religion, actively defending its right to own an ultimate truth, makes the Bible, as the "Book of Revelation," the main source of the latter. So it is the Bible that appears at the center of the attention of intellectuals of this time, while Nature as such is supplanted into the background; by analogy with the "Book of Revelation" it is now interpreted as just a "Book of Nature." According to this metaphor, introduced by one of the Church Fathers, ex-neo-Platonist Aurelius Augustine, the human does not investigate Nature; instead, he "reads" the "Book of Nature" and delves into its text guided by the meanings that the "Book of Revelation" carries [Gaydenko 1980; Lyubarsky 2015]. This view of Nature gives rise to a specific *semiotic* cognitive model; on its basis, at the turn of the 16th–17th centuries, Galileo Galilei, one of the founders of European science, will formulate his aphorism "The Book of Nature is written in the language of mathematics," which will become a kind of symbol of all contemporary mathematized natural science. Its notable manifestation in systematics of the 19th century will become the following aphorism: "Species are the letters whereby the Book of Nature must be read" [McOuat 1996: 473].

All this leads to a loss of interest in the "collectionist" program (gathering facts about phenomena), but gives rise to a powerful intellectual work within the framework of the "methodical" program (substantiating judgments about phenomena). Due to this, Medieval scholasticism plays a very important role in the formation of the *analytical* cognitive model of modern science. The basic problems it tries to solve, among most closely related to the subject of this book, were originally defined by neo-Platonists: (a) do genera and species exist in reality (objectively) or only in thought (subjectively)? (b) if they really exist, are they material or insubstantial? (c) if they are immaterial, do they exist outside of things or in the things themselves? These questions relate to the most fundamental issues of ontology and epistemology, in which contexts the whole of natural science, including systematics, will be developing [Gaydenko 1980; Gaydenko and Smirnov 1989]. These categories are realism, nominalism, conceptualism, and rationalism; they are briefly considered in the theoretical part of this book (see Section 3.3.3).

For the formation of natural-philosophical foundations of early systematics, an idea of the levels of the hierarchical structure of natural and super natural worlds, developed by scholastics, is of great importance. In the teachings of Aurelius Augustine and Anselm of Canterbury, the Platonic *eide* and the Aristotelian "genera" and "species," embodied in the biblical context, become hierarchically arranged Divine *prototypes* of the material world contained in the Divine plan of creation [Amundson 1998; Sokolov 2001]. Supplemented with Anselm's concept of the *principal essence* (*essentia principalis*) [Holopainen 1996], it will become an important part of the substantiation of the natural method of scholastic systematics.

The generic–species classification scheme, a key for rationalization of the future systematic, after its reworking by scholastics, acquires the following principal features. It is construed as a sequential deductive division of notions by which "genera" of different levels of generality are distinguished, with "species" being designated at its final step. It begins with the highest "genus" (*genus summum*), proceeds with recognition of intermediate "genera" (*genera intermedia*); the last of the latter (*genus proximum*) is divided into "species" (*species infima*). At each step, this division follows the *principle of dichotomy*, according to which all "genera" are divided strictly into two "genera" of the next lower rank (and eventually into two "species"). All these "genera" are primarily logical, their levels of generality are not fixed, so this scheme yields rankless generic classification. Sequential division is carried out on a certain *basis of division* (*fundamentum divisionis*). To the extent that "genera" and "species" should reflect certain natural phenomena, the basis of division is to be shaped by an *essential character* (*essentia*), while non-essential characters (*accidentiae*) should not be taken into account. The logical consistency of the entire generic–species scheme is ensured by the *principle of single basis of division*: it prohibits the use of different characters in the given scheme if they yield different divisions, since (according to the *principle of the excluded third*) only one of its possible outcomes (hierarchies of "genera") can be true. Accordingly, of prime importance is the selection of a "true" essential character at the very first step of division and, once selected, it should determine the entire sequence of division steps. With such an embodiment of the generic–species divisive algorithm, a hierarchy of essences of different levels is first revealed (partonomic division), and then, on this basis, classes of objects endowed with these essences are distinguished (taxonomic division). Thus, in modern terms, this scholastic scheme presumes rather strict *taxon–character correspondence*, in which "character precedes taxon"; it will be accepted in scholastic systematics, typology, and biomorphics [Pavlinov 2018] (see Section 6.2).

Concluding this briefest review, once again the very significant contribution of general principles of onto-epistemology developed by scholasticism to the emergence of early systematics should be emphasized. Adoption of the generic–species scheme of division of notions ("genera") endows natural history with the *classification* cognitive program which will define the principal way of comprehending Nature as its classifying [Stafleu 1971; Rozova 1986; Wilkins and Ebach 2014]. Considered from this program, the general trend in the conceptual history of systematics over the next several centuries will be shaped by a combination of two different vectors [Pavlinov 2018]. One is set by an adaptation of systematics to scholastic rationality, whereas the other is set by an adaptation of this rationality to the complexity of plant and animal diversity. At first, biological systematics will be developing following the formal classification principles (early scholastic systematics), and later by overcoming them (late scholastic and nearly all post-scholastic systematics).

2.2.3 THE HERBAL EPOCH

The Renaissance, unlike the Medieval age with its focus on the rational interpretation of the Book of Genesis, shifts from God to man: among the revived attributes of Antiquity, an important place is occupied by Protagoras' aphorism "Man is the measure of all things." With this, Nature becomes perceived partly in a mystical and partly in a utilitarian manner, rather than in a rational cognitive one. For proto-systematics, this turn means the growth of an interest in plants and animals not so much as a field of application of classification techniques but as a means of satisfying various human needs requesting more tangible knowledge of their properties and uses [Ogilvie 2006; Pozdnyakov 2015]. This causes a revival of the Late Ancient "collectionist" program laid down by the above-mentioned works of Theophrastus, "Natural History" by Pliny the Elder, and especially "De Materia Medica" by Pedanius Dioscorides. During the 12th–15th centuries, these works are reproduced many times and serve as a model for new manuals.

Such interest in plants and animals gives rise to the *Herbal epoch*, the main concern of which is to gather and systematize various data about them and how they are used [Sachs 1906; Arber 1938; Larson 1971; Kupriyanov 2005; Pavlinov 2018]. This results in specific incunables and books called *herbaria*, hence the name of this epoch. Plants become their main object due to their being an important source of a variety of medicines; "bestiaries" dealing with animals are few and replete with descriptions of curiosities [Gould 1886; Ivanova-Kazas 2004]. This focus on botanical objects will continue by and large in scholastic systematics.

> It should be noted that herbals are considered here as belonging to the field of botany, according to the established natural science tradition. This reservation is necessary because there is another historiography focusing not on cognitive but on the utilitarian aspirations of the herbalists, so it considers herbals not botanical but rather medical references. Thus understood, herbals have a very long documented history dating back to the third millennium BC [Arber 1938].

European Renaissance herbalistics, marked by "science-centered" historians as an important stage in the development of botanical science [Sachs 1906; Arber 1938], is distinguished within the general herbal tradition mainly by its manner of presenting information about plants. In the most advanced herbals, plants (and occasionally animals) are often ordered in hierarchical systems in which categories are denoted as "genera," "species," and occasionally by other terms [Anderson 1977; Perfetti 2000; Pavlinov 2015, 2018]. Due to this, disparate information about different organisms is combined into certain natural history compendia, so herbalistics is a forerunner to scientific systematics; from a more general perspective, this epoch can be designated, with certain reservations, as the beginning of "natural science" [Foucault 1970; Pratt 1982; Lyubarsky 2015; Pozdnyakov 2015].

For the Herbal epoch, the following features seem to be the most significant to justify its characterization as part of proto-systematics [Ogilvie 2006; Pavlinov and Lyubarsky 2011; Pavlinov 2018]. Since herbalistics grew up out of the patristic tradition of biblical scholars, the authors of herbals refer to the ancient authorities with whom they agree in their observations on plants and animals [Greene 1909]. This lays the foundation for the practice of references to sources (manuscripts, books, letters), which is so characteristic of contemporary science. Thanks to the Great Geographical Discoveries, knowledge of plants and animals becomes significantly expanded; this prompts the compilation of herbal catalogs in a form of rather advanced classification. This stimulates improvement of the means of describing and naming organisms; hence, there are other more important features. On the one hand, the very form of presentation of materials gradually becomes more complicated by employing the hierarchical generic–species scheme borrowed from scholasticism. At the same time, herbalists make first attempts to fix and designate taxonomic categories in a different way than scholastics: these are *Order* in Gesner, *Classis* in l'Ecluse, and *Section* in Bauhin, and they will be adopted subsequently by scientific systematics. In those herbals which are inclined towards the "methodical" cognitive program, typical "trees of Porphyry" appear in the second half of the 16th century [Pavlinov 2018], i.e., a hundred years before those works of the 17th century (Rivinus, Morison, Ray), which are usually given priority [Voss 1952].

The use of predominantly Latin names of organisms in the herbals, instilled by the university education of their authors, will become an important part of the professional language of systematics. It is very important that, probably under the influence of scholasticism, an attitude to "test" Nature becomes more rational and "mechanical." Accordingly, organisms turn into analytically studied objects torn from their natural environment and described according to certain fixed rules [Lyubarsky 2015]. In particular, herbalists of the late period customarily base their descriptions on observations of dead derivatives of once-living beings: in this way, the "collectionist" program of herbalistics stimulates the initial development of herbaria and museum collections, which will subsequently become the factual basis of scientific systematics [Ritterbush 1969; Pavlinov 2016].

The Herbal epoch covers the period from the 13th to the 16th centuries, i.e., about 400 years in total. Its beginning falls in the late Medieval age, its main development coincides with the Renaissance, and it ends at the very beginning of the New Age. Following the development of the ways in which organisms are described in herbals, this epoch can be divided into three main stages [Pavlinov 2018].

The *first stage* (13th–14th centuries) belongs to the Medieval to Renaissance transition; it deals mainly with rewriting, translation, and commentaries on the ancient authors. These works first appear in Italy, so the history of herbalistics, like the entire Renaissance, begins there [Janick 2003].

The *second stage* (15th to early 16th centuries) is associated with the beginning of book printing in Europe, which, thanks to new technical capabilities, stimulates the rapid development of printed herbals. It not only simplifies their production but also partly changes their structure: multilevel sections, indexes, and cross-references appear in them, making it easier to compare different sources in search of particular plants and authors. Among the herbalists, the most active and well-known figures of this stage are the "German fathers of botany" [Sprengel 1808; Sachs 1906]. Of them, Otto Brunfels and Hieronymus Bock (Tragus) make the most noticeable contribution to the initial development of herbalistics towards future systematics [Greene 1909; Larson 1971; Ogilvie 2006]. Thus, O. Brunfels in his three-volume "Images of living plants..." widely uses the notions of "genus" and "species" in their general meaning to designate botanical categories [Brunfels 1530]. H. Bock in his "New book of plants..." rejects the alphabetical arrangement of plants as preventing understanding of their natural order. He emphasizes in the Preface: "in describing things, I strive to get as close as possible to how the plants themselves, apparently, were united by Nature by similarity in their form" [Bock 1546: xiv]. Very significant is the contribution of Tragus to the initial development of organography, i.e., the anatomical nomenclature of plants: he identifies and designates more than a hundred elements of the plant body, in the first attempt to introduce rather a strict descriptive terminology in botany. Thanks to this work, analytical "construction" of the botanical object begins: a plant appears as a combination of explicitly identified and named parts of an organism [Arber 1938, 1950; Pavlinov and Lyubarsky 2011; Lyubarsky 2015, 2018].

The *third stage* (the mid to late 16th century) can be characterized (with reservations) as more scientific rather than purely applied; one of the principal figures is Konrad Gesner [Fisher 1966; Gmelig-Nijboer 1977; Ogilvie 2006]. His early botanical works are in the herbal tradition, but his encyclopedia "History of Animals..." (*Historiae animalium*, 1551–1587) differs significantly from them in its structure. The latter pretends to reflect something like the natural system of the animal kingdom: it is organized into five books largely corresponding to the main Aristotelian "higher genera" and partly coinciding with the division of plants and animals according to the biblical days of their creation [Gmelig-Nijboer 1977]. In a separately published "List of aquatic animals..." [Gesner 1560], the chapters are called "Orders": for example, "Ordo III. De Pisciculus," "Ordo XII. De Cetis": this is how this important taxonomic category first appears in the biological literature. A particularly noteworthy detail of Gesner's work as a systematist is that in this "List...," classification of the Order Mollibus is illustrated by the "tree of Porphyry": as noted above, this is probably the first instance of this type of presentation of a taxonomic system for living beings.

The entire Herbal epoch is finalized by the works of Gaspar Bauhin: their general style is transitional between the herbals and early systematical books [Arber 1938, 1950; Larson 1971; Kupriyanov 2005; Pavlinov 2018]. Considered from the conceptual history perspective, the most significant among these works is the "Illustrated exposition of plants...," whose priorities are underlined by its subtitle: "Methodical description of plants according to their genera and species [...] providing correct names and differences" [Bauhin 1623]. All this work is organized systematically: it is divided into "books," which are further divided into "sections"; they can be correlated by their ranks with later orders or families. Categories of lower ranks (designated in fonts without specific terms) do not have any fixed meaning, therefore Arthur Cain, in his analysis of this work, will prefer to denote them not with standard but with modified terms: not "genus" but *generoid* ("as if genus"), not species but *speciate* ("as if species") [Cain 1994]. The main content of the "Illustrated exposition..." is quite rational: unlike most herbals, it does not include illustrations with explanations, but gives brief diagnostic descriptions of plants based on herbarium specimens. And yet G. Bauhin, like most of his predecessors and contemporaries, is mostly a "collector" and not an advanced "methodist." Therefore, he hardly deserves the laurels of a forerunner of the future "Linnaean reform," which will be sometimes awarded to him [Sachs 1906; Kamelin 2004]; actually, his work is but a hint of the coming rationalization of systematics [Whewell 1847; Bartlett 1940; Larson 1971; Cain 1994; Pavlinov 2018].

2.3 THE BEGINNING OF SYSTEMATICS: SCHOLASTIC REVOLUTION

The beginning of the formation of scientific systematics coincides with the beginning of modern natural science, which marks (together with the technological revolution) the transition from the Renaissance to the New Age at the turn of the 16th and 17th centuries. It inherits basic ideas from Antiquity: (a) the all-unity of Nature driven by a universal law of the Universe; (b) the unity of Nature and Method; and (c) the rational nature of the Method as a logical tool. The key innovation becomes an understanding of Nature as a rationally arranged "mechanism," which should be investigated by an analytical method reducing everything complex to simple components [Foucault 1970; Gaydenko 1987, 2003; Pozdnyakov 2015]. The well-developed foundations of this new rationality are attributed mainly to the natural philosopher and mathematician René Descartes, so it is usually called *Cartesian*. Direct references to Descartes can be found in the key works completing this stage of the conceptual history of systematics, viz. in the "Elements of Botany…" by Tournefort and in the "Philosophy of Botany…" by Linnaeus.

Since rationality presupposes a direct proving, the development of science from the very beginning is associated mainly with the mathematical method of describing phenomena of "mechanical" Nature. In this regard, the rationalist philosopher Francis Bacon divides the entire corpus of natural sciences into analytical *natural philosophy* (in a strict sense) and descriptive *natural history*. In the former, which includes "exact sciences," the method is based on a quantitative mathematical apparatus linking quantitative variables in a single formula; this will become an ideal of science (*scientia*) and a measure of scientificity of research approaches and their results. In contrast, the main method of natural history is a qualitative generic–species scheme that links different groups to each other in a hierarchical classification based on similarities in qualitative characters. However, a traditional opposition of these two methodologies is not entirely correct: the development of sufficiently rigorous analytical tools for a qualitative (categorical) description of the diversity of objects makes the classification method of natural history to a certain extent comparable with the mathematical apparatus of natural philosophy [Lesch 1990].

2.3.1 Major Features

The key ideas of the rationality that provides a powerful stimulus to the initial development of systematics at the end of the 16th century can be summarized as follows.

At an ontological level, an idea of the unity of rationally arranged Nature receives a symbolic incarnation of the *System of Nature* [Foucault 1970; Pratt 1985; Lesch 1990; Lefevre 2001]. This is illustrated by a short essay under the iconic title "A new system of nature…" issued by the philosopher and mathematician Gottfried Leibniz at the end of the 17th century [Leibniz 1900]. This natural-philosophical idea of a rationally arranged System of Nature (*Systema Naturae*) is embodied in systematics by two basic concepts: the hierarchical *Natural system* (*Systema naturalis*) and the linear *Natural order* (*Order naturalis*); the former dominates in the period under consideration. An understanding of Nature as a specific "mechanism" consisting of interacting "details" inevitably entails the same understanding of the organism. It means the latter's "decontextualization," i.e. its "pulling" out of its natural environment and its representation as a set of anatomical parts using formalized diagnostic characters [Slaughter 1982; Lyubarsky 2015]. With this, under the influence of Aristotelian ousiology, organismal inner properties become thought of as essences expressed by some of these characters. Going back to Antiquity, an idea of the unity of Nature and the Method yields the concept of the *Natural method* (*Metodus naturalis*), which allows the System of Nature to be revealed and leads to the development of the *Natural classification* (*Classificatio naturalis*). Because of the natural-philosophical idea of the unity of the Natural system and the Natural method, they become "the same" in a sense in the minds of early taxonomists [Larson 1971].

Thus, systematics begins to take shape at the end of the 16th century based on the development of the rational natural method which includes a fairly formalized algorithm for classifying organisms by their essential properties. In a general sense, the method of early systematics is defined as a means of "combining similar and separating different [things] and arranging them by genera and species according to how the nature of things indicates" [Cesalpino 1583: 28]. From a contemporary standpoint, this idea looks rather banal and quite empirical; however, from the previous herbalistics perspective, it is quite non-trivial: actually, it denotes a radical shift from classifying organisms according to their essences instead of their "uses."

This methodological accent becomes the "key innovation" in the initial formation of systematics, which separates it from the "collectionist" tradition of herbalists [Slaughter 1982; Kupriyanov 2005; Pavlinov 2018]. This leads to the formation of the first research program in systematics and marks the first scientific revolution in its conceptual history, making it quite an advanced natural science discipline for its time. Its method allows the elaboration of rationally organized classifications designed to represent the Natural System in its essentialist understanding. Since the classification method in question is borrowed from scholasticism, both early systematics and its research program and the respective scientific revolution should be synonymously called *scholastic* [Pavlinov and Lyubarsky 2011; Pavlinov 2018, 2019]. From an actualism standpoint, there is nothing derogatory in such a designation: this program fully corresponds to the standard process of development of the entire natural history of this epoch. Therefore, a widely accepted presentism-guided consideration of this important initial stage in the conceptual history of systematics as a "stagnation" and to postpone the beginning of this discipline to the second half of the 18th century [Sachs 1906; Hull 1965; Mayr 1969; Uranov 1979], seems to be largely incorrect.

The "genera" and "species" of different levels of generality, distinguished by the classification method, remain initially logical predicates; their biological content begins to manifest itself at a later stage. As far as possible, the *principle of single basis of division* is fulfilled by focusing on the above-mentioned *principal essence*: the latter manifests itself in a certain single character, which forms the sought basis of division, other characters being discarded. Thus, the fundamental *principle of differential weighting of characters*, which in this case means their treating as essential (*essentia*) and accidental (*accidentia*), is introduced in systematics. The resulting hierarchical classifications are illustrated by the "trees of Porphyry," in which "genera" and "species" are designated by the respective diagnostic characters.[1] The latter's

---

[1]   It is worthy of note that the influential phytographer Robert Morison, in his "New arrangement of umbellifers...," along with a lot of "trees of Porphyry," provides (probably for the first time in natural history) several tree-like schemes indicating a kind of kinship relation between genera and species and thus antedating future genealogical trees [Morison 1672].

descriptive designations become verbose taxonomic names—a very characteristic feature of scholastic systematics of the 16th to 18th centuries.

Starting with the development of the classical genus–species scheme, scholastic systematics begins to recast it intentionally to make it more consistent with the real diversity of organisms and at the same time rationalize the natural method in its own way. A very important innovation becomes the replacement of the (quasi)continuous rankless hierarchy of this scheme with one with a limited number of fixed ranks. By its meaning and significance, the introduction of a ranking scale in classifications can be compared to an introduction of the Cartesian coordinate system into descriptions of the imaginary spatial relations between objects [Lyubarskiy 2018]. It is note-worthy that this innovation designates a rationalization of early systematics in a direction opposite to the natural-philosophical one, where the development of the analytical method is associated with its continualization (differential calculus, etc.). The general idea of the ranked organization of the System of Nature is seemingly borrowed from that of society and especially the army, where this form of rational-ization is pushed to the limit (this analogy is mentioned by Linnaeus). The taxonomic ranks are partly borrowed from herbalistics and partly from other sources: these are *Section* (Magnol, Tournefort), *Order* (Rivinus, Linnaeus), *Familia* (Magnol), *Classis* (Tournefort, Linnaeus), and *Regnum* (Linnaeus); their standardization is established by Tournefort and Linnaeus in the early to mid 18th century. The introduction of these ranks abolishes the "higher" and "intermediate" genera of scholasticism, while the "closest" genus becomes the only genus that is characteristic of contemporary systematics; a fixed rank is finally also assigned to species. Thus, "invention" of the ranked hierarchy becomes a significant new step in the development of the natural method of systematics, breaking its connection to previous scholastic tradition [Pratt 1985; Lyubarsky 2018; Pavlinov 2018].

Although the system of fixed ranks developed by later scholastic systematicians is undoubtedly a departure from scholasticism, it undoubtedly follows from it, so it is predominantly logical [Larson 1971; Pavlinov 2018]; this is unequivocally pointed out by Carl Linnaeus in several of his works [Linnaeus 1751, 1766]. A close correl-ation between an old (scholastic) and a new (systematic) hierarchy is visible from the following direct correspondence

Genus summum = Classis
Genus intermedium = Sectio/Order/Familia
Genus proximum = Genus
Species infima = Species.

This evident scholastic background of the originally fixed categories, later called "Linnaean," means that attributing to them some special ontic status [Ereshefsky 1997, 2001a; Shatalkin 2012] has scarcely any serious natural justification.

In the hierarchical method of scholastic systematics, the notion of genus becomes of key importance. Such a position is predetermined by a principal character of the scholastic generic–species scheme, in which genera are fundamental, while species just designate the last steps of the generic sequential divisions. The first to express this attitude is A. Cesalpino, asserting (quite in the style of Boëthius) that "if genera are confused, everything will inevitably be confused" [Cesalpino 1583: 4]. He is echoed by one of the luminaries of the scholastic systematics, J. Ray, who in his work "On Plant Variations…" emphasizes that "to be one in nature and to be one in genus are the same" [Ray 1696: 13]. Thus, this first stage in the conceptual history of scientific systematics can be regarded primarily as a "classification of genera" [Cain 1959a], which will be partly retained by early post-scholastic systematics.[2]

In connection with the fixation of ranks, an old problem of realistic vs. nominal-istic interpretation of the categories of "genus" and "species" becomes of particular importance; not only naturalists but also prominent philosophers such as Locke and Leibniz become involved in the discussion around it [Larson 1971; Goodin 1999; Jones 2006]. This problem appears especially acute in the second half of the 18th cen-tury because of opposition of the supporters of two natural-philosophical concepts, the discrete hierarchical System of Nature and the continuous linear Ladder of Nature; the former presumes and the latter denies discrete categories in Nature. In the sub-stantiating reality of these categories, a reference to their Divine creation is usually thought of as a kind of crucial argument, which seems quite natural under the strong impact of biblical mythology on the religious minds of systematicians of this time [Wilkins 2009; Richards 2010, 2016]. However, it is to be noted that, in this argumen-tation, the focus is not on genera, but on species: at any rate, none other than J. Ray, an author of the pamphlet "The Wisdom of God…" [Ray 1714], in his "History of Plants…" says that "God interrupted on the sixth day His great work, the creation of new species" [Ray 1686: 40]; apparently, there is no talk of genera here, as if the Creator did not care about them.[3]

With the development of scholastic systematics, a noticeable element of nov-elty is introduced in the interpretation of the single basis of division, in which two ways of considering this basis loom, first ontic and then epistemic [Winsor 2003; Pavlinov 2018]. The first is present in the works of early taxonomists in two slightly different versions. In one of them (from Cesalpino to Rivinus and Morison, "early" Ray), a single principal essence appears as such a basis, manifested in a certain single part of the organism. In botanical classifications, either flowers or fruits are most often considered in this capacity, so scholastic phytographers are divided into either "corollists" or "fructists." In another version ("late" Ray, Magnol), the essence is understood as multiple, i.e., manifested in a combination of several vital structures. An epistemic understanding of the classifying characters is rather typical of late scho-lastic systematics (Tournefort, Linnaeus): they are selected as the basis of division according to their importance not for organisms themselves, but rather for classifiers according to their diagnostic values.

---

[2] This is illustrated by a fundamental work of one of its founders, A.L. de Jussieu, which is entitled, in the Linnaean style, "Genera plantarum…" [Jussieu 1789].

[3] Here, it should be borne in mind that in the English-language editions of the Bible, creations are designated as kinds which may correspond to both the genus and the species of naturalists [Wayne 2000].

Despite an obvious accentuation of early scientific systematics on "methodism," the "collectionist" program also founds a specific and quite important place in its development. It involves an active formation of the botanical herbaria and zoological collections begun in the Herbal epoch [Impey and MacGregor 2001; Gritskevich 2004] and leads to the emergence of professional collection activity and the transformation of systematics into a "museum science." Surely, such a literal implementation of the "collectionist" program is part of the growing rationalization of the scientific systematics. On the one hand, the aforementioned alienation of organisms from their natural habitats turns them into exhibits, whole or torn apart, which are placed on sheets, and in boxes and cabinets. Such exhibits, for all their shortages, are closer to their living prototypes and reflect more accurately their anatomical features as compared to the artistic illustrations in books replete with factual errors [Nickelsen 2006]. On the other hand, these systematic collections and herbaria are thought of by researchers as visual embodiments of the arrangement of Nature, according to which they are placed in their respective order [Ogilvie 2006]. Thanks to this, a reliable factual basis for scientific systematics begins to take shape.

One of the characteristic features of scholastic systematics is the close connection between classifying and naming organisms [Pavlinov 2015]. Pitton de Tournefort writes that "knowing plants is tantamount to knowing their names [therefore] the study of plants should begin with their names" [Tournefort 1694: 1]; Linnaeus recognizes in systematics two "foundations" of equal importance, viz. classification ("disposition") and naming [Linnaeus 1736, 1751]. Taxonomic names are interpreted in a strictly essentialist way: "the idea of a trait that essentially distinguishes some plants from others must be invariably associated with the name of each plant" [Tournefort 1694: 2]; according to Linnaeus, "the specific name is the essential definition" [Linnaeus 1751: § 257].

2.3.2 Major Stages

The conceptual history of scholastic systematics begins at the end of the 16th century and reaches a conclusion at the end of the 18th century. Based on the changes in its natural method characterized above, the development of this research program can be quite clearly divided into early (16th–17th centuries, from Cesalpino to Ray) and mature (the 18th century, from Tournefort to Linnaeus) phases. The former retains the classic rankless hierarchical system combined with the ontic (essential) interpretation of the basis of division. The latter, culminating in the first phase of "Linnaean reform," builds the taxonomic hierarchy in the form of several fixed ranks; the basis of division is interpreted mainly epistemically.

The starting point of the scholastic stage in the development of systematics is identified quite clearly: it is the work of the Aristotelian philosopher and naturalist Andrea Cesalpino "16 Books on Plants" [Cesalpino 1583]. In the conceptual history of systematics, he appears as a "methodist" reformer purposefully introducing

Aristotelian cognitive categories into the foundations of the rational method of classifying plants [Whewell 1847; Bremekamp 1953; Larson 1971; Pratt 1982; Pavlinov and Lyubarsky 2011; Lyubarsky 2015, 2018; Pavlinov 2018]. For this, Cesalpino was awarded by Linnaeus the honorary title of "the first true systematician" (*Primus verus systematicus*) in the "Philosophy of Botany..." (Linnaeus 1751, § 54]; Vernon Pratt [Pratt 1982] calls him "the herald of the new century," while Georgy Lyubarsky honors him as the "Galileo of biology" [Lyubarsky 2015]. Accordingly, the scholastic revolution may be personalized as *Cesalpinean*, and the first of the two stages in the conceptual history of scholastic systematics deserves being called the "*Cesalpinean epoch*" [Whewell 1847].

Cesalpino's classification method is mainly based on an analysis of the principal essence. For him, the latter is identical to the "soul" of plants (with a slightly different meaning as compared to the Aristotelian version) and is manifested in their fruits: it is through them that some organisms give rise to others, providing inseparable chains of plant beings. Obeying the emerging general trend of mechanistic understanding of Nature, Cesalpino appears the first among systematicians to declare that descriptions of anatomical structures should be based on their "number, position, shape" (*numerus, situs, figura*; see [Cesalpino 1583: 29]); "Cesalpino's formula" will be repeated later by Jung and Linnaeus. At the same time, his natural method is not entirely formal: like Aristotle, he declares rigid logical rules for deductive classification but distinguishes the main groups of plants on the basis of a conjoint analysis of their various features [Stafleu 1969; Sloan 1972]. Therefore, Cesalpino, following Theophrastus and the most advanced herbalists, begins with the recognition of the most "obvious" plant life forms (herbs, trees, etc.) and then divides them based on the characters of fructification alone.

Cesalpino's scholasticism is manifested in the fact that the natural method itself interests him much more than a result of its application, viz. the natural system of plants; therefore, his work lacks any explicit botanical classification in the form of a list of "genera" and "species." In this respect, his "16 books on plants" falls out of the herbal tradition that still continues to dominate, and, in a sense, turns out to be premature. Because of this, adherents of the empirical history of botanical systematics believe that, despite considerable intellectual efforts, Cesalpino's contribution to its development is minute [Sachs 1906; Miall 1912]; of course, such an assessment contradicts the predominating trend of the conceptual history of this discipline.

Attention to the rational foundations of the emerging scientific systematics at the turn of the 16th and 17th centuries is an evident imperative of the times. This is evidenced by the fact that, almost simultaneously with Cesalpino and, apparently, independently of him, a similar logical method of plant classification is developed by Adam Zaluzhiansky in his treatise "Three Books on the Method ..." [Zaluziansky 1592]. He applies a strictly dichotomous deductive division, which gives a detailed hierarchical classification of plants in the form of identification keys. The latter are illustrated by the "trees of Porphyry," in which the corresponding diagnostic features are indicated for all division steps.

A very significant contribution to early systematics is made by the philosopher and mathematician Joachim Jung. The main content of his method, set out in his posthumously published book *Brief Botanical Physics...* [Jung (1662) 1747], is recognition of "genera and species on a consistent base [...] in accordance with the rules of logic." To this end, Jung undertakes the first (after H. Bock; see Section 2.2.3) serious and quite successful attempt to develop clear formulations of *differentiae* of the plant "genera," seemingly understood in a logical sense. Believing (following Galilei) that "the book of Nature is written in the language of numbers and geometrical figures," he describes in detail a rigorous method of recognizing individual parts of plants and reducing their diversity to several strongly fixed combinations that allow clear descriptions akin, to a degree, to mathematical formulas. With reference to Cesalpino, Jung discusses in sufficient detail the issue of distinguishing essential and accidental features of plants, referring to the first their inner characters (leaves, flowers, fruits, etc.), while to the second the outer ones (place and time of growth, meaning for humans, etc.). Although Jung, like Cesalpino, does not elaborate precise botanical classifications, his approach will significantly influence the further research of the botanical "methodists" of the 17th–18th centuries [Sachs 1906; Arber 1950].

Among the brightest and most famous "methodists" of scholastic systematics of the second half of the 17th to early 18th centuries is John Ray (initially Wray), mainly a botanist, to a lesser extent a zoologist, also interested in the ideas of natural theology and partly in the "philosophical language" of natural sciences [Greene 1909; Oliver 1913; Bryan 2005]. Ray starts as a herbalist, but quickly realizes that, within the herbal tradition, one cannot get a truly "philosophical" system of plants reflecting truly their nature. In the 1660s, he gets to know the classification methods of Cesalpino and Jung and becomes their follower, quoting them abundantly in his books. Unlike his predecessors, Ray focuses not only on the method but equally on elaboration of the hierarchical classifications called by him the "Ariadne thread" [Ray 1686: Prefatio, p. 2]. Believing that species are "Divine creations," he considers them as certain "natural" units, and not just the final steps of logical division; by this, he seems to make the first important step towards the fixation of the ranked hierarchy at its lower level. At the same time, his general approach to plant classification is quite scholastic. Indeed, Ray says that "the complete definition [of species] consists of the closest genus and essential difference" [Ray 1733: 6], and denotes all supraspecific categories as "genera," hierarchizing them as "primary" (*Genus primum*), "secondary" (*Genus secundum*), and "tertiary" (*Genus tertium*). His classifications are presented consistently in the form of "trees of Porphyry" with diagnostic characters indicated at their branching nodes.

It should be noted that, in Ray's method, both Aristotelian understandings of species are present, i.e., as a group of organisms and as its distinguishing essence. In particular, in the above-cited work "On Plant Variations...," Ray, in the style of Aristotle's "On the Parts of Animals," writes about the species of seeds, species of parts of a flower, etc. [Ray 1696]. From this one may conclude that the plant and animal species recognized by Ray are largely Aristotelian "forms," and not groups of organisms in their modern biological understanding

[Cain 1999]. However, when considering such groups, Ray emphasizes, referring to the ancient formula "the generation of like by like," that individuals of
one species always arise from the seeds of the same species, but never from the
seeds of other species [Ray 1686]. On this basis, he is often considered a predecessor (at least within the framework of the New Time) of the contemporary
generative concept of species [Skvortsov 1967; Wilkins 2009].

Developing the natural method of scholastic systematics, Ray enriches it with rather
deep empirical natural philosophy, which is largely borrowed from J. Locke with his
idea of separating real and nominal essences [Sloan 1972; Pratt 1982; Slaughter 1982;
Cain 1999]. This influence is clearly seen from the following passage: "The essence
of things is not known to us. Since all our knowledge comes from sensations, we do
not know anything about things that exist outside of us, except for the effects that they
have on our perceptions [...] and reflection on them" [Ray 1696: 5]. Developing this
idea, Ray argues that:

> since from the same essences involve the same qualities, functions, and also
> other secondary features [of organisms], there can be no more reliable indicator
> of the essential, and thus generic affinity than agreement on many characters in
> common, that is, by similarity in many parts and properties.
>
> Ray 1733: 6

In this regard, in his later zoological classifications, Ray abandons the basis of
division corresponding to a single essence and compounds it from the observations
on different anatomical features of animals. A peculiar combination in this method of
the ideas of Aristotle and Locke points to Ray as actually the first herald of the future
natural systematics, operating with different categories of characters.

Another important departure from the purely scholastic method of classification
is made by the botanist Pierre Magnol, whose understanding of the natural method,
like that of Ray, is influenced by the philosophical ideas of J. Locke [Stearn 1961;
Alello 2003]. In his "Introduction to General History of Plants..." he follows Ray in
pointing out that the Natural System, being a natural affinity of organisms, can only
be comprehended by the study of several anatomical structures [Magnol 1689]. So
he rejects the idea of a sole principal essence, on which the natural arrangement
of plants should be based, by emphasizing that "all parts [of plants] which do not
serve the fruit are no more accidental than the arms and legs are accidental parts of
animals" [Magnol 1689: *Prefatio*]. For him, "there is a certain likeness and affinity
in many plants which does not rest upon parts taken separately but in their total
composition, which strikes the sense but cannot be expressed in words" [Magnol
1689: *Prefatio*]. Magnol becomes the first to lay down fixed suprageneric categories in systematics by introducing *Section* and *Family* in his system of the plant
kingdom. He borrows the first from the herbalists and does not attach much importance to it, while the second is introduced as the most important: his entire botanical
system is titled "The Arrangement of Families." The designation of this category is
intended to emphasize its natural rather than logical status: Magnol emphasizes a

close analogy between these fundamental units of the plant kingdom and the families of mankind [Sloan 1972; Alello 2003].

The mature phase in the conceptual history of scholastic systematics begins with Joseph Pitton de Tournefort, who is a Cartesian by his philosophical sympathies and an opponent of Ray and Magnol in his interpretation of the natural method of systematics [Becker et al. 1957; Larson 1971]. His main work is "Elements of Botany..." [Tournefort 1694], which in some important points marks a significant departure from the scholastic tradition in the direction that finished with "Linnaean reform" [Pavlinov 2018].

The first important innovation in the Tournefortian method is a clear structuring of the entire taxonomic hierarchy developed for the first time, though without special substantiation. He uses four fixed and terminologically designated major categories with as fixed subordination: *Class, Section, Genus,* and *Species*; as noted above, such a four-level ranked hierarchy corresponds to the main categories in the genus–species scheme of scholastics: *Genus summum, Genus intermedium, Genus proximum,* and *Species infima*. With this, genus and class are basic for Tournefort: he considers the principles of their recognition in greatest detail in the introductory section of his book, and their definitions are provided in the Dictionary (see below). He considers that it is "necessary to establish two kinds of genera, those of the first order and those of the second order": the first-order genera are distinguished by the structure of flowers and fruits, the second ones by less significant characteristics [Tournefort 1694: 30]. Accordingly, Tournefort is often called the "father of the generic concept" in botany [Bartlett 1940; Stuessy 2008]. Species for him are simply "plants of the same genus differing from each other by some peculiarity [...] besides the generic character" [Tournefort 1694: 13]; as to Sections, he hardly mentions them in the Introduction; and neither species nor section is mentioned in the Dictionary.

Another important innovation of this method is its evidently inductive character: Tournefort says that "To make a system of botany, it is not enough to know the characters of the genera of plants, it is also necessary to arrange these genera in certain classes, and then to arrange these classes in a simple and natural order" [Tournefort 1694: 40]. The same inductive character seems to occur in the way the main categories are defined in the Dictionary at the end of "Elements...": "A class of plants is the cluster of several genera of plants, all of which are suitable in that they have certain common marks which essentially distinguish them from all other genera of plants"; "A genus of plants is the cluster of several plants which have a common character established on the structure of certain parts, which essentially distinguishes these plants from all others" [Tournefort 1694: 525, 542]. As can be seen, here the classes are not divided into genera but represent clusters (*l'amas*) thereof. Classes are characterized by the same features of flowers and fruits as the genera they unite, so the principle of single basis is obviously followed, though it is not division but rather agglomeration. Strangely enough and unlike most of his colleagues, Tournefort does not use the "tree of Porphyry" to illustrate his system of plants.

Tournefort analyzes in detail the characters that should be used to classify plants, based on the belief that the author of nature [...] imprinted a common character on each of its species, which should serve as our guide to put them in their natural place. We cannot change these marks of distinction without deviating too evidently from the truth.

Tournefort 1694: 20

It is to be stressed that his choice of characters is based not so much on their importance for the execution of the vital functions of plants as their stability and reliability in distinguishing genera. So he says that:

There is no question about the goals of nature or about the exclusive qualities of the parts; rather, it is a matter of finding the means to differentiate the plants as clearly as possible; and if the smallest of their parts would be more suitable than those that are called the most outstanding, they should be given preference.

Tournefort 1694: 6

He examines in detail anatomical structures of plants that can be used for their classification and comes to a conclusion that "it is absolutely necessary to have regard only to the true relations which exist between them, that is to say, to the relationships drawn from the structure of their essential parts, which are the flower and the fruit" [Tournefort 1694: 40].

The most significant and influential figure in systematics of the middle and second half of the 18th century is the naturalist Carl Linnaeus (originally, Linné). His numerous works become the pinnacle of the scholastic stage in the conceptual history of this discipline and create prerequisites for the formation of some ideas of its early post-scholastic development. The great merit of Linnaeus is that he generalizes, partly supplements, and, most importantly, systematizes the previous developments of many of his predecessors in numerous works that become famous among naturalists of this epoch. His repeatedly republished "System of Nature..." [Linnaeus 1735, 1766, 1767] is the most complete, for this time, encyclopedia of the "three kingdoms of Nature" studied by natural history (in Bacon's sense). His "Philosophy of Botany..." [Linnaeus 1751] is a follow-up of the advancements in the scholastic stage of the development of systematics; it is quoted below in paragraphs based on its 2009 translated reprint.

According to the established tradition, Linnaeus equates the System and the Method (in its general sense), therefore, "The natural method is the ultimate purpose of botany" [§ 163]. Like other scholastic systematicians, his Method is loaded philosophically: he assures that "The philosophers have turned botanical knowledge to the form of a science [...] The rules and regulations in botany are due to them" [§ 19].

An important part of the natural method of Linnaeus is his proposed distinction between natural and artificial systems [Sachs 1906; Daudin 1927]. In the "Philosophy of Botany...," he calls the natural system the *System*, while the artificial one is the *Synopsis*; it may be that this distinction reflects a peculiarity of Linnaeus' worldview, in which his belief that "Nature is the law of God" (*Natura est lex Dei*) is combined with a certain amount of agnosticism [Hofsten 1958; Lindroth 1983; Petry 2001; Harrison 2009]. From this point of view, the System of Nature, as an attribute of Divine providence, is overall and therefore the one, it is an incomprehensible and unattainable ideal. Based on the latter, Linnaeus suggests, in the first edition of his

"System of Nature…," that it is needed to develop "artificial systems [which] are absolutely necessary" [Linnaeus 1735: § 12], like "Ariadne's thread" (J. Ray's metaphor), which allows us to orient ourselves in the labyrinth of the diversity of Nature.

Linnaeus borrows the basic ranked hierarchy from Tournefort, replacing his "Section" with "Order" and directly correlating his ranks with the respective categories of "philosophers" (= scholastics) [§ 155]. However, in Linnaeus' writings, these categories are not applied consistently: for example, the largest botanical groups (fungi, algae, mosses, ferns, grasses, etc.) are called "species" in the "Foundations of Botany…" [Linnaeus 1736], whereas they are called "families" in the "Philosophy of Botany…" [§ 78], and "tribes" in the 12th edition of the "Systems of Nature…" [Linnaeus 1767]. Moreover, in the latter edition, in a summary of his "theory of botany" [*Delineatio Plantae*], Linnaeus recognized only two categories, "species" and "genus," and included class and order in the latter in its broad scholastic understanding.

In general, how Linnaeus arranges the plants among these categories is in part deductive and in part inductive. The former is evident from the following passages: "A system separates the classes by 5 appropriate divisions" [§ 155] and "An order is a subdivision of the classes" [§ 161]. Following scholastic tradition, he uses quite actively the "tree of Porphyry" as an illustration of many of his classifications, not only of plants but also of botanists. An inductive element of his method is evident from his saying that "there are as many genera as there are similarly constructed fruit-bodies produced by different natural species" [§ 159], and "a class is the agreement of several genera" [§ 160]. In fact, in the first edition of the "System of Nature…" he objects to those who "descend according to the law of division from classes to orders and further to species. By this hypothetical and artificial principle, they destroy and divide natural genera, thus exercising violence against nature" [Linnaeus 1735: § 8 *Ratio operis*]. Thus, the opinion that Linnaeus strictly follows the scholastic deductive scheme [Cain 1958; Ereshefsky 1997] is not true, and this is noted by many researchers [Whewell 1847; Larson 1967, 1971; Winsor 2001, 2006a; Pavlinov and Lyubarsky 2011; Lyubarsky 2015; Richards 2016; Pavlinov 2018].

The categories of "species" and "genus" are basic for Linnaeus, and in a realistic (natural) sense: for him they "are always the work of Nature," while "class and order are the work of Nature and Art" [§ 162]. And yet, he indicates "That natural classes exist so created" [§ 160], while his "work of Nature and Art" refers to "artificial classes." At any rate, he begins the botanical section of the first edition of the "Systems of Nature…" with the statement that "the foundation of botany lies in the systematic arrangement [of plants] according to genera and species" [Linnaeus 1735: § 2], and invokes "enlightened botanists" to recognize their natural status, "for without accepting this principle, it is impossible to comprehend the art" of systematization [Linnaeus 1735: § 6 *Ratio operis*].

However, one should not be under the delusion that Linnaeus' realistic interpretation of species and genus in plants and animals is biological in the modern sense. Indeed, his System of Nature also encompasses the kingdom of minerals, and the respective section on them of the "System of Nature..." includes their classification by classes, orders, genera, and species, with distinguishing minerals through generic commons and species particulars. Thus, for Linnaeus, these categories are seemingly of the same nature in all three Kingdoms of Nature, and their naturalness seems to be determined by their Divine creation in the case of species and by the combinatorics of species characters in the case of genera.

Linnaeus pays great attention to the characters, dividing them into three categories—natural, essential, and artificial, but in this division, there is no direct correspondence with the interpretation of *essentia* and *accidentia* of Aristotle and scholasticists. For him, *natural characters* define genera (they are *definitio*) and *essential characters* discriminate them (as *differentia*) [§ 258] in the System, while *artificial characters* simply outline them in the Synopses. In this respect, his attitude to the choice of the basis of division is quite inconsistent. In some places, he shows his adherence to the scholastic tradition: an essential feature "remains unchanged, even if unlimited new genera should be discovered" [§ 189] and "The character should be kept without change in all systems, even the most diverse" [§ 202]. Elsewhere, however, Linnaeus states something opposite: "those [characters] that are effective to establish one genus do not necessarily produce the same effect in another" [§ 169], and "a character does not make a genus, but the genus makes the character" [§ 169]. Thus, contrary to a recognized viewpoint [Cain 1959b; Hull 1965, 1985; Mayr 1982], Linnaeus does not follow strongly the essentialism doctrine in his treatment of characters [Winsor 2006a, 2006b; Richards 2016; Pavlinov 2018].

Taking into account the overall contribution made by Linnaeus to the development of the natural method of systematics, he is most often referred to in the contemporary literature as the "founder of the science of systematics," almost the entire classical systematics of the 19th–20th centuries is called "Linnaean," and his achievements are declared the "Linnaean reform."[4] As can be seen from the above, none of these epithets seems to be fully correct: the "science of systematics" began two centuries before Linnaeus, and his "reform" consists mainly of the generalization of ideas that have been developing by the mid-18th century. What is especially important in this respect is that the subsequent development of the natural method of systematics will become much more "non-Linnaean" and partly essentially "anti-Linnaean" (see the next section).

2.4 EMERGENCE OF BIOLOGICAL SYSTEMATICS: ANTI-SCHOLASTIC REVOLUTION

In the second half of the 18th century, a decisive role in the conceptual history of systematics is played by a new turn in the development of natural science, largely

[4] It is noteworthy that Linnaeus himself is the first to designate his achievements as "Linnaean reform" in his autobiography [Bobrov 1970].

predetermined by a combination of the ideas of English philosophical empiricism and French Enlightenment. In the branch of natural history associated with the study of living nature and of which systematics is an important part, a new "biophilosophy" begins to emerge. It is based on a fundamentally new understanding of the organism—not as a set of essential or diagnostic characters, but as a complexly organized developing living being interacting with its natural environment. Detailed anatomical, histological, and physiological studies of animals and plants, including comparative embryological ones, become of great importance as suppliers of new kinds of data. The emergence of the term "biology" (its authors are G. Treviranus and J.-B. Lamarck) signifies an acknowledgment of the important border between living and inanimate natures, so the old scholastic universal systematics gives a way to a new, actually *biological* one [Stafleu 1969; Wilkins 2009; Pavlinov 2018].

All this means nothing more than the "biologization" of systematics, and its new content and tasks can be summarized in the following main positions. First, to reveal the natural orderliness of living nature, it is necessary to recognize the natural groups of plants and animals as dictated by their nature instead of construing artificial "Ariadne's threads" following a formal classification algorithm. Secondly, since the System of Nature is a plurality of interrelations of "all with all" (one of Leibniz's ideas), it is necessary to distinguish natural groups by many characters. Further, the latter should be uncovered as a result of empirical research instead of pre-assigning them certain "weights" based on ideas about essences. Finally, the classification algorithm should be not divisive (deductive), but agglomerative (inductive): one should not divide higher "genera" into lower ones, but start at the species level and then sequentially unite species into assemblages of higher ranks. All this together provides a new understanding of the natural method of systematics implementing, with significant delay, one of the key conditions of natural science established at the end of the 16th century: it should be empirical in its base.

New ideas set a significantly new trend in the conceptual history of systematics. The shift from an "old" to a "new" systematics in the late 18th and early 19th centuries deserves the designation *anti-scholastic revolution* dividing scientific systematics into scholastic and post-scholastic stages [Pavlinov 2018, 2019]. The classic writer on botany of the early 19th century, A.-P. de Candolle, equates it with the revolution in chemistry made at the end of the 18th century by the analytical chemist Antoine Lavoisier [de Candolle 1819]. In botanical historiography, this transition is usually designated as the end of the epoch of "artificial systems" and the beginning of the epoch of "natural systems" [Lindley 1836; Sachs 1906; Starostin 1970; Uranov 1979; Stevens 1994, 1997]; in fact, all post-scholastic systematics should be called "natural" in the most general sense.

The second scientific revolution in systematics, being anti-scholastic, becomes to an extent also "anti-Linnaean." Leading experts of systematic botany of the first half of the 19th century, evaluating its first results, accentuate enthusiastically that systematics did not remain limited to the Linnaean (actually, scholastic) formal method of identifying species and genera with a few diagnostic features, but went further to become the science of "exhaustive characterization" [Sprengel 1808; Brown 1810; de Candolle 1819; Lindley 1836]. Later, the historian of botany Julius von Sachs will join this critical assessment by emphasizing that if Linnaean systematics had triumphed, "the inevitable result was that botany ceased to be a science," while the truly natural systematics that began to take shape in France "proved to be the only one endowed with living power, the true possessor of the future" [Sachs 1906: 108, 110]. The method of natural systematics will be hailed as a new "philosophy of botany" [Drouin 2001].

The empiricism of post-scholastic systematics develops not in a strict Lockean, but no less in the rational Cartesian style, i.e., it emerges as *rational-empirical* systematics. The latter means that the basis of the new rationality, like the previous one, is the belief in the Method (Organon), i.e., the very natural method that allows "methodists" to reveal a sought "nature of things." In its new interpretation, such a method "must be based on the nature of organisms, which includes a totality of all their features and structures" [Adanson 1763: clv], and this method "must be universal, or unified, that is, there should be no exceptions for it" [Adanson 1763: civ]. An important part of this rationality becomes a separation of the "System" from the "Method": the first is equated with the scholastic (Linnaean) "artificial system," while the second is acknowledged in a new empirical meaning as the "natural order" corresponding to the true "nature of things" [Stevens 1994; Pavlinov 2018]. The need for a new conceptual design of biological systematics and its natural method, as opposed to the previous scholastic one, stimulates a purposeful interest in the development of its theoretical foundations. So, at the beginning of the 19th century, a theoretical section of this discipline, *taxonomy*, is delineated and denoted [de Candolle 1819; DeCandolle and Sprengel 1821], followed by articles especially devoted to the illumination of new methodological principles of systematics [Jussieu 1824; Strickland 1841].

> This new rationality of biological systematics, with its emphasis on the use of a large number of characters, becomes fixed terminologically. The botanist Charles-François Brisseau de Mirbel, in the first volume of his "Natural History of Plants" [Brisseau-Mirbel 1802], suggests calling the new multi-character method *polytypic*, in contrast to the *monotypic* single-character method of scholastics; later they will be denoted as *polythetic* and *monothetic*, respectively [Sokal and Sneath 1963].

The beginning of the post-scholastic stage in the conceptual history of systematics dates back to the mid–end of the 18th century, with representatives of the French Enlightenment playing the most active role. Its initiation is undoubtedly marked by the publication of the two-volume "The Families of Plants" by botanist Michel Adanson: the above quotation from him most clearly expresses, for the first time, a fundamental idea of the natural method in its new understanding, supplemented by a new "natural order" of plants [Adanson 1763]. Thus, an initial phase of the anti-scholastic revolution deserves being called *Adansonian*, by analogy with the beginning of the scholastic revolution, called above Cesalpinean. However, the novelty of the technical implementation of Adanson's method turns out to be too radical: it is based on the analysis of all available characters and justifies natural groups by specific combinations of most compatible ones. Therefore, in the early post-scholastic systematics of the late 18th to early 19th centuries, a more conservative approach of Jussieu, rooted in rather traditional ideas about essential features, will prevail over that of Adanson (see Section 2.4.2).

The post-scholastic revolution in systematics, freeing it from the dictate of scholastic procedures, leads to the emergence of a bunch of new research programs, each with its own understanding of the subject, tasks, and the natural method of systematics [Pavlinov 2018, 2019]. Thus, at the turn of the 18th and 19th centuries, two programs are almost simultaneously being shaped to take leading positions: natural systematics in its narrow, mostly botanical capacity (Jussieu, Candolle, Lindley, etc.) and typological systematics of zoologists (Vic d'Azir, Cuvier, Baer, etc.). A more peripheral position is taken by natural-philosophical concepts of a rather esoteric make-up, these are "biblical Platonism" (Agassiz, Owen), organismism (Oken), and numerism (MacLeay); to them undoubtedly should be added the idea of transformism, appearing ahead of its time (Lamarck). Most of the "taxonomic esotericism" will leave almost no trace in the subsequent conceptual history of systematics (albeit a noticeable though indirect impact of organismism on classical phylogenetics), typological and natural schools systematics will fade into the background, while transformism will take a leading position half a century later.

It is worth mentioning separately a version of natural systematics in its most general (natural-philosophical) understanding which focuses on "apparently natural" groups of plants. An idea of naturalness of such groups dating back to Theophrastus, in the period under consideration, is declared by one of the leaders of the Enlightenment, naturalist G.-L. Buffon. It is defended most insistently by the natural philosopher and naturalist Alexander von Humboldt, who believes that the Natural System cannot be built without taking into account the inseparable connection of organisms with their environment, which is revealed by detailed investigations of their natural history. Accordingly, the Natural System of plants must include, first of all, obvious and clearly delineated ecological groups, such as grasses, bushes, trees, vines, etc.: such groups are designated by Humboldt as the "basic forms of life" [Humboldt 1806]. Active assimilation of this idea will begin at the end of the 19th and the beginning of the 20th centuries when a more developed concept of life forms will be elaborated followed by the appearance of systematics of life forms (see Section 5.5).

2.4.1 MAJOR NON-SCHOLASTIC MOTIVES

The problematic character of the new cognitive situation, taking shape as a result of the antischolastic revolution, is primarily determined by the multiplicity of the natural-philosophical world pictures interpreting the orderliness of the System of Nature in different ways. One of the deepest of them includes different natural-philosophical comprehension of fundamental properties of Nature, viz. its continuity ("nature does not make leaps") and its "unity in the plural." For systematics proper, the core of this difference constitutes a contradiction between the acceptance or denial of the natural (real) status of discrete groups of organisms (taxa) and their ranks. Within scholastic systematics, this problem was not seriously considered: a generic–species scheme implementing discrete Aristotelian logic inevitably generated discrete taxa;

in a similar, mostly logical, way, discrete ranks were fixed. Both these discretenesses were taken by scholastic systematics as "metaphysical given," as an evident and therefore undisputable manifestation of the universal System of Nature. However, a transition to post-scholastic systematics makes this a central problem; it is considered in the context of several basic concepts, viz. the hierarchical Natural System, the linear Natural Order, the two-dimensional "map of Nature," and the branching "Tree of life." Some natural philosophers oppose these concepts as suggesting fundamentally incompatible properties of Nature (Buffon, Lamarck), whereas others consolidate them in different ways within specific natural methods (Jussieu, Cuvier).

With regard to ranked hierarchy, its growing fragmentation becomes an important novelty, making it "less discrete" to a degree. Expanding knowledge of taxonomic diversity reveals a more complex structure of mutual affinity of organisms and makes the canonical ranks of Tournefort–Linnaeus insufficient for its description. Accordingly, the system of ranks is developed first by the invention of more categories (i.e., cohort, series, tribe, etc.) and then by the introduction of auxiliary ones, *subordinate* and *superordinate* (i.e., subfamily and superfamily). At the same time, a skeptical attitude toward the meaningfulness of ranked hierarchy is expressed regarding the non-equivalence of similarly denoted categories in different sections of the Empire of Live Nature [Brown 1810]. As a consequence of all these disagreements, the ranked system established by the beginning of the 19th century loses its certainty and stability and becomes really chaotic. In the second half of the 19th century, a new very fractional ranked hierarchy is stabilized [Gray 1858], making it, contrary to recognized opinion, non-Linnaean to a certain extent [Needham 1911; de Queiroz 2005; Pavlinov 2015, 2018]. In the 20th century, skepticism about the system of fixed ranks will revive and lead to a proposal to abandon it (see Section 6.3).

An idea of the continuous *Natural Order*, as an alternative to that of the discretely fragmented *Natural System*, is an embodiment of the natural-philosophical understanding of Nature as the unbroken Great Chain of Being. It goes back to Aristotle's idea of the Ladder of Perfection, which at the time under consideration is better known as the *Ladder of Nature* (*Scala Naturae*) [Lovejoy 1936; Rieppel 2010a; Archibald 2014]. This "Ladder" (or "Chain") implies three interrelated fundamental properties of the Universe: continuity, linear ordering, and polarity. This means that the diversity of essences is ordered into a single gapless sequence according to a certain gradient interpreted as either a regression or progression. Regression corresponds to the ideas of the emanation of the One (Platonism) or embodiment of the Divine plan of creation (bibleism): the further a certain essence "falls away" from its source, with which gradation begins, the less perfect it is. Progression is more consistent with the understanding of Aristotle's Ladder of Perfection as a sequential transformation of beings from less to more developed: so it leads from inert matter through living entities (plants and then animals) to mankind and further to supernatural creatures (angels, etc.).

An important part of the "Ladder" natural philosophy is the conception of Nature as a whole interconnected by a single chain of affinities due to the unity of its integrative creative principle, viz. a prototype or archetype [Kanaev 1963; Hammen 1981;

Pozdnyakov 2015]. One of the most peculiar and important features of this natural philosophy is that it implicitly contains a general idea of the development of Nature, which is not only the world of being but also the world of becoming [Rieppel 1985; Hopwood et al. 2010]. According to this idea, diversity of organisms appears due to the sequential "unfolding" (*evolutio*) of the prototype, just as a simply organized embryo "evolves" into a much more complex and perfect adult organism. This "Ladder" worldview gains very great influence among natural philosophers of this time; it contains important prerequisites for the formation of both early typological and evolutionary conceptions usually placed in opposition [Hammen 1981; Richards 1992; Pavlinov 2018].

Among "ladderists" of the second half of the 18th century involved in the conceptual history of systematics, the most notable is Georges-Louis Leclair de Buffon. At the beginning of his career, he is unconditionally committed to the idea of the continuity of Nature: in his early treatise "Preliminary Reasoning..." he argues that "only individuals really exist in nature, while genera, orders, classes exist only in our imagination" (cited after [Buffon 1835: 44]). However, later Buffon abandons such an absolute nominalism and acknowledges the reality of species: he begins a small opus "On Nature..." with the statement that "an individual [...] is nothing in Nature; a hundred and a thousand individuals are still nothing in Nature. Species are the only creatures of Nature, eternal and unchanging like itself" (cited after [Buffon 1843: 52]). His argument in favor of species reality originates from the ancient generative species concept and repeats the viewpoints of Ray and Linnaeus: each species at its beginning had a prototype molded by the Creator, based on which all other organisms of this species are reproduced as its copies [Farber 1972; Bowler 1973; Sloan 1979, 1987; Wilkins 2009; Richards 2010; Pavlinov 2013a]. This marks a radical break with a pure classification interpretation of species and very soon becomes a dominant concept, leading directly to its historical interpretation.

To understand the route of the conceptual history of systematics, it is important to keep in mind a deep natural-philosophical background of the species-generative concept, which attracts the attention of not only naturalists but also philosophers. Thus, G. Leibniz, despite his adherence to an idea of the continuity of Nature, believes (before Buffon) that the existence of generative chains of ancestors and descendants witnesses the real division of Nature into species [Look 2009]. This idea is taken up by I. Kant (with reference to Buffon) and included in the general natural-philosophical principle of the historical development of Nature [Mensch 2013].

The Ladder of Nature is most often represented graphically as a linear scheme of the arrangement of organisms in (usually) ascending series. Obviously, such a representation (as a kind of cognitive model) is more than schematic by showing only principal stages of the prototype's "evolution," so it is far from being an adequate reproduction of the idea of gapless Nature. Thereby, elements of discreteness are introduced indirectly into the representation of the Ladder, while its true continuity is only implied. Another serious problem is that the real diversity of living beings does not fit into such a one-dimensional ordering; therefore, linear diagrams are usually supplemented with short lateral branches showing variants of the realizations of particular stages of a progressive perfection. This indicates a smooth transition from the linear Natural Order to the branching "Tree of Life"; it is worth noticing that the latter is mentioned by the most devoted "ladderist" Charles Bonnet [Archibald 2014].

In systematics, an idea of the fundamental continuity of the Ladder of Nature leads to an unconditional nominalism: living nature is a continuous series of organisms without any noticeable gaps in it, so there is no discretely ranked hierarchy of discrete groups (species, genera, etc.). Thus, if "gaps in Nature do not exist, it obviously follows that our classifications do not describe it. The classifications we create are completely nominal" [Bonnet 1769: 28]. The first true evolutionist and convinced "ladderist" Jean-Baptiste de Lamarck fully agrees with this conclusion and claims in his "Zoological Philosophy…" [Lamarck 1809] that all taxonomic categories "commonly used in natural sciences, are purely artificial aids […] Nature has made nothing of this kind [so it] has not really formed either classes, orders, families, genera or constant species but only individuals" (cited after [Lamarck 1963: 20–21]). In other words, though hierarchical classifications are permissible and even useful, they are arbitrary: as the continuous Chain of Being has no joints, it can be cut at any of its "links" by constructing any arbitrary taxa and categories which are of practical value only.

It should be noted that the general idea of the Ladder of Perfection is quite organic not only to the Natural Order but also to the Natural System. In the latter, it is embodied in its simplest form by the *rule of progression*, which determines the arrangement of taxa of the same rank according to the levels of advancement (progressiveness) of the respective organisms. It occurs in the classifications in two main versions: some implement it as a regression series with the most advanced organisms being listed first (i.e., Linnaeus, Cuvier, etc.), whereas others do that as a progression series beginning with the most primitive organisms (i.e., Adanson, Lamarck, almost all classifications of the 19th and 20th centuries).

The *metaphor of taxonomic map* represents the structure of taxonomic diversity in a two-dimensional planar scheme similar to a geographical map [Stevens 1984a, 1994; Lesch 1990; Pavlinov 2018]. This metaphor is most consistent with the "network" picture of the world, which focuses on the totality of multilateral connections "all with all" without any preferential axis [Barsanti 1992]. In it, natural groups of organisms are likened to the territorial units of different levels of generality (islands, archipelagos, mainlands, etc.) to reflect a close mutual affinity between "neighbors"; so, both taxonomic hierarchy and discreteness are clearly presumed by this metaphor. Within the realm of systematics, such a metaphor was mentioned by Linnaeus in his "Philosophy of Botany…," and the respective scheme appears (probably for the first time) in the work of his disciple Paul Giseke as a "map of geographic genealogical affinity" [Giseke 1792]. A half of a century later, the geologist and zoologist Hugh Strickland directly indicates that "the true order of affinity can only be exhibited (if at all) by a pictorial representation on a surface [...] illustrated by a series of *maps*" [Strickland 1841: 192; italics in the original]. Bertrand de Jussieu in Paris and A.-P. de Candolle in Geneva bring this metaphor to reality by planning botanical gardens under their curation so that the distribution of plants over the territory would reflect their systematic affinity [Stevens 1994]. It is noteworthy that, for one of the founding fathers of botanical natural systematics, A.L. de Jussieu, chain and map are not alternatives: the natural method

> connects all forms of plants into an indissoluble whole and follows step by step from simple to complex [...] in an unbroken series, like a chain whose links represent countless species [...] or like a geographical map, in which the species are distributed over territories, provinces and kingdoms.
>
> Jussieu 1789: xxxiv

This metaphor loses its popularity from the middle of the 19th century due to growing interest in tree-like genealogical schemes, and yet its indirect influence on taxonomic ideas of post-scholastic systematics is noticeable. It explicitly introduces into systematics the concept of hiatuses, which separates taxa just like physical space separates islands and archipelagos on a map [Stevens 1994]. In this way, this metaphor indirectly anticipates the modern concept of quantitative taxonomic distance: Strickland believes that, on such a map, "the distance from each species to every other is in exact proportion to the degree in which the essential characters of the respective species agree" [Strickland 1841: 185].

Considered epistemically, the metaphor of a taxonomic map is remarkable in two respects. Firstly, it represents not the actual, but an "imaginary" reality, showing what is important to a cartographer, affecting (by feedback) the latter's perception of the reality it displays [Winther 2020]. Secondly, it implies a substantially different logical procedure as compared to the generic–species classification. The latter is based on a logical division of notions and thus divides a set into its subsets. In contrast, mapping implies territorial zoning, which is a kind of partonomic division of the whole into parts; this important distinction will receive special attention in the 20th century [Meyen 1977; Rodoman 1999; Chebanov and Martynenko 2008; Lyubarsky 2018].

The *tree-like representation* of relations between objects (organisms, ideas, etc.) goes back to neo-Platonism and Medieval scholasticism, where it is realized in the form of the classification "tree of Porphyry." The latter is more than popular in scholastic systematics and is partially preserved until the middle of the 19th century. With an assimilation of the evolutionary idea by post-scholastic systematics, this classification tree is replaced by another, genealogical one. It is important to keep in mind that there is no direct logical link between these two categories of trees; a fundamental difference between them is as follows [O'Hara 1991, 1992; Pavlinov 2007a, 2015,

2018; Podani 2013]. The classification tree is a *dividing tree*: it shows sequential logical partitions of general notions into particular ones; in systematics these are taxa divided into subtaxa. In contrast, the genealogical tree is a *connecting tree*: it shows connections between objects according to the degree of their affinity determined by their origins (say, species from species). However, operationally, the latter tree is converted into a classification in the same manner as the "tree of Porphyry," i.e., by dissecting it "branches" in a descending (deductive) manner.[5]

Genealogically interpreted tree diagrams have been known since at least the Middle Ages; their branching order could be either ascending or descending depending on the position of the respective "starting points." In the earliest versions, they illustrate genealogical relations between biblical characters, then between representatives of the noble families; starting from the 16th century, such diagrams illustrate genetic links between human tribes and languages [Gontier 2011; Archibald 2014]. The possibility of representing the Natural System in a form similar to the family tree is noticed by the naturalist Peter Simon Pallas in the middle of the 18th century [Barsanti 1992; Kolchinsky et al. 2004; Archibald 2014]: in his "Index..." he writes that "the system of organic bodies is best represented in the form of a tree that comes directly from the root from the most simple plants and animals and takes shape of closely adjoining double animal and plant trunk" [Pallas 1766: 23–24].

In scholastic and early post-scholastic systematics, connecting trees of affinity are most often represented as undirected networks [Stevens 1994; Rieppel 2010a]. The first such schemes were published by R. Morison in the second half of the 17th century (see Section 2.3.2). The natural systematician John Lindley likens complex networks of affinity of plants "to rays drawn from the center of a sphere, which spread in all directions, and impinge upon the affinities of other spheres in their neighborhood" and concludes from this that "all attempts at discovering a lineal arrangement are chimerical" [Lindley 1835: 42]. Such network diagrams are sometimes combined with the above taxonomic maps, as in one of the books by A.-P. de Candolle [Stevens 1994]. The earliest rooted (vertical) genealogical schemes appear in the early 19th century; the first of them are occasionally descending (for example, in Lamarck's "Zoological Philosophy") but ascending trees soon become most popular, and nowadays they are usually called *phylogenetic trees*.

Another "hot point" of early post-scholastic systematics is shaped by the discussion of choice of classifying characters; two general approaches are considered and substantiated. In one of them, dominating throughout the 19th century, it is asserted

---

[5] From this viewpoint, linking "tree thinking" exclusively to genealogy [O'Hara 1997; Baum and Smith 2013] is incorrect: actually, "tree thinking" is no less characteristic of scholastics than of phylogenetics. Moreover, considered from a taxonomic perspective, the genealogical tree does not differ greatly from the classification tree. To emphasize the fundamental difference between these two metaphors, it is better to speak of "genealogical" or "evolutionary" thinking rather than "tree thinking" [Pavlinov 2005, 2007a].

that these characters should reflect certain most significant vital functions of organisms and/or some of their fundamental structural features (body plans). This approach, whose foundations are laid by botanists A.L. de Jussieu and A.-P. de Candolle and zoologist G. Cuvier, continues to a certain extent an essentialist tradition in the Ray–Magnol version; it is characteristic of natural systematics of botanists, early typology, and some "esoteric" theories. In another approach, characters are mainly assigned a diagnostic function: they serve as indicators ("marks") of natural groups. This position may be considered a continuation of the development of the method of scholastic systematics by Tournefort and Linnaeus, and is characteristic of the natural method of M. Adanson and, subsequently, of evolutionary interpreted systematics.

An important part of the anti-scholastic motivation of the emerging "new systematics" becomes the shift from essentialist to nominalist interpretation of taxonomic names [Pavlinov 2015, 2018]. M. Adanson appears the first here too: he affirms the key principle of a new understanding of taxonomic nomenclature, according to which "names denote objects [...] and do not express their nature or at least their most essential features" [Adanson 1763: cxxii–cxxiv]. At the beginning of the 19th century, his main thesis "a name is just a name" becomes a symbol of the all contemporary nomenclature.

2.4.2 THE NATURAL SYSTEMATICS

As was noted at the beginning of this chapter, botanist historians usually designate a shift from the scholastic to post-scholastic stage of the conceptual history of systematics as a transition from the epoch of artificial to the epoch of natural classifications, with their development becoming the main task of natural systematics. If the latter is understood not in a general sense (as post-scholastic systematics in general) but as a particular taxonomic theory, its main content can be formalized as an aspiration to reveal the Natural System by analyzing a large number of characters following the inductive argumentation scheme.

The first attempts to develop such an understanding of the natural method of systematics, making it possible to get as close as possible to the Natural System, were made by J. Ray and P. Magnol in the 17th century. Since the middle of the 18th century, it is especially actively defended by representatives of the natural-historical wing of the French Enlightenment. Thus, Louis Daubenton, in his article "Botany" for the "Encyclopedia" of Denis Diderot, asserts that the natural order is to be "grounded on a complete description of 'all the relations' of resemblance, rather than on one concentrating on similarities in a single part" (cited after [Sloan 1979: 121]). In a similar way, this general idea is declared and embodied by many theoreticians of natural systematics: in developing natural classification, "attention is paid as much as possible to all and each of the parts, and no preference is given to one above the rest" [DeCandolle and Sprengel 1821: 311], it "is an accumulation of facts which are to be arrived at only by a slow inductive process" [Strickland 1841: 185], by which "organisms are united based on a similarity in the greatest number of their parts or characters [...] starting by combining individuals into species and species into genera, then genera into families and families into classes" [Jussieu 1824: 9].

The general idea of this method is expressed, perhaps in its most refined form, by the natural philosopher and anthropologist Johann Blumenbach: in his extremely popular "Guide to the Natural History," he writes that "we should consider not just a few selected features, but all external characteristics [so] the animals that are similar in 19 structures and differ only in the twentieth should [...] be grouped together" [Blumenbach 1782: iii–iv]. Thus, if anyone could be considered a forerunner of the modern phenetic idea, then it is not M. Adanson, as is argued mistakenly [Sokal and Sneath 1963; Hull 1988; Lyubarsky 2018], but rather J. Blumenbach [Pavlinov 2018].

Although general foundations of natural systematics in its narrow (mostly botanical) sense are first outlined and embodied by M. Adanson in his "Families of Plants," Antoine-Laurent de Jussieu is usually considered its "founding father," referring to his "Genera of Plants..." [Sachs 1906; Carr 1923; Daudin 1927; Uranov 1979]. This is partly true, indeed, since the method of Jussieu becomes the most influential in the period under consideration: the typologist G. Cuvier writes that "his natural method forms perhaps as important an epoch in the sciences of observation, as the *Chimie* of Lavoisier does in the sciences of experiment" (cited after [Carr 1923: 63]). Such significant popularity of the method of Jussieu and his followers is due to its noticeable conservativeness, as it incorporates certain elements of old essentialism. The latter means prior "weighting" of characters by their significance for organisms as a basis for classification: Jussieu and de Candolle consider reproductive organs [Jussieu 1789, 1824; de Candolle 1819]; Lindley thinks axial structures and leaves are most significant [Lindley 1836]; while Strickland emphasizes a certain "physiological importance" in general [Strickland 1841]. According to the particular interpretations of this general idea, three more particular taxonomic theories can be quite clearly recognized within its scope: Adanson's method is purely rational, that of Jussieu includes noticeable elements of essentialism, and de Candolle's one gravitates towards typology.

A specific feature of the *method of Adanson* is that it does not presume any prior recognition of characters as significant or insignificant in any sense [Adanson 1763]. Instead, all available structures of plants are first investigated, preliminary classifications are then construed based on each of these structures, and next the groups of genera most often recognized in these pre-classifications are treated as natural, so the final classification (called "natural method" by Adanson) is elaborated to include these groups [Stafleu 1963; Guedes 1967; Pavlinov 2018]. Thus, the greater the number of genera that are marked by particular characters, the more the posteriorly assessed input of the latter is in the resulting classification; this inductive method resembles that developed by contemporary *compatibility analysis* [Pavlinov 2018]. In Adanson's method, the main taxonomic category is family; he borrows it from Magnol and criticizes Linnaeus as he "could neither substantiate nor give an exact definition of the natural genus using artificial one" [Adanson 1763: cv]. In substantiating his method, Adanson asserts that "it is nature that prescribes to the 'author-methodist' the steps to be taken [to] distinguish with certainty the true from the false" in the arrangement of plants [Adanson 1763: xciii].

The *method of Jussieu* is originally outlined in his "Sketch of a New Order of Plants..." [Jussieu 1773], then is more fully developed in his "Genera of Plants..." [Jussieu 1789] and later detailed in a paper "Principles of the natural method..." [Jussieu 1824]. As a natural philosopher, Jussieu believes that the task of "the science of botany" is to comprehend "the unchanging laws that Nature has imprinted on plants and which are revealed to any attentive researcher" [Jussieu 1789: xxxiv]. Proceeding from this, he suggests "to carefully study and reveal the entire organization of plants [...] and all characters [...] and not to miss anything that would allow to reveal the mutual affinity of all plants and achieve possession of a complete knowledge about them" [Jussieu 1789: xxxv]. This natural method "ties all plant forms into an indissoluble whole and moves step by step from simple to complex [...] in a continuous series, like a chain whose links represent countless species" [Jussieu 1789: xxxvi].

In Jussieu's method, in contrast to Adanson's, classification begins with ranking anatomical structures of plants according to the following basic principles. The *principle of functional significance* defines an essentialist core of this method, though significance of characters is substantiated inductively: according to Jussieu, "it is necessary to study all parts of the plant to know their functions, so that to better determine their significance" [Jussieu 1824: 46]. The *principle of constancy* means that the importance of features should be inferred from the number of genera of which they are characteristic; this is a certain analogy with the method of Adanson. To these, Jussieu adds the *principle of character ranking*, according to which plant features are divided into *primary* (most significant and permanent), *secondary*, and *tertiary*. Jussieu argues that "common and permanent characters cannot be obtained except from the organs that are most essential for life, namely for the reproduction of species" [Jussieu 1773: 183]; in this way, characters should not only "be counted as units, but each is added according to its relative value, one permanent character being equal to or even superior to many non-permanent ones" [Jussieu 1773: 196; Jussieu 1824: 27]. In developing the guiding ideas, J. Lindley emphasizes that "affinity is an accordance in all essential characters" and "the characters by which natural affinities are ascertained, are valuable in proportion to their importance to the existence of a plant" [Lindley 1835: 40]. In a similar way, H. Strickland writes that "the natural system [is] the arrangement of species according to the degree of resemblance in their essential characters," with the latter being selected in agreement with their "physiological importance" [Strickland 1841: 184, 185].

The *method of Augustin-Pyramus de Candolle* is developed in a markedly different way. Its first version is set out in his fundamental work "Initial Theory of Botany..." (first published in 1813; a more developed second edition of 1819 is cited here: [de Candolle 1819]); its revised version is published in German in collaboration with Kurt Sprengel in 1820, its English translation entitled "Elements of the Philosophy of Plants..." is cited here [DeCandolle and Sprengel 1821]. De Candolle bases his method largely on Jussieu's principle of character ranking (de Candolle calls it *subordination*) according to their functional significance: "the value of a character stands in a compound proportion to the importance of the organ" [de Candolle 1819: 172; DeCandolle and Sprengel 1821: 137]. This *principle of subordination* is complemented by the *principle of symmetry*, according to which the basic diversity of plants is a diverged

manifestation of an initial symmetry of a certain "original type," or a "regular primitive form" [*ibid.*], so the natural system ("method") is designed to reflect the diversity
of symmetry patterns. In substantiation of this principle, an important role is played
by ideas developed in crystallography [Stevens 1984a]. In this case, de Candolle acts
as a structural typologist; therefore all botanical systematics grown from his method
are sometimes qualified as typological [Ruzhentsev 1960].[6] De Candolle supplements
these principles with three "theorems," one of which asserts on an *a priori* basis that
"really natural classes based on one of the main functions of a plant will, of necessity,
be the same as based on another function" [de Candolle 1819: 79]. This assumption
serves for de Candolle as a theoretical basis for the possibility of classifying plants
according to only one of the most significant anatomical structures; this means an evident return to the one-character ("monotypic") method of essentialists, rationalized
by him in a new way. This "theorem" will later be formalized as the *principle of character interchangeability* [Meyen 1978].

The founders of natural systematics (in its narrow understanding by botanists) pay
very serious attention to its theoretical outlining in a rational manner. As noted above,
de Candolle combines the set of his key principles in the *taxonomy*, thus delineating
and denoting the theoretical section of systematics [de Candolle 1819]. Formalization
of the classification method by scholastic systematicians, especially by Linnaeus,
in the form of "canons" and "aphorisms," is continued by the leaders of natural
systematics: examples are the already-mentioned "theorems" of de Candolle; Lindley
presents the principal theses of his method in the form of "axioms" [Lindley 1836].
Guided by the principles of taxonomy, de Candolle divides classifications into *empirical* and *rational*: the former "do not depend on the nature of the object, [the latter] are
associated with the true nature of objects, only they deserve attention" and the name
of scientific ones [de Candolle 1819: 28].

The research program of natural systematics (as presented here) is one of the most
notable and influential of the first half of the 19th century, along with classification
typology. In the second half of the 19th century, it will partially cede its leading position to evolutionary interpreted systematics, but will also partially merge with it.

2.4.3 THE ORIGIN OF TYPOLOGY

Typology is a semantic and etymological derivative of the general concept of *type*
(Greek τυπος, example, mold), which has very ancient roots. It was put forward by
ancient thinkers in two main interpretations, epistemic and ontic, i.e., either as a
model (standard) of comparison or as an ideal form ("matrix") that gave rise to certain real forms. The second interpretation was reinforced by the notion of *arche*type
(Greek ἀρχή) as the "beginning" [Hammen 1981; Chebanov and Martynenko 2008].

The epistemic interpretation of the type permeates implicitly folk systematics;
it is very characteristic of herbalistics and scholastic systematics. According to the
empirical rule called the *type method* [Whewell 1847], classification is built by first

---

[6] Based on the principle of symmetry, E. Haeckel, by the end of his life, will elaborate a version of the
natural classification of animals guided by his *promorphology*, i.e., by the laws of transformation of the
types of symmetry of their body structure [Haeckel 1917].

recognizing a certain typical example and then uniting other elements of diversity by their similarity with the type. Based on the ontic interpretation of the type, a fundamental typological concept begins to take shape in post-scholastic systematics in the late 18th and early 19th centuries as the natural-philosophical one, with the archetype (or prototype) being considered as one of the organizing principles of all Nature [Amundson 2005; Pavlinov 2018].

Typology in its ontic interpretation is usually derived from essentialism [Hull 1965; Mayr 1982; Shatalkin 1993, 1996, 2012; Ereshefsky 2001a]. One of the reasons for this is a certain similarity of the structural understanding of type to the Cesalpinean principle of describing organisms based on the "number, position and shape" of their anatomical structures; therefore it is sometimes argued that the "Aristotelian essence is the structure of an object" [Shatalkin 2012: 129]. However, this linking is incorrect [Hammen 1981; Pratt 1982; Grene 1990; Winsor 2003, 2006a; Levit and Meister 2006; Lewens 2009a; Pavlinov 2018]. Indeed, the core of essentialism is the conception of essence in its predominantly Aristotelian teleonomic, i.e., mainly functional understanding. According to this, in scholastic systematics of the 16th–18th centuries and in one of the sections of natural systematics (in its narrow sense) of the late 18th and early 19th centuries, taxa are distinguished by essential properties most important for the functioning of organisms. Unlike this, the basis of the typology developed by structuralist anatomists is an idea of proto- or archetype shaped by the ratio of parts (organs etc.) of organisms [Meyer-Abich 1934; Hammen 1981; Card 1996]. Thus, these two fundamental natural-philosophical concepts are not very close, although they both refer to certain inner fundamental properties of organisms, without which they just cannot exist.

Considered in the context of the conceptual history of systematics, typology develops from the very beginning in three main versions, viz. stationary, dynamic, and epigenetic [Pavlinov 2018, 2020]. Common to them is an understanding of the (arche)type as an attribute of an ideal (imaginary) superorganism. An important difference is that the (arche)type is thought of as either a stationary structural plan of that superorganism or as a result of transformations of its parts; these transformations are thought of as either certain mental operations or real (more precisely, ideal objective) processes.[7]

Stationary typology is based on the concept of an ideal *structural type*; its implementations in particular organisms produce an overall hierarchical pattern of their typological diversity. It is originally developed by zoologist-anatomists, largely under the influence of the "Ladder" natural philosophy. In its origins lies the "Treatise on Anatomy..." by Felix Vicq d'Azyr; according to him, "nature follows apparently a type, or a general model, not only in the structure of various animals but also in the structure of their various organs" [Vicq d'Azyr 1786: 12]. Thus, this concept appears initially in two basic interpretations of the type, viz. (a) as a prototype of the variants of the same structure in different organisms or (b) as a prototype of different

---

[7] Outside biology, the mathematician and sociologist Adolphe Quetelet develops in the 1830s a significantly different version of *empirical* typology based on a statistical understanding of the type. This concept is not popular among researchers during the 19th century, but ethnographers, anthropologists, and some biologists will accept it later [Smirnov 1924; Klein 1991].

structures in the same organism. This marks the beginning of the formation of typological thinking, which has a direct bearing on biological systematics [Voigt 1973;
Stevens 1984a; Lewens 2009a].

This general idea is further developed by Georges Cuvier to make it the basis of
his version of the natural method of systematics. His concept is initially outlined in
his "Lessons in Comparative Anatomy" [Cuvier 1801] and then in the introductory
section of "The Animal Kingdom..." (of 1813); the latter is reprinted several times
in different languages, and here its 1840 edition is referred to [Cuvier's 1840].
The key for it is the concept of the stable *general plan* of an ideal superorganism
determined by the spatial relations of its parts;[8] it manifests itself in certain forms
characteristic of particular organisms denoted by Cuvier as *types* [Cuvier 1801].
Cuvier recognizes four basic general plans, each serving as a prototype for specific manifestations. Following Vicq d'Azyr, Cuvier substantiates the stability of
the general plan by the *principle of correlations of parts* supplemented by the *principle of the conditions of existence*, according to which there is a limited number
of certain viable combinations of parts of organisms, each constituting a particular
general plan, while others are incompatible within a certain plan, thus making it
non-viable (so functionality is yet present). Any significant change in the general
plan due to miscorrelations of its parts makes it non-functional (non-viable) under
certain conditions of existence. Substantiating his typological concept empirically,
Cuvier refers to the real anatomical organismal types as "kinds of experiments
ready prepared by Nature" [Cuvier's 1840: 15]. In addition, he adopts the *principle
of subordination of characters*, which means dividing the latter (more precisely, the
parts of organisms expressed by them) into *dominant* and *subordinate* according to
their (a) significance for the existence of organisms and (b) constancy due to their
mutual correlations. This principle is borrowed from the natural method of Jussieu
(see Section 2.4.2).

Based on these principles, Cuvier elaborates his natural method of classification,
which he thinks of as "the ideal to which Natural History should tend" [Cuvier's
1840: 16]. It is shaped by a specific combined implementation of two general natural
philosophies, "System" and "Ladder"; with this, his typology-inferred method becomes
very similar to that of Jussieu. Based on the "System" conception, Cuvier elaborates
a hierarchical classification of animals reflecting the hierarchy of implementations of
four general plans into respective types based on ranked characters. As he says, "from
their influence and from their constancy, result equally the rule, which should be
preferred for distinguishing grand divisions [of the system], and in proportion as we
descend to the inferior subdivisions we can also descend to subordinate and variable
characters" [*ibid.*]. Based on the "Ladder" conception, he arranges each respective
division, called *branches*, according to the *principle of progression* so as to reflect a
gradual perfection, in a descending order, of the body types of animals in each branch.
Cuvier's colleague Henri de Blainville denotes these branches as *types* in a taxonomic

---

[8]   Cuvier's term "general plan" will gain popularity in various refinements (body plan, common plan,
ground plan, *Bauplan*).

meaning [de Blainville 1816], thus making Cuvier's natural method a core of the *classification typology* [Farber 1976; Pavlinov and Lyubarsky 2011; Pavlinov 2018].

The main ideas of *dynamic typology* are laid down by the poet and naturalist Johann Wolfgang von Goethe under the influence of the ideas of organismism and the "Ladder" natural philosophy, complemented by some artistic and linguistic images [Arber 1950; Kanaev 1970; Hammen 1981; Steigerwald 2002; Zabulionite 2011; Riegner 2013]. The first sketch of Goethe's concept is presented in his "The Experience of Plant Metamorphosis" [Goethe 1790]. Central to this version of typology are the concepts of *archetype* and *metamorphosis* (transformation). In his natural-philosophical worldview, Goethe proceeds from likening Nature to a superorganism: its parts, undergoing mutual metamorphoses, give rise to the whole variety of real forms; so this version of typology is sometimes called *organismic* [Arber 1950; Steigerwald 2002]. Such a typological construction can be represented as some kind of an imaginary structural prototype, a totality of mutual transformations of its parts represent a single *dynamic archetype* [Naef 1931; Meyer-Abich 1949; Hammen 1981; Ho 1988; Beklemishev 1994; Levit and Meister 2006; Riegner 2013]. The latter, according to Goethe, is a certain law of mutual transformations (general metamorphosis) of parts of this superorganism; the archetype thus understood cannot be watched embodied in a particular genus or species: they "relate to it just like particular instances relate to a law; they are contained in it but they do not contain or give it" (cited after [Goethe 1957: 76]). In the 19th century, Goethe's concept appears to be one of the sources of two other very important concepts, viz. Owen's concept of homology and Haeckel's concept of phylogenesis (see next sections), but it is mostly disregarded by taxonomists. In the 20th century, it will become rather popular among morphologists and typologically minded taxonomists [Naef 1919, 1931; Kälin 1945; Troll 1951; Beklemishev 1994; Lyubarsky 1996], when it will be designated as *transformational typology* [Zakharov 2005]; eventually, Goethe's typology will be declared one of the forerunners of the newest concept of "evo-devo" [Riegner 2013].

At the beginning of the 19th century, the general typological idea receives a very significant development in the works of the German natural-philosophical school, based on the doctrine of vital materialism [Lenoir 1988]; its result is *epigenetic typology* that combines the main features of the two versions described above [Pavlinov 2018, 2019]. Its initiator and most prominent exponent is the zoologist embryologist Karl von Baer, who is fascinated by the organismic natural philosophy with elements of Platonism and transformism [Raikov 1969; Rieppel 1985; Lenoir 1988]. He defines the *type* as an arrangement of the parts of an ideal organism, which corresponds structurally to the body plan of Cuvier; the fundamental difference is that, in Baer's concept, the type is developing. Both general types and variants (subtypes) detailing them at different levels of generality are ideal prototypes of certain stages of individual development of living bodies arranged in an inclusive hierarchy. Interpreted natural-philosophically, the latter is a single organizing force that directs the real development of animals in two parallel "streams," viz. individual development as a transition from one stage of ontogeny to another and historical development as transformations (transmutations) of some organisms into others [Baer 1828]. Thus, Baer extrapolates ontogenetic transformations to all living nature as a developing superorganism, endowing them partly with a genealogical meaning [Raikov 1969; Lenoir 1988]. On this basis, the Baerian type may be called with good reason the *developmental type* [Lenoir 1988; Amundson 2005; Pavlinov 2018], although its hierarchy of particular types seems to be a stationary construct [Rieppel 1985].

According to Baer's natural method, transformations of a single structural plan in the course of the development of organisms make it possible to uncover a natural affinity among them. For this reason, Baer's typological method very soon becomes consolidated as one of the guiding principles for elaborating a version of the natural classification of animals [Milne-Edwards 1844]. A general idea of parallelism between ontogeny and phylogeny, developed by Baer and some other zoologist typologists (Serre, Meckel, Agassiz), will become one of the cornerstones of classical phylogenetics (Haeckel). In the 20th century, this idea will be recalled by theoreticians of structural cladistics [Nelson 1978]; besides, basic elements of Baerian epigenetic typology will be incorporated by *ontogenetic systematics* [Ho 1992; Pavlinov 2018].

In a work especially devoted to taxonomic issues (published in our days), K. von Baer emphasizes that the Natural System should be based on the hierarchy of developmental types, with the proximity of organisms in it reflecting their affinity expressed by a similarity in their ontogenetic patterns [Baer 1959]. Each natural group thus defined can be represented as a sphere, with the animals in it being placed to reflect their embodiment of the respective type: more typical organisms constitute the core (center) of this sphere, while less typical ones are allocated along its periphery. These spheres touch each other (as in the quinary system; see the next section) and may even overlap with their peripheries: this is probably the first case of a fuzzy interpretation of taxa in systematics; it will be used in one of the contemporary typological approaches [Tchaikovsky 1990].

2.4.4 "TAXONOMIC ESOTERICISM"

Among the natural-philosophical teachings that are very popular at the beginning of the 19th century, there are several that are usually qualified as esoteric. On their basis, several classification theories are formed in systematics that deserve the name of "*taxonomic esotericism*" [Pavlinov 2018, 2019]. Each of them is based on a specific natural-philosophical world picture constituting an ontic basis of the corresponding cognitive situation. All of them, in one way or another, implement the cognitive program of onto-rational systematics, according to which the general properties of classifications (hierarchy, etc.) are deduced from certain general laws of the structure of Nature (see Section 5.2.1). This basic natural philosophy presumes a realistic worldview, from which follows the "reality" of classifications derived from it, with all their taxa and ranks. These classifications are elaborated in a deductive manner, just as in scholastic systematics and in typology: the higher-rank taxa are first recognized and subsequently divided into the lower-rank ones. In contrast to those of natural (in its narrow sense) and typological versions of post-scholastic systematics, these classifications do not imply any "external" verification by empirical data: they remain unchanged insofar as their theoretical foundations remain steady, but changing the latter usually entails serious modifications in the respective classifications.

Such teachings are called "esoteric" as they are based on unverified "inner knowledge." They are considered non-scientific from a presentism standpoint developed by the physicalist philosophy of science and thus usually omitted or just casually mentioned in historical reviews as "delusions." However, at the time under consideration, they are quite influential and make their input in the formation of certain "respected" taxonomic ideas and concepts. Therefore they deserve closer consideration, from an antiquarism standpoint, as noticeable parts of the conceptual history of systematics.

To begin with, let us consider a peculiar combination of natural philosophy of Platonism and Bibleism ("Christian Platonism"), which in one form or another occurs in the ontology of natural science in the 15th–18th centuries [Gaydenko and Smirnov 1989; Harrison 2006; Lindberg 2007]. Among taxonomic theories of the 19th century, the most vivid example of the implementation of this doctrine is the theory of the paleontologist Louis Agassiz [Winsor 1979; Rieppel 1988a; Stamos 2005; Pavlinov and Lyubarsky 2011; Pavlinov 2018], who in his "An Essay on Classification" substantiates the view, according to which Nature is a result of the embodiment of the Divine plan of Creation that had been being coherently incarnated at different stages of the history of the Earth [Agassiz 1859]. This "plan" includes "prophetic divisions" of different levels of generality as "categories of thinking [...] of the Divine Intelligence" [Agassiz 1859: 8], which were materialized sequentially in taxa of different ranks, starting with the highest and ending with the lowest. This viewpoint goes back to that of Aurelius Augustine and Anselm of Canterbury about Divine prophetypes (archetypes), and through them to Plato's idea of the emanation of the One. According to Agassiz, all taxa, regardless of their position in the taxonomic hierarchy, are endowed with a dual realism: they exist both in the world of ideas (as "prophetypes") and as their incarnations in the created world: "Species [...] exist in nature in the same manner as any other groups, they are quite as ideal in the mode of existence as genera, families, etc., or quite as real" [Agassiz 1859: 176]. The taxonomic concept of Agassiz will remain in the future without apparent continuation, if not to mention contemporary *baraminology* [Wayne 2000; Todd 2006].

An essentially similar natural philosophy is implemented in a particular natural history theory by the zoologist-anatomist Richard Owen based on an idea of the hierarchy of archetypes, understood partly in the sense of Platonic *eide*, implying Divine Intelligence as their primary source [Rupke 1993; Amundson 1998, 2005; Camardi 2001]. For Owen, the general archetype is a really (objectively) existing eternal ideal form; through its organizing force formless matter becomes successively incarnated and embodied in the anatomical features of particular organisms in specific environments [Owen 1848]. According to this, the unity of archetype is what unites organisms into the Natural System; variations of this archetype are what provide subdivisions of this System. In Owen's view, the general archetype embodies the entire construction of an ideal super-organism, therefore the main task of comparative anatomy is to analyze this construction in both its general and specific manifestations. This analysis is based on the *concept of homology*: R. Owen is credited with its initial development in an idealistic typological manner based on ideas borrowed largely from Goethe and partly from geometry [Blacher 1976; Hall 1992; Rupke 1993; Panchen 1994; Amundson 1998, 2005]. A preliminary version of this concept is presented in his *Lectures on the Comparative Anatomy* and fully completed in a special work *On the Archetype and Homology* [Owen 1843, 1848]. In this, homology is defined theoretically through the archetype as the structural correspondence of its parts, *homotypes*; this concept will become of prime importance in the 20th century (see Section 6.6).

A peculiar natural-philosophical worldview, traced throughout the history of the development of the European cognitive program, is expressed by the Pythagorean aphorism "everything is Number." This is based on the postulate that the world is the Cosmos subject to the laws of harmony of numbers, therefore everything that exists can be expressed and measured by numbers; comprehension of the Universe as a whole and of any of its parts is equal to comprehension of respective numerical relations [Zarenkov 2009]. In one of the versions of natural theology, the ideas of Pythagoras are attributed to the biblical Creator having shaped Nature in accordance with this harmony, in which the numbers 5 and 3 occupy the central place; this, in particular, is the basis of Johann Kepler's "numerical astronomy" [Gaydenko and Smirnov 1989].

A taxonomic theory based on this numerology is developed in the first half of the 19th century by the naturalist William MacLeay in his "Peaks of Entomology..." [MacLeay 1819]. Named *quinary* for its being based on the number 5 (*quinta*), it is shaped by the following principles [Panchen 1992; Williams and Ebach 2008; Pavlinov and Lyubarsky 2011; Pavlinov 2018]. According to the *principle of hierarchy*, the Natural System of organisms is hierarchical, and its hierarchy is deductive (from higher to lower categories). The *principle of quinarity* states that, at each level of the hierarchy, five subgroups should be recognized in a given group to reflect the numerical harmony of Nature. The *principle of affinity* means that these subgroups should be ordered into a single series according to their *essential* characters, with their similarity in these characters reflecting their close affinity. Other characters not associated with this affinity are called by MacLeay *analogous*: they are of secondary significance and reveal distant affinity. Finally, the *principle of circularity* states that the entire series of subgroups within a given group should be organized so that its completion is adjacent to its beginning: this gives a cyclical arrangement of these subgroups, and the result is a kind of *ring classification*. In it, the groups within each circle (series) are placed according to their close affinity, while different circles are placed relative to each other according to their distant affinity. With this, a typical subgroup is located in the center of the circle, evaders are located along its periphery, and different circles contact each other via their respective evading subgroups according to their analogous characters. The naturalist Johann Kaup in his "Classification of Mammals and Birds" takes quinarity to extremes by representing classifications of natural groups with five-pointed stars arranged in circles [Kaup 1844]. A follower of MacLeay, ornithologist William Swainson adds trinarity to quinarity by dividing five groups of the same series based on their closest and distant affinities as follows: one typical, one semi-typical, and three deviating [Swainson 1835]. The idea of quinarism still retains some influence at the end of the 19th century, and its traces can be found in the second half of the 20th century [Coggon 2002; Pavlinov 2018].

The most influential and rather original doctrine of the esoteric kind is an idea of organismism developed by the natural philosopher Friedrich Schelling and his followers. In it, Nature is likened to a living super-organism and is endowed with many fundamental properties inherent in biological beings. Among such properties are the integrity of Nature, its continuity in being and becoming, and its objective and regular subdivision into parts. Now almost forgotten among natural scientists, in the 19th century this idea plays an important role in the formation of transformism in general and of modern evolutionism in particular [Gould 1977; Richards 1992; Pavlinov 2009a]. In particular, the evolutionist Ernest Haeckel imagines phylogeny as the historical development of a "genealogical individual" [Haeckel 1868]. Besides, organisms underlines typological concepts considered above.

In biological systematics, organismic natural philosophy, supplemented with ideas of alchemy and numerology, is most consistently embodied by Lorenz Oken (Okenfuss) [Raikov 1969; Richards 1992; Breidbach and Ghiselin 2006]. He opposes the natural system built on the principles of organismism with an artificial one based on Linnaean principles, which is only suitable, in his opinion, to memorize names and diagnoses [Oken 1847]. The backbone of Oken's taxonomic theory arising from organismic natural philosophy is shaped by the following principles. The key *principle of likeness* means the likeness of the structure of the respective taxonomic system to that of the super-organism of Nature. According to this, each taxon of plants and animals is likened to a certain part of that super-organism: the correspondence between them is inferred from certain speculations as to which part of organisms belonging to the respective taxon is supposed to be most significant for their life. The *principle of hierarchy* establishes a ranked hierarchy of taxa corresponding to the subordination of parts of the prototypical super-organism. Oken's adherence to the idea of the harmony of numbers that governs the world yields the hierarchy of five ranks with three taxa in each of them. The *principle of parallelism* states that in different sections of the natural system, groups of organisms should "repeat" each other in their essential features, just as different "analogous" parts of a single super-organism are similar to each other, an example being animal paired limbs (not to be confused with the "triple parallelism" of Agassiz and Haeckel). In systematics, Oken's ideas have not left noticeable trace. At the same time, the principle of parallelism will appear of great importance in biology: its embodiment is the theory of analogs by E. Geoffroy de Saint-Hilaire, in which this principle is affirmed as one of the basics for patterning the entire diversity of organisms [Geoffroy Saint-Hilaire 1830]. Here lies the origin of one of the versions of classical phylogenetics, which includes the concept of parallel homologous series as the main evolutionary trend [Cope 1887].

2.5   A STEP FORWARD: EVOLUTIONARY REVOLUTION

Historically (and maybe conceptually), the primary basis of the world picture of mankind is *fixism*, asserting the immutability of forms after their appearance (creation, etc.). Such is, for example, the biblical concept of the Divine creation, after which the System of Nature remains unchangeable. This natural philosophy forms an ontic core of those taxonomic theories (primarily scholasticism, stationary typology, etc.), in which "genera" and "species" are constant and unchanging universals or essences.

The fixist standpoint is opposed by the idea of development, or *transformism*, which means transformation (transmutation) of some forms into others. Transformism has two main versions, atemporal and temporal [Grushin 1961]. According to the first version, development does not imply its actual duration in time: this is an "ideal" transformation exemplified by "evolution" of the prototype giving rise to the Ladder of Perfection, metamorphoses of Goethe's archetype, etc. According to the second version, development is a material process that has both temporal extension and sequence, and within its route some forms actually turn into others; transformations of parts of an organism during ontogenesis provide a clear example. This second version is part of the materialistic natural philosophy: it is centered around the idea of the self-developing Nature, with cosmogonic conceptions of Descartes, Leibniz, and Kant playing a key role in its formation in the epoch under consideration [Gaydenko 1987; Hoque 2008]. On this basis, Kant (with reference to Buffon) differentiates between the fixist "descriptions of nature" (*Naturbeschreibung*) and the transformist "history of nature" (*Naturgeschichte*).

In the 19th century, the notion of "evolution" begins to be used to denote real historical transformations, thus emphasizing a certain analogy between historical and individual development [Chambers 1860]; this is what Haeckel means and emphasizes by coining the new terms "phylogeny" and "ontogeny" [Haeckel 1866]. In the middle of the 19th century, in the works of philosopher Herbert Spencer, an iconic notion of the "theory of evolution" appears and then becomes widely acknowledged [Bowler 1975; Richards 1992]. In biology, the general idea of transformism is embodied in a more particular idea of the origin of some species from others, so the transformist (evolutionary) concept, as applied to living matter, is often called the *theory of origin* [Haeckel 1866; Cope 1887].

The first historically interpreted transformist ideas relevant to systematics appear at the beginning of the 19th century based on different natural philosophies: J.-B. Lamarck thus interprets the Ladder of Nature, while Gottfried Treviranus and Friedrich Tiedemann infer an evolutionary worldview from organismism [Raikov 1969; Richards 1992]. Their adoption leads to the "historicization" of systematics, making it *evolutionary interpreted* [Lenoir 1988; O'Hara 1988; Pavlinov 2009a, 2018; Pavlinov and Lyubarsky 2011]. This gives a significantly new understanding of the Natural System, in which the unity of organisms is based on the unity of their historical (evolutionary) origin and blood (genealogical) kinship relations. This idea forces taxonomists to reconsider the major old problems of systematics in a new way: how to define natural groups of organisms, whether they are real or nominal, how they can be recognized, etc. The main illustrative representation of the Natural System in its historical understanding becomes the genealogical tree, which refers to the timing of the origins of the respective groups, so it is interpreted as the *tree of history* [Barsanti 1992; O'Hara 1992, 1996]. All this makes up the content of the next scientific revolution in systematics—the *evolutionary revolution*, which begins in the 1860s.

2.5.1 FIRST IDEAS

Lamarck appears the first taxonomist to integrate his evolutionary concept directly into the respective taxonomic theory and classification [Lamarck 1809]; his natural method becomes the first to be evolutionarily interpreted, and it can be designated as *classification Lamarckism* [Pavlinov 2009a, 2018; Pavlinov and Lyubarskiy 2011]. Its main ideas are briefly as follows: the historical development of organisms is represented as a continuous (without breaks) time-extended chain of live beings; any gaps separating its fragments are due to extinct or unknown intermediate forms; the taxa corresponding to these fragments are arbitrary, and are introduced as discrete classification units to indicate the main stages of the general unidirectional trend of evolution from lower to higher organisms; this trend determines an arrangement of taxa in the classification according to the above-mentioned *rule of progression*. As a "ladderist," Lamarck clearly distinguishes between the natural order (or arrangement), which reflects the evolutionarily interpreted Ladder of Nature, and artificial classifications developed according to "systemic" principles.

Early attempts to master the evolutionary idea by systematics during the first half of the 19th century have no success due to their probable prematurity because of the domination of the theories of natural systematics (narrowly understood) and typology. In a full-fledged manner, evolutionary (in the contemporary sense) natural philosophy penetrates systematics in the second half of the 19th century thanks mainly to the works of Charles Darwin and Ernest Haeckel. They both are "systemists," so one of their tasks is to justify evolutionarily the tree-like representation of the Natural System [Winsor 2009]; they are unanimous about a fundamentally new understanding of the Natural System as genealogical. These two persons become the authors of two interrelated, yet different, evolutionary concepts, which lay the foundations for two respective research programs in systematics. Darwin is mainly interested in the origin of species, so his theory refers (in modern terms) to *micro*systematics; Haeckel develops a theory of phylogeny of the animal kingdom relating to *macro*systematics.

Charles Darwin's contribution to the development of the evolutionary interpreted systematics is fundamentally important: as a matter of fact, he appears the first to express clearly the general idea that the Natural System is the genealogical one. His taxonomic conception, which is sometimes called Darwin's "central discovery," at least as far as systematics is concerned [Reif 2006, 2007], can be outlined as follows. According to him, "propinquity of descent,—the only known cause of the similarity of organic beings,—is the bond, hidden as it is by various degrees of modification, which is partially revealed to us by our classifications" [Darwin 1859: 413–414]. Therefore, "the arrangement of the groups within each class, in due subordination and relation to the other groups, must be strictly genealogical in order to be natural" [Darwin 1859: 302], the latter statement can be formalized as the *principle of genealogical arrangement*. With this, Darwin provides a very important justification for the selection of characters for the reconstruction of genealogies, rejecting their essentialist interpretation. He emphasizes that "the characters [...] showing true affinity between any two or more species, are those which have been inherited from a common parent" [Darwin 1859: 420], therefore, "the less any part of the organization is concerned with special habits, the more important it becomes for classification"

[Darwin 1859: 414]; this statement is referred to as *Darwin's principle* [Mayr 1965]. This principle in its expanded form includes several important evolutionary criteria for assessing taxonomic significance (weighting) of the characters. One of them goes back to the ideas of Jussieu and Cuvier about character ranking based on their constancy; it can be designated as the *criterion of commonality*; the second resembles partly Adanson's method based on the *principle of congruence*; the third repeats in many respects the principles of parallelism of Oken, although in Darwin's interpretation it is the *principle of evolutionary parallelism*; finally, according to Darwin, "amount or value of the differences between organic beings all related to each other in the same degree in blood" [Darwin 1859: 421] should not be taken into consideration if organisms are to be grouped according to their genealogical relationships; this is the *principle of irrelevance of differences*. Most of these principles (except parallelism) constitute the foundation of the modern cladistics, so the latter's adherents are better to call themselves Darwinists rather than Hennigians; the principle of parallelism is especially significant in the evolutionary taxonomy of Simpson (on these schools of taxonomic thought, see Section 5.7).

An important part of Darwin's natural method is a denial of the particular importance of the species category; the latter should be emphasized—not species as a natural unit (as is often misbelieved), but just a certain fixed category [Komarov 1940; Zavadsky 1968; Stamos 1996, 2007a; Ereshefsky 2011; Pavlinov 2013a]. This position is substantiated by reference to the continuity of the microevolutionary process, which is a gradual transformation of local races into different species, so a distinction between categories of race and species is not qualitative but quantitative. Indeed, based on his observations supporting this evolutionary model, Darwin becomes convinced how "entirely vague and arbitrary is the distinction between species and varieties" [Darwin 1859: 48]. So he iterates that "species [are] only strongly-marked and well-defined varieties" [Darwin 1859: 55], while "a well-marked variety may be justly called an incipient species" [Darwin 1859: 52]. All this urges him to call against "a vain search for the undiscovered and undiscoverable essence of the term species" [Darwin 1859: 485]. Referring to this thesis, supporters of the extraordinary importance of species in both evolution and systematics accuse Darwin of "eliminating the species as a concrete natural unit" [Mayr 1963: 14], although this is hardly true. At any rate, it is this view of the gradual transition between species and infraspecies units that constitutes an ideological core of what was called above *classification Darwinism*.

The background of Ernst Haeckel's consideration of biological evolution, as noted above, is clearly natural-philosophical: in one of his lectures he refers to Kant as the person who was the first to give mankind a comprehensive transformist doctrine [Haeckel 1868]. For Haeckel, the history of living matter, i.e., *phylogeny*, is the development of a "genealogical individual," by its fundamental properties being similar to the development of particular individuals, i.e., *ontogeny* [Haeckel 1868]. These general ideas go back to the natural philosophy of organismism of the beginning of the 19th century, in particular, to the transformist ideas of Treviranus and Tiedemann [Richards 1992; Rieppel 2016]. In this regard, it is pertinent to note that Haeckel is a great admirer of Goethe: quotations from him are placed as epigraphs to all chapters of his fundamental "General Morphology" [Haeckel 1866].

The macroevolutionary character of Haeckel's transformism is clearly seen from his genealogies. In contrast to the well-known Darwinian schemes, in which races and species are represented, Haeckel's scale of consideration of the history of organisms is set mainly by the level of classes and types. With this, Haeckel understands phylogeny as genealogical history reconstructed on the basis of the study of fossils: for him, "phylogeny includes paleontology and genealogy" [Haeckel 1868].

Haeckel in all his works follows Darwin in insisting that "the tree-like form of the Natural System can only become understandable when we acknowledge it as a real genealogical tree of organisms." However, generally speaking, he is more a phylogenetic morphologist rather than a systematician; at the center of his attention are the problems and laws of historical changes in animal body plans, which determine the groups of different levels of generality. Accordingly, he designates his approach as *systematic phylogeny*, as is evident from the title of his three-volume book thus entitled [Haeckel 1894–1896]; it will only turn out to be *phylogenetic systematics* in the works of his adherents. Studying this systematic phylogeny, he simply "applies" genealogical interpretations to ready-made classifications, mainly borrowed from Cuvier and Baer, turning them into phylogenetic schemes [Remane 1956; Williams and Ebach 2008; Rieppel 2016].

For Haeckel, the ascending genealogical tree, with its main trunk and lateral branches, is a standard representation of phylogeny as a "genealogical history." The predominating orientation of the trunk, as in the Ladder of Nature, is based on Aristotle's principle of perfection, from lower organisms at its base to higher ones at its top; each of its branches is "an aggregate of all those organisms, the common origin of which from one ancestor we cannot doubt" [Haeckel 1868]. Haeckel calls such an "aggregate" *monophyletic* and refers to it as *phylon*; if a group combines descendants of different ancestral forms, it is *polyphyletic*. The main task of systematic phylogeny is to identify monophyletic groups of organisms; together they constitute the Natural System, while polyphyletic groups cannot be considered elements of the Natural System. This general provision is known as the *principle of monophyly*; it is implemented by the *principle of phylo-taxonomic correspondence*, according to which a certain (in this case, non-strict) isomorphism should be established between a genealogical tree and a classification reflecting it.

Another phylogenetic concept is developed in the same years by the zoologist paleontologist Edward Cope: unlike Haeckel's, it is based on an assumption of unidirectional parallel macroevolution, in which dominate certain general trends (scales) of transformations of organisms not tied by close genealogical relationship [Cope 1887]. Each trend is manifested in many parallel phyletic lines: they originate from different "roots," and in all of them evolutionary transformations obey the same trend in both the general direction and successive passing through the same phases. So, not a tree but a phylogenetic "lawn" serves as an adequate representation of this phylogenetic model. According to Cope, the natural system is a hierarchy of polyphyletic groups, in which each "class is a scale of orders, the order of tribes [...] the family composed of one or more scales of genera" [Cope 1887: 45]. Each of these scales forms a *homologous series*; each successive phase passed by organisms in parallel scales forms a *heterologous series* (both terms borrowed from organic chemistry).

The groups encompassing these scales are distinguished by characters that are ranked into species, generic, familial, etc., though the criteria for their ranking are not specified. Characters of different ranks are mutually independent in their evolution: "the process of development of specific and generic characters does not proceed *pari passu* [...] therefore, species may be transferred from one genus to another without losing their specific characters, and genera from order to order without losing their generic characters" [Cope 1887: 123]. The last passage paradoxically repeats the idea of Theophrastus that plants can "pass" from one species to another because of their becoming modified under varying environmental conditions (see Section 2.2.2).

2.5.2 FIRST DEBATES

The proposal to treat the Natural System genealogically immediately sparks a lively debate. Many taxonomists take this proposal very enthusiastically, and the most radically minded phylogeneticists decisively break with "pre-Darwinian" systematics; this attitude will be summarized later as follows: "for phylogenetics, the centuries-old work of the natural systematics is of little value. For it, everything that had been done before Darwin needs to be redone" [Kozo-Polyansky 1922: 8]. With this, disagreements among phylogeneticists arise about how to interpret phylogeny, how to reconstruct it, and how to represent it in classifications. Thus, paleontologists object to using the phylogenetic approach in the systematics of extant organisms: in this, they refer to Haeckel's idea that phylogenetic reconstructions must be based on paleontological data, without which they remain purely speculative [Scott 1896]. Many taxonomists insist that phylogeny should not be interpreted simplistically as just a genealogy; they emphasize, with reference to Darwin, that in recognizing closely related groups, it is necessary to take into account the "internal" parallelisms between their members [Scott 1896; Engler 1898; Osborn 1902]. Devoted "Haeckelians" (such as William Scott) insist that genealogy should be reconstructed based only on the characters inherited from a common ancestor, while those inclined to natural systematics (such as Adolf Engler) believe that it should be supported by the largest number of characters. All these debates will be echoed in modern cladistics by disputes between adherents of the principles of synapomorphy and total evidence (see Section 5.7.3).

At the same time, the relation of systematic phylogeny to two other dominant approaches, namely natural systematics and typology, is hotly discussed. Early attempts to elaborate evolutionary classifications are criticized for their authors not actually reconstructing phylogenies but simply "attaching" phylogenetic interpretations to the preceding classifications without any significant changes in them [Bessey 1915; Kozo-Polyansky 1922]. For this reason, Haeckel's systematic phylogeny is sometimes referred to as a "hidden" typology [Naef 1919; Meyer 1935] and resulting systems are estimated as "pseudo-phylogenetic" [Kozo-Polyansky 1922].

There are many who sharply criticize an equation of the natural and phylogenetic classifications. From a philosophical background, this is objected to by insisting that classifications based on similarity as such are primary, while their phylogenetic interpretations are secondary, so the latter are redundant [Gray 1876; Caruel

1883; Driesch 1899]. Proponents of the natural systematic criticize the Haeckelian approach in that it is based on analysis of only a few characters as evidence of genealogical relationships, while actually natural classifications should be based on a large number of characters. Typologically minded researchers, even admitting the evolutionary idea, consider it inapplicable in systematics because the latter must be "an exact and logical ordering of verifiable facts," therefore the natural system should be developed without reference to unreliable genealogical schemes [Huxley 1864]. At the same time, many taxonomists remain apart from these discussions, considering the genealogical idea as just one more (along with many others) speculative and therefore redundant interpretation of the fundamental concept of affinity.

A very important consequence of mastering the evolutionary idea becomes the phylogenetic interpretation of the originally typological concept of homology: it leads to a revision of the basic criteria for distinguishing homologous and analogous structures and to the introduction of a new terminology. Thus, Haeckel calls Owen's special homology *homophyly* [Haeckel 1866]. Zoologist Edward Ray Lankester suggests eliminating Owen's notion of homology altogether as "idealistic" and replacing it with the notion of *genealogical correspondence*, with its two distinguishing forms [Lankester 1870]. One of these is *homogeny* as a correspondence of structures inherited from a single ancestral form; it is close to Haeckel's homophyly. All other correspondences are *homoplasies*, in their wide understanding including parallelisms, analogies, and also serial homologies. A very detailed system of gradations between homologies and analogies is proposed by George Mivart, who includes genealogical and ontogenetic criteria in their definition [Mivart 1870]. He identifies more than two dozen variants of correspondences of anatomical structures: for example, according to Mivart, there are "homological analogy" and "analogical homology."

At lower levels of taxonomic hierarchy, the basic idea of classification Darwinism provides a strong impetus to a nominalistic understanding of species: since species are constantly changing, they are "a human contrivance [...] made simply for convenience's sake" [Bailey 1896: 457, 459]; "species and variety are [...] abstract concepts, they do not exist in nature" [Timiryazev 1904: 81]. However, more moderately minded taxonomists, along with rejecting species as real natural units, ascribe a real status to local geographic races. This viewpoint leads to a "species splitting": in the second half of the 19th century, races and subspecies are frequently denoted by binomens, i.e., they are actually assigned a formal species status in classifications. In the first half of the 20th century, an emphasis on intraspecific units will be actively developed in biosystematics (see Section 5.6).

2.6   A STEP ASIDE: POSITIVIST REVOLUTION

At the end of the 19th and in the first half of the 20th centuries, natural science is developing under a strong influence of positivist philosophy, dating back to the philosophical empiricism of the 16th century (F. Bacon, J. Locke, etc.) and finally taking shape in the second half of the 19th century (O. Comte, E. Mach, etc.). The central idea is that true scientific knowledge can only be achieved by the inductive scheme of argumentation: at the heart of everything is empirical data, generalizations arise on their basis, all kinds of *a priori* judgments about the "nature of things" must be discarded, everything complex must be decomposed into elementary "bricks," and the relations between them serve as the basis for explaining complex things. An ideological core of this philosophy of science is so-called *physicalism*, according to which any judgments about natural phenomena deserve to be considered scientific if they are substantiated by means and expressed in the language of physics, including employment of experimental evidence and mathematical apparatus.

Against this background, new biological disciplines make it possible to take a significantly different look at organisms and at relations between them; with this, they correspond, to a greater or lesser degree, to the physicalist understanding of the content and principles of natural science. First of all, genetics should be mentioned here, in which an organism is reduced to a "sum of genes" and thus the evolutionary issues are transferred from the level of organism to the level of genetic units. This reductionist trend is reinforced by biochemical studies making it possible to compare organisms not by anatomical (macrolevel) but by biochemical (microlevel) features. Finally, the emergence of ecology, in which the main emphasis is placed on the processes occurring in local populations, also becomes an important novelty of the "new biology."

Because of all this, classical biology associated with the study of the diversity of organisms at just the macrolevel appears to be outside the realm of science thus understood. This concerns first of all classical systematics, which is assigned the role of the Cinderella of the "new" natural science. Thus, Sinaï Tschulok divides biology into biophysics (exploring "real" relations) and biotaxonomy (exploring "imaginary" relationships) [Tschulok 1910]; similarly, the ecologist Eugene Odum divides biological disciplines into fundamental and taxonomic [Odum 1953]. Classical systematics is criticized as purely descriptive by the founders of the onto-rational doctrine [Driesch 1908; Lyubishchev 1923]; developers of new taxonomic ideas and approaches (biosystematics, numerical phenetics) reject it as "morally obsolete" [Turrill 1940; Gilmour and Turrill 1941; Mayr 1942; Heslop-Harrison 1960; Michener 1962; Sokal 1966]. The growing influence of the Darwinian micro-evolutionary theory also plays a significant role in the decline of interest in classical systematics: its interpretation of the causes of the diversity of organisms appears to be more compatible with physicalist philosophy and makes the general concept of the Natural System completely redundant.

Progressive rationalization of systematics in a new guise, trying to respond to the challenges of the new philosophical scientific context, becomes the most noticeable trend of its conceptual history during the first half of the 20th century. The emergence of this "new rationality" marks a drastic shift from the old to a new cognitive situation in systematics, which can be called the *positivist revolution* [Pavlinov 2018, 2019]. Accordingly, as a result of the latter, something like *positivist systematics* is shaped; this term designates a certain cognitive program of a physicalist kind, and within its scope several particular research programs are formed. The article "Taxonomy and Philosophy" by the botanist John Gilmour appears in a festschrift with the iconic title *The New Systematics* [Huxley 1940a] to become a kind of "positivist mani-festo" for this revolution [Gilmour 1940]. The symposium "Philosophical Basis of

Systematics" organized by the Society of Zoological Systematics (USA) in 1961 (materials published in *Systematic Zoology*, vol. 10, issue 4) clearly shows the dominance of positivist ideas in the theoretical systematics of this time.

This movement of systematics towards a new rationalization is most clearly evident from the attempts to re-formulate its onto-epistemic foundations. On the one hand, the task of its nomotetization is designed to make classifications by status similar to physical or chemical laws [Driesch 1908; Lyubishchev 1923, 1972, 1975; Ho 1990, 1992; Ho and Saunders 1993, 1994; Webster and Goodwin 1996; Zakharov 2005]. On the other hand, attempts at axiomatizing systematics and building its foundations by employing more rigorous formalisms become more accentuated [Woodger 1937, 1952; Thompson 1952; Gregg 1954; Mahner and Bunge 1997; Pavlinov 2011a, 2018].

An active assimilation of research methodologies of a physicalist kind becomes another important response of systematics to these new challenges: this includes issues in experimental (reliance on experiments with organisms or their derivatives), numerical (reliance on the aphorism "mathematics is the queen of all sciences," by Carl Gauss), and phenetic (substantiated by direct reference to the elementarist background of positivism) systematics. One of the first becomes *biosystematics*, which grows out of classification Darwinism [Hall and Clements 1923; Camp and Gilly 1943; Camp 1951]. In it, all discussions about species and supraspecific taxa, phylogeny, and homology are discarded as "unscientific"; the main emphasis is placed instead on classifications of intraspecific categories. A prerequisite for this is the interpretation of populations as the key units of the ecological communities and the real actors of the evolutionary process. Among its important parts are new approaches to solve systematic tasks based on ecological, hybridological, and immunogenetic experiments. Penetration to the subcellular level gives rise to several branches of biosystematic studies that can be called "*character-based*"; these are chemosystematics (analysis of chemical compounds, including macromolecules), karyosystematics (analysis of chromosomes), etc.

One of the most important parts of the positivist revolution in systematics becomes a combination of phenetic and numerical ideas: this gives rise to *numerical phenetics*, which ties biological systematics most closely with the philosophy of physicalism. It is the maturation of these ideas that their supporters declare to be another revolution in systematics estimated as almost greater than the previous "Darwinian" one [Sneath 1995]. Quantitative methods for demarcating populations developed by biosystematics are the first step in this direction, though elaborating full-fledged classifications on a quantitative basis is not presumed at this step [Simpson and Roe 1939]. Such a task for "*exact systematics*" is set for the first time by the zoologist Evgeny Smirnov in the 1920s–1930s [Smirnov 1923, 1938], but his approach does not become popular because of its typological terminology. Actually, the beginning of this "revolution" is laid by a series of works by zoologist Robert Sokal and microbiologist Peter Sneath, and the appearance of another "new systematics" is immediately announced [Michener 1962], with the book *Principles of Numerical Taxonomy* becoming a kind of "gonfalon" [Sokal and Sneath 1963]. In the second half of the 20th century, the development of computer technology, including ready-to-use computer programs with standard statistical and classification methods, leads to a nearly total "numericalization" of most of systematics. New high-performance computers make it possible to process large amounts of data, which causes the announcement of the emergence of yet another "new systematics" stimulated by quantitative processing of "big data" [Schram 2004]. These newest instrumental developments in systematics are considered to mark the early stages of a technology-driven revolution [Wilson 2005], resulting in *cybertaxonomy* that deals with processing data and metadata on biodiversity using standard electronic tools [Smith 2013; Wheeler and Hamilton 2014].

The biosystematic research program is being developed mainly by botanists and retains a fairly large influence on botany to this day [Lines and Mertens 1970; Takhtadjan 1970; Hedberg 1997; Feliner and Fernandez 2000]. Numerical phenetics in its pure form turns out to be very limited in its effective application to microsystematic research only and, because of this, fades away from the 1970s. However, its principal ideas are in demand in the currently dominant "new phylogenetics," which flourishes within the framework of post-positivist systematics.

2.7 HOMAGE TO METAPHYSICS: POST-POSITIVIST REVOLUTION

In the middle and second half of the 20th century, the "historical pendulum" in the development of natural science begins to move in the opposite direction. The influence of positivist (first of all, physicalist) ideas, putting a hard boundary between "science" and "stamp collecting" and founding all natural science on a reductionist counter-metaphysical basis, is noticeably weakening. Instead, the ideas of the post-positivist philosophy of science come to the fore, forming the basis of its *non-classical* cognitive paradigm [Popper 1959; Kuhn 1962; Lakatos 1978; Quine 1996; Ilyin 2003; Stepin 2005]. An understanding emerges that the general cognitive situation of natural science is structured much more complexly than is assumed by both the classical and positivist philosophies of science. With this, a key place in its ontology is given to a rather rich scientific metaphysics shaping the background knowledge of natural science research. Besides, scientific pluralism is proclaimed, according to which cognitive activity in science is structured by particular research programs and paradigms interacting in a complex way. The fundamental cognitive status of the diversity of biological phenomena, which is in no way inferior to the search for unifying laws and constants, is acknowledged [Rosenberg 1985; Mayr 1988; Chebanov 2016]. This results in a revival of interest in the Natural System as a manifestation of general regularities supposedly occurring in the structured diversity of any natural phenomena [Rozova 1980, 1986; Tchaikovsky 1990; Rozov 1995].

Discussion of new ideas of onto-epistemic foundations of natural science, with a return of interest in metaphysics and respective loss of interest in positivist elementarism, has a noticeable impact on theoretical systematics and marks the beginning of a new phase in its conceptual history. Biological systematics begins to return to its former prestige and gradually frees itself from the "Cinderella syndrome" [Rosenberg 1985]. This leads to a loss of attractiveness of the research programs that emerged in the context of the positivist paradigm (phenetics, partly biosystematics), and creates certain preconditions for a revival of interest in the macro-scale historical and structural manifestations of taxonomic diversity. According to how philosophers of science call the new cognitive paradigm, this shift of systematics in a new direction can be designated as a *post-positivist revolution*, the last (at the present moment) in its conceptual history [Pavlinov 2019]. From the conceptual historical perspective, its main content is substantive, so it does not include the above-mentioned "technological" revolution continuing the physicalist trend.

At a philosophical scientific level, it is important, although hardly fully realized so far, that this revolution, generated by the influence of the non-classical scientific paradigm, stimulated the beginning of the formation of a *non-classical* frame for systematics [Pavlinov 2006, 2013b, 2018]. It legitimizes not only metaphysics as an important part of the cognitive situation of this discipline, but also taxonomic pluralism as an inevitable consequence of the diversification trend in its conceptual history.

Among particular research programs that begin to develop most actively in the context of the post-positivist history of systematics, phylogenetics in its new guise occupies first place. It is shaped by a combination of three independently emerging components, viz. cladistic methodology, molecular factual basis, and numerical technology. Together, they shape the *new phylogenetics* [Pavlinov 2003, 2005]; its taxonomic application suggests the name *genosystematics* [Mednikov 1980; Antonov 2006]. The changes it makes in practical systematics are so significant that they are sometimes equated with a scientific revolution [de Queiroz 1988], which is called *cladogenetic* [Pavlinov 2019, 2020]. However, this "revolution" is not so much a conceptual as a technological one, connected with the development of new approaches to phylogenetic reconstruction; therefore, in this book its formation is presented as part of the post-positivist revolution.

The basic ideas of a cladistic version of phylogenetics were foreseen by C. Darwin (see Section 2.5.1), and in the 20th century they are formalized by the botanist Walter Zimmermann and the zoologist Willie Hennig [Zimmermann 1931, 1934; Hennig 1950, 1966]; from the 1970s, it becomes increasingly popular, and manuals are published one after another [Nelson and Platnick 1981; Wiley 1981; Ax 1987; Shatalkin 1988; Pavlinov 1990, 2005; Wägele 2005; Williams and Ebach 2008]. This version differs from Haeckel's mainly by a refined theoretical and operational definition of monophyly. This refinement is supplemented by a proposal to classify only strictly monophyletic (holophyletic) groups, which causes the radical restructuring of many "classical" phylogenetic classifications. The molecular component of the new phylogenetics is developed by *phylogenomics* (genophyletics): it gives the greatest importance to DNA and RNA macromolecules and ignores other organismal features, so it is a version of the above-mentioned chemosystematics. It is of importance that molecular genetic data remove many restrictions that did not previously allow analysis of the diversity of prokaryotes and their direct comparison with eukaryotes, so an idea of the global "Tree of Life" is revived, now built on a unified molecular basis [Cracraft and Donoghue 2004]. An active development of the numerical component of the "new phylogenetics" is predetermined by its cladistic and molecular genetic components. Numerical phenetics and phyletics develop almost synchronously in the

1960s–1970s in acute competition, but numerical phyletics wins this "struggle for existence" [Hull 1988] and now dominates in taxonomic research. Inspired by this, the statistician Joseph Felsenstein, one of the leaders of molecular phyletics, announces in a somewhat joking manner that he "founded the fourth great school of classification, the It-Doesn't-Matter-Very-Much school" [Felsenstein 2004: 145]; its methods are considered in a number of manuals [Hillis et al. 1996; Nei and Kumar 2000; Felsenstein 2004; Albert 2005]. With this, the methodology of this "great school" is largely guided by the same reductionist positivist philosophy that earlier gave rise to numerical phenetics, so "de-physicalization" of the most recent sytematics appears to be only partial [Pavlinov 2018, 2019].

A shift from a classical to cladogenetic version of phylogenetics provides an interesting example of the so-called "butterfly effect" in the conceptual history of systematics. Indeed, it begins with just a narrower definition of the monophyletic group as the only allowable element of the phylogenetic classification. Then it becomes enthusiastically accepted and massively applied by phylogeneticists of the new generation. This leads to significant changes in most traditional phylogenetic classifications, at first without any "molecular" interference.

In the recent conceptual history of systematics, a noticeable place is occupied by structuralist theories that gravitate towards typology in its various manifestations; they compose *onto-rational systematics* [Pavlinov 2011b, 2018; Pavlinov and Lyubarsky 2011]. The latter is aimed at uncovering structural causes of the morphological disparity of organisms; its basic task is to identify law-like regularities in this disparity and to present them in the form of natural classifications of their own.

At present, it is realized that "cladogenetic revolution" in systematics appears to be too radical and this leads to its significant "de-biologization." In this regard, an impression arises that a new revolution in systematics is slowly brewing, promising a *post-cladistic* stage in its conceptual history [Wheeler 2008; Williams and Knapp 2010; Zander 2013; Pavlinov 2019, 2020]. Its driving force seems to be a very interesting research program, which begins to emerge within the framework of evolutionary interpreted systematics. It links the structural consideration of the disparity of organisms with evolutionary developmental biology ("evo-devo") [Minelli 2015; Pavlinov 2019, 2020], which focuses on the evolution of epigenetic regulators of ontogenetic patterns [Minelli 2003; Minelli and Pradeu 2014]. This program, in its own evolutionary way, revives a significant element of epigenetic typology and provides a new stimulus for the development of the above-mentioned ontogenetic systematics. Whether or not this revolution will happen and whether it will actually lead to the emergence of a new research program of *evolutionary ontogenetic systematics* remains to be seen (see Section 5.8).

Currently, however, a new and more practical challenge seems to rise that may seriously affect the nearest future of biological systematics. It is caused by a certain gap between the needs of the biodiversity issues and the capabilities of biological systematics, which is designated as *"taxonomic impediment"* (e.g., [Godfray 2002; Wheeler et al. 2004; Agnarsson and Kuntner 2007; de Carvalho et al. 2007; Raposo et al. 2020]; etc.). It remains just to guess at the moment what might be its consequences for the conceptual development of systematics to make it another "new" one.

3 Some Philosophical Considerations

> Science is what you know, philosophy is what you don't know.
>
> Bertrand Russell

One of the main motives of this book is the idea that any scientific discipline, including systematics, is impossible without its own theory—otherwise, it turns out to be "wandering by touch" within an unknown. Another motive, less obvious but just as important, is recognition that the theory of science is impossible without the philosophy of science, or *metascience*: it is the philosophy that lights the lantern illuminating the paths of scientific knowledge. As a matter of fact, it is the philosophy of science that poses and tries to answer the questions of paramount importance for systematics, such as: how can its studied object be outlined? How can we be sure that our classifications actually reflect something of interest to us? What makes the research method and the knowledge it provides scientific?—etc., etc., etc. Therefore, if science does not exist without scientific theories, then the latter do not exist without a certain philosophy of science [Heisenberg 1959].

In systematics, as E. Mayr emphasized, "every basic biological concept is also a philosophical concept [...] they run into each another" (cited after [Greene 1992: 259]). So, the scientific status of both systematics and the taxonomic knowledge developed by it cannot be comprehended outside a certain philosophical and theoretical context that deserves being termed *philosophy of systematics*. Therefore, before considering the theory of systematics, it is necessary, very briefly and in quite a simplified form, to touch upon some principal issues of the philosophy of science that concern our discipline most closely.

3.1 CLASSICAL AND NON-CLASSICAL SCIENCE

From the beginning of the formation of European science in the 16th century and up to the end of the 19th century, all natural sciences developed within the framework of the classical philosophical canon. Throughout the 20th century, a movement shifted slowly from the classical towards a non-classical version of the philosophy of science. They differ in their understanding of the ways of defining objects, tasks, and methods of natural sciences [Lektorsky 2001; Gaydenko 2003; Ilyin 2003; Stepin 2005].

The *classical philosophy of science* is basically monistic. It acknowledges that Nature in all its manifestations is organized by a unified global cause—the overall universal Law of Nature—so it is possible in principle to explain all phenomena of Nature by its direct or indirect action. If Nature is unified in its foundations, then scientific knowledge about it should also be unified in its content and form. Such knowledge must be strictly objective and reflect what is "in reality," and to do this it must be induced impartially and consistently from observations of objective facts. Such knowledge can be achieved with the help of the universal Scientific Method that is equally applicable to all phenomena of Nature; what is subject to such Method is the science—everything else is "non-science." For this Method to be "objective," any influence of both prior ideas about the "nature of things" (metaphysical background knowledge) and any subjects' opinions and preferences should be excluded from it. This makes the Method—now a method (with a lowercase letter) as a set of technical devices—objective, and its application by itself should warrant scientific knowledge to be objective. Such method must certainly be mathematical, for "The Book of Nature is written in the language of mathematics" [Galilei]. The method is intended to link all conceivable variables that characterize the diversity of natural phenomena into a single "formula." The latter "shapes" a unified and therefore the only "general theory of everything," and if it actually explains "everything at once," then this is the "final theory." This is how the Nobel laureate physicist Steven Weinberg, in his book under a rather symptomatic title *Dreams of a final theory* defined a goal towards which natural science should strive [Weinberg 1992]. A movement towards this goal, albeit slow but steady and unidirectional, subordinates the development of science to a unified trend of achieving the only possible true knowledge—from less to more complete, from less to more objective, from less to more precise, etc.

For *classical systematics*, a single universal Law is given in the form of the System of Nature: let us recall that, according to Linnaeus, "Nature is the Law of God" [Breidbach and Ghiselin 2006]. Accordingly, this System is reflected by a unified natural classification, which is achieved by means of a unified natural method: as M. Adanson once assured, such a method "should be universal, or unified, that is, there should be no exceptions for it" [Adanson 1763: civ]. This general idea dominated throughout the conceptual history of systematics and remains highly respected to this day, regardless of how natural classification and natural method are understood. So, with regard to such a claim to possessing a "final theory," the newest cladistics hardly differs from, say, the quinarism of the early 19th century.

In the *non-classical philosophy of science*, understanding of how Nature and knowledge about it are arranged is fundamentally different—it is pluralistic. Nature is organized quite complexly; its organizing principle is a combination of different causes, local in their actions and not reducible to some unified "cause of everything"; therefore, it is impossible to encompass it with a single "general theory of everything." As a result of cumulative actions of such causes, Nature breaks up into various local manifestations—fragments, aspects, levels, etc., each with its own local cause or a local set of causes. Therefore, non-classical science focuses on the very diversity of manifestations of Nature: this means that natural (as well as social, cultural, etc.) diversity is investigated not to reduce it to a single law, but in all its particular manifestations that are actual knowable objects. In such a "dismemberment" of overall Nature, an active role is played by the cognizing subject: without the latter, cognition is not feasible, which means that "absolutely objective" knowledge is impossible. Scientific knowledge cannot be simply inferred from direct observations: the latter become scientific facts only after they are comprehended in the context of some theoretical background knowledge that forms the metaphysical (eventually natural-philosophical) context of cognitive activity. The latter is organized as a set of research programs, each concentrating on a certain manifestation of Nature outlined by the respective metaphysical context, and a particular scientific theory is developed for it. Accordingly, scientific knowledge is an organized set of "local" theories with their specific objects, aims, and methods. However, such scientific pluralism, contrary to the widespread misconception, does not imply a denial of some "unified" truth: it only means that there is no single "classical" truth in the form of a "general theory of everything," but there is a multiple "non-classical" truth encompassing all "local" truths elaborated by the respective theories that complement each other in a "whole." With this, a plurality of such "local" truths means that scientific knowledge inevitably includes some unavoidable amount of uncertainty.

The following provisions are of fundamental importance for the development of what might be called *non-classical systematics* [Pavlinov 2006, 2012, 2018]. First, focusing on the structure of diversity and corresponding means of studying it, non-classical science allows systematics to assert its right to be a "normal science" with its specific subject area (taxonomic reality; see Section 4.2.1), specific research method adequate to it (classifying), and specific generalization reflecting it (natural classification). Second, non-classical systematics acknowledges the recognition of particular manifestations of the taxonomic reality that deserve to be considered respected objects of research to the extent that these manifestations are individualized naturally (by presuming certain natural causes generating them). It follows from the above, third, the reasonability of developing particular taxonomic theories dealing with these manifestations, along with research programs implementing the respective theories. And lastly, such theories and programs and, of course, the "locally" natural classifications reflecting these manifestations, can all be ascribed the same scientific status.

3.2   COGNITIVE SITUATION

Any research activity is conducted within the framework of a specific *cognitive situation* generated by some general cognitive goal and particular cognitive tasks formulated within the latter's context [Barantsev 1983; Yudin 1997]. In natural science, such a goal is determined by an aspiration to comprehend a certain phenomenon of Nature; in systematics, this is taxonomic reality. Under the provisions of non-classical science, this general goal cannot be reduced to specific tasks in a single trivial way, so a *cognitive problem* arises [Prytkov 2013]. Therefore, research activity is designed to resolve the respective cognitive problem within the framework of the respective cognitive situation. For example, in systematics, among them are the problem of the Natural System, the problems of homology and species, etc.: they collectively shape the overall cognitive situation this discipline deals with.

Both the particular research problem and the cognitive situation constructed upon it are conditioned by certain philosophical-scientific and historical contexts [Miller 1996; Rozov 2002; McCray 2006]. The variety of these contexts entails the respective variety of cognitive situations, while the historical changes in the contexts entail the respective changes of situations; such changes constitute the content of the conceptual history of systematics.

3.2.1 Cognitive Triangle

The structure of the cognitive situation is formed generally by its three basic components: ontic (= ontological), epistemic (= epistemological), and subjective.

Ontology considers the question *"what?"*—what is the studied object? With this are defined the object itself, its main properties, the processes associated with it, etc.—certainly not all of them but those considered significant within a particular cognitive situation. Here belong issues concerning the correct ways of defining both taxonomic reality and its manifestations, speculations about their causes, real *vs.* nominal status of the taxa and their ranks, etc.

Epistemology addresses the question *"how?"*—how research should be conducted to obtain the knowledge sought about the object defined by the ontic component. It concerns the scientific status of knowledge, how it could be obtained and substantiated, etc. An important part of epistemology is *methodology*: it substantiates research methods and develops the criteria for their scientific consistency, suitability for taxonomic research, etc.

The *subject component* is about *who* decides what and how should be investigated; this "who" may be a scientific community, a particular scientific group (school, lab, etc.), and eventually a researcher. The role of the subject (in a broad sense) in a cognitive situation is rarely taken into account, but it is very significant. At a personal level, it manifests itself in a subjective preference for a holistic or reduction perception of the cognizing reality, for an intuitive or analytical way of knowing it. At a higher level, it manifests itself, for example, in the social regulation of preferences in the choice of tasks, methods, etc. on the grounds not related to science itself (see the beginning of Chapter 5). So the most important issues here concern the character of the subject's influence on the structure of a cognitive situation.

Generally speaking, it is the subjective component that shapes the entire cognitive situation and establishes mutual interrelationships between all its basic components: due to this "subject effect," they do not occur in isolation from each other but interact in a complex manner. Both their triplicity and interdependence make it possible to represent the entire cognitive situation metaphorically as a *cognitive triangle*.[1] Its vertices correspond to its basic components, the edges reflect their interrelationships, and its center corresponds to a certain concept (taxonomic theory). So the latter appears as a result of the complex interaction of all three components and should be considered taking this interaction into account.

The interrelation of the first two components of the cognitive situation is fixed by the *principle of onto-epistemic correspondence*; it goes back to the ancient ideas of unity of Nature and Method, and in terms of the contemporary philosophy of science it means the following [Pavlinov 2018]. On the one hand, the object under

---

[1]   There are several versions of the triadic representation of things related to cognitive activity. Some of them ascend to Pierce's semiotic triade, some (for instance, in cognitive psychology, behaviorism, etc.) are more original [Surov 2002; Mechkovskaya 2007; Atkin 2013; Innis 2020]. The concept of cognitive triangle does not claim to be quite original; it is introduced to emphasize interrelations between the three basic components of a cognitive situation.

study (a part of ontology) is outlined depending on certain epistemic considerations (first of all, on supposed cognizability by scientific means). On the other hand, some important methodological principles (epistemology) are developed based on certain fundamental properties attributed to the object under study. For instance, in systematics, a purely empirical view of Nature, devoid of any metaphysics, reduces it to "physically" perceived organisms; from a more holistic standpoint, certain speculations about causes of the diversity of organisms constitute an important part of the ontology. In cladistic systematics, specific methods of phylogenetic reconstructions are substantiated by reference to the divergent character of phylogenesis [Hennig 1966; Wiley 1981; Pavlinov 2015]. The onto-epistemic correspondence in numerical phyletics is opposite: based on the properties of statistical method of maximum likelihood, phylogeny is presumed to be a stochastic process [Felsenstein 1982, 1983, 2004].

The fundamental significance of the onto-epistemic correspondence for systematics is quite obvious. Since biodiversity does not lend itself to direct manipulations in experiments, and since phylogeny, as one of its supposed causes, is not reproducible and observed, some logical arguments are needed to substantiate certain correspondence of the results obtained by a particular analytical method to what is assumed to be the studied reality. And the principle in question is a key part of these arguments.

The *problem of instrumentalism* [Rieppel 2007; Pavlinov 2018] is to be mentioned in this regard. It means that the epistemic assessment of taxonomic knowledge prevails over the ontic, which is characteristic of the approaches that substantiate consistency of the research methods with reference to their logical (axiomatic, etc.) foundation. As a result, something like *inverse correspondence* appears: the method as such dictates how the reality should be comprehended and studied, so the properties of the method indirectly shape the properties of the reality.

3.2.2 CONCEPTUAL SPACE

The cognitive situation is "enveloped" by a *conceptual (notional) space*, or a *conceptual framework* [Botha 1989; Gärdenfors 2000; Efremov 2009; Pavlinov 2010b, 2011a, 2018]. It is formed by the basic *thesaurus*—those concepts and respective notions with which an object to be studied is defined; in our case, it is the taxonomic reality with all its properties and elements and occasionally causes. Metaphorically, the thesaurus can be called a *conceptual (notional) model* of this reality: the more complete and detailed the thesaurus is, the more adequately such a model reflects the reality in question, so if something in the latter is not reflected in the thesaurus by respective notions, this "something" does not exist in the cognitive situation [Kuraev and Lazarev 1988; Margolis and Laurence 2011]. From this, it follows by a "reverse reading" that each conceptual (notional) model shapes the studied reality in a specific way: what this model is, so also is the reality implied by it. This is just what is presumed by concepts of linguistic relativity and linguistic world picture developed by cognitive semantics [Gumperz and Levinson 1996; Talmy 2000].

The main elements of the thesaurus are *concepts, notions,* and *terms* [Voyshvillo 1989; Murphy 2002; Margolis and Laurence 2011]; the first two are semantic (substantive), whereas the third is semiotic (symbolic). They all relate in some way to an object being studied: the notion designates it, the concept provides a substantive interpretation of the notion, and the term denotes it with some symbol. A connection between a concept and an object is established by a *definition* of the notion; there are two main methods of definition. The *intensional* definition indicates the intrinsic properties of the object, while its *extensional* definition indicates its elements. In systematics, for example, the intensional definition of a taxon includes a list of the essential properties of the organisms allocated to it, while the extensional definition includes a list of these organisms.

The definitions of notions/concepts that compose a thesaurus should be complete, explicit, and strict for the conceptual space to be properly formatted: this is one of the key conditions of the *principle of constructiveness* [Voyshvillo 1989]. *Completeness* means that the definition provides exhaustive meaningful characteristics of the biological phenomenon reflected by the respective notion. *Strictness* presumes one-to-one correspondence between the notion and the object under a variety of possible conditions supposed by the cognitive situation. *Explicitness* means that the concept/notion should be not presumed but clearly defined, either directly in the given thesaurus or with reference to a more general thesaurus that includes this one. Indeed, in order to discuss effectively (constructively) the problems of species or homology, all concepts related to them must be defined in the manner just outlined; in ordinary language, this means to "agree on words." Exceptions are undefined concepts introduced in the cognitive situation as basic; in systematics, for example, these are concepts of biota, biodiversity, organism, for which it is just assumed that researchers using the respective terms mean (approximately) the same natural phenomena.

However, in the natural sciences, these requirements are never met for several reasons, of which the main one is the fundamental inaccessibility of one-to-one correspondence between a natural phenomenon and its conceptual reflection in a cognitive situation. The boundaries of inaccessibility are established by the *principle of the inverse relationship between rigor and meaningfulness of notion* [Kuraev and Lazarev 1988; Voyshvillo 1989]: the more strictly the latter is defined, the less likely there is something in nature to which it may exactly correspond. The reason is that the rigor of definition is in close conjunction with formalization, and the latter is the reverse of substantiveness; mathematical notions are most strictly defined, but they are pure abstractions, which are not supposed to relate to something in Nature [Perminov 2001]. In systematics, this contradiction occurs, for example, in an aspiration to define the concept of monophyly or species as strictly as possible: intuitively more or less obvious, they appear less and less applicable in practice as they become more and more rigid [Pavlinov 1990, 2007b, 2018; Hull 1997; Hołyński 2005].

One of the important causes of the non-rigid character of taxonomic notions/concepts is that their definitions are context-dependent. This means that a concrete definition of a particular notion is given not for all "possible worlds," but for a particular cognitive situation, and it depends on the entire contents of the basic thesaurus "enveloping" it. Therefore, some general notion (phylogeny, monophyly, species, homology, etc.) is split into particular interpretations whose meanings depend on the context of the particular thesauruses.

It follows from the above that definitions of notions/concepts shaping the conceptual space in systematics are inevitably fuzzy, contextually dependent, and therefore metaphorical to an extent [Murphy 2002; Pavlinov 2018]; therefore, the notions elaborated by systematics cannot be considered rigid designators in the sense of S. Kripke [Kripke 1972]. Thus, the semantics of systematic thesaurus is adequately described by a probabilistic model of the natural science language [Nalimov 1979], with proper formalizations provided by fuzzy logic [Zadeh 1992; Kosko 1993]. This means that, for each concept, it is possible to fix more or less strictly only its "core" using the logical relation "A is B," while its periphery remains fuzzy, presuming multiple contextual interpretations of the type "A may be $B1, B2, B3...$" As a result, each notion/concept in systematics occurs as a set of particular context-dependent interpretative definitions; with this, the latter should be strict only to the extent it is really needed within particular cognitive situations [Hołyński 2005; Pavlinov 2010b, 2011a, 2018].

3.2.3 Conceptual Pyramid

A fundamental property of the conceptual space is its hierarchical structure, which is determined by how concepts are introduced into the cognitive situation through their respective definitions [Hempel 1965; Quine 1996; Hacking 1983]. This hierarchy is explained by the epistemic *principle of theory incompleteness*,[2] according to which no particular theory (as a conceptual system) can be exhaustively defined within its own thesaurus: for such a definition, a certain metatheory is needed, in terms of which the basic concepts of this particular theory are interpreted [Antipenko 1986; Perminov 2001]; in this case, theory and concept are equivalent as definable elements of the general conceptual space. This principle follows from the logical genus–species scheme, in which each particular concept is defined consistently as "species particular" in the context of the "generic common."

For a formal representation of this hierarchy, the so-called *conceptual pyramid* can serve as a suitable metaphor; its forerunner is the "pyramid of concepts" of Medieval scholasticism [Makovelsky 2004; Pavlinov 2018]. The top of this "pyramid" corresponds to a framework theory (concept) of the most general order, lower levels correspond to more particular theories and concepts—and so on down to the lowermost levels of operational concepts. The "pyramidal" character of such a hierarchy is a consequence of the fact that, moving from initial to final links of the interpretation chain, the number of concepts gradually increases at each step. The reason is that the number of particular interpretations of each concept is always more than one at any level of generality; therefore, there are always less general concepts exceed in number more general ones. For example, in systematics, the general notion of affinity can be

---

[2] The epistemic principle of theory incompleteness originates from the incompleteness theorem deduced within one of the versions of formal arithmetic by the mathematician Kurt Gödel.

interpreted as either similarity or a kinship relation, while the latter can be defined in different ways depending on how the phylogeny is interpreted.

The metaphor under consideration presumes the following important features of the hierarchical structure of the conceptual space to be emphasized.

Firstly, concepts of lower levels can be meaningfully defined only in the context set at higher levels of the "pyramid" [Carrier 1994; Quine 1996]. This, in particular, is true for operational concepts and classification algorithms: neither they themselves nor the classifications elaborated with them have a fixed biological meaning, unless it is specified by some biologically sound concept (theory) of higher levels. This observation concerns classifications based on strictly empirical (in particular, phenetic) approximations, i.e., on similarity relations as such. This is because, outside of the context set by certain biologically sound concepts, in which these relations are substantively interpreted by indicating their biological causes, their specific biological meaning remains undefined. Indeed, operationally defined groups (phenons) can correspond to species or to biomorphs or to certain intraspecific groups (such as age stages or castes in insects). So these phenons as such can hardly be considered biologically meaningful without an indication of the particular elements of the structure of biota they correspond to, which requires prior substantive definitions of these elements.

Secondly, various particular (subordinate) theories and concepts that figure in systematics turn out to be interconnected through more general theories/concepts that unite them into a single conceptual system. This is provided by their meaningful cross-interpretations in terms of certain metatheories of higher levels of generality. For example, in phylogenetics, the possibility of cross-interpretation of particular meanings of kinship is provided by setting particular definitions of phylogeny, referring to the latter's general understanding as one of the manifestations of the evolutionary development of biota.

3.3 SOME COGNITIVE REGULATORS

In cognitive activity, the key role is played by basic cognitive categories (concepts, principles, etc.) acting as its general regulators. They "dictate" to researchers how to perceive and describe natural phenomena, so their consideration is one of the most important sections of the philosophy of science. Among them, there are regulators of the most general order (analysis *vs.* synthesis, abstraction *vs.* concretization, generalization, etc.) and more specific ones involved in the formation of specific cognitive situations. This section discusses those regulators of the second group that are most relevant to biological systematics.

3.3.1 BETWEEN *UMGEBUNG* AND *UMWELTS*

All natural science is aimed at comprehending Nature in all its contents, manifestations, and details—that is, what it is "in reality." However, this ultimate end is principally unattainable: Nature is so global and diverse that it is impossible to comprehend it exhaustively because of the limited character of human cognitive means. For this reason, any cognitive activity is inevitably selective and reductionist: a cognizing subject selects something in Nature as a studied object guided by certain reasons. This is the principal cause of the differentiation of natural science into disciplines, each with its own object and adequate tools—physics, chemistry, geology, biology— and in biology—ecology, physiology, genetics, anatomy, and, of course, systematics.

This circumstance is regulated by the generally valid *principle of ontic reduction*; its important epistemic addition is the already-mentioned *principle of constructive- ness*. With their effect in action, cognitive activity is actually addressed not to the *objective reality* (Nature in its entirety) but rather to a certain *cognizable reality* that corresponds to a certain manifestation of the former. It is "construed" by a cognizing subject so that, on the one hand, it corresponds to a certain manifestation of Nature, and on the other hand, it becomes accessible for research on some operational (con- structive) basis [Hayes and Oppenheim 1997; Devitt 2005; Stepin 2005; Knyazeva 2006; Riel and Gulick 2014]. This position is generally known as *ontological rela- tivity* [Quine 1996; Dupré 1993; Mahner and Bunge 1997], developed by contem- porary conceptualism [Swoyer 2006].

A peculiar justification for such an ontic reduction is the general scheme of two- level division of overall reality into the objective world as such, termed *Umgebung*, and its actually knowable manifestation, termed *Umwelt*; it was proposed at the beginning of the 20th century by the zoopsychologist Jakob von Uexküll [Kull 2009; Uexküll 2010]. Although Uexküll himself meant selective "biological" perceptions of the environment by particular organisms, his metaphoric concept appeared to be widened to include human cognitive activity, which is no less selective with respect to Nature being cognized [Knyazeva 2015]. As a result of this reduction, a natural phenomenon as part of objective reality (*Umgebung*) turns into an investigated object constituting a particular *cognizable reality* (*Umwelt*). The fundamental difference between them is that the *Umgebung* exists outside and independently of a subject, while an *Umwelt* is "constituted" by the latter's cognitive activity and thus does not exist outside the respective cognitive situation. So, this metaphor is an illustration of the above-stressed irremovable occurrence of the subject component into any cogni- tive situation.

In fixing an *Umwelt*, only those properties and relations of a cognizable phenom- enon are taken into account which are considered significant for a particular research task within the framework of a particular cognitive situation. Other properties are discarded, which means "cutting out" the respective *Umwelt* from the general context of the *Umgebung*. Thus, the delineation of an *Umwelt* is a conceptual operation of the ontic reduction of the *Umgebung*, which turns a certain manifestation of objective reality into a particular *conceptual reality*. The latter is separated from the *Umgebung* by an *ontic break*: the wider the latter is, the more properties and relations of a natural phenomenon are "cut out" by individuation of the respective *Umwelt*. The latter can be further reduced to more specific "*sub-Umwelts*" down to some purely operational concepts.

In systematics, the ways of delineating particular *Umwelts* of various levels of generality and content are quite diverse depending on the ways biota (*Umgebung*) and its manifestations can be considered. Examples of higher-level *Umwelts* are phylogenetic pattern and typological universum; examples of lower-level *Umwelts* are cladistically and biosystematically defined aspects of the diversity of organisms;

cognitive situations shaped by the problems of species and homology can also be thought of as based on particular *Umwelts*.

For the construction of an *Umwelt*, conditions of rationality are of great importance: they presume that operating with analytical methods is more effective in the case of a simply construed cognized object. Therefore, simplification of *Umwelt*, by its "cutting out" from the complexly organized *Umgebung*, makes it more operational. As a result, losing in one (complexity), one gains in another (operability); the only question is whether the substantial losses are covered by formal acquisitions.

In folk systematics, particular *Umwelts* are isolated implicitly based on specific mythologies (ontology) or pragmatic considerations (epistemology). In scientific systematics, its *Umwelt* is rationally (explicitly) delineated by certain conceptual means, so *taxonomic reality* thus appears as the basic conceptual reality that constitutes its specific *Umwelt* [Zuev 2002, 2015; Pavlinov 2010b, 2011a, 2018; Pavlinov and Lyubarsky 2011]; for more details of this concern, see Sections 4.2.1 and 3.3.2.

3.3.2 BETWEEN HOLISM AND REDUCTIONISM

Holism and reductionism interpret admissible ways of understanding and studying complex objects in an opposite manner. The former implies consideration of such objects as integrated wholes not reducible to a sum of their constituents. The latter, on the contrary, means an obligatory reduction of such objects to their elementary components or manifestations.

The *holistic approach* justifies its position by supposing that a complexly organized natural phenomenon of whatever level of generality is endowed with a certain intrinsic essence, which makes it what it is "by its nature." At a global level, this ontology is most evidently expressed in some natural-philosophical concepts: according to one of them, Nature is likened to an integrated super-organism (Oken); according to another, phylogeny is the historical development of the "genealogical individual" (Haeckel). At the local level, it is believed that, when an organism is described as just a sum of its characters, its integrity is lost, so a classification resulting from such an approach ceases to be biological to a certain extent.

The *reductionist approach* is an important part of the epistemic component of the cognitive situation and therefore, in one form or another, is always present in rationally organized natural science, including systematics [Rosenberg 2020]. As noted above, this is inevitable because of certain epistemic restrictions; one of its results, important from a rational epistemology perspective, is such an optimization of the cognitive situation, in which analytical methods could work most effectively.

Though reductionism is usually considered as an intrinsic part of the rational way of knowing, it is also built into intuitionism as a specific cognitive program. In this case, it means reducibility of the cognizable world to personal perceptions and reflections about it. The phenomenology of Edmund Husserl works in just this way: in it, comprehension is addressed not to an object itself, but to its itsintuitive grasp, so reality "disappears" behind certain subjective reflections [Oskolski 2011].

Multiple and multistep ontic reduction can be represented as a sequential expanding *reduction cascade* [Pavlinov 2011a, 2018]. The multistep character means that the entire top-down reduction is carried out as an ordered sequence, with each step involving the next "cutting out" of a certain fragment from the previously delineated *Umwelt*. Plurality means that, at each step of the reduction, a previously delineated *Umwelt* can be reduced in more than one way. As a result, this cascade yields something like a "prototype" of the conceptual pyramid considered above. Such a conceptualist interpretation of the "pyramidal" hierarchy of taxonomic concepts, each referring to a certain manifestation of the diversity of organisms, leads to an inevitable recognition of their plurality and shows that the monistic consideration of the taxonomic reality [Lam 2020] is untenable.

Since ontology and epistemology are interconnected (see Section 3.2.1), the entire reduction cascade, though addressed to the cognizable object, can be considered a form of *epistemic reduction*. The latter means reducibility of the ontic component to the epistemic one; accordingly, the object itself is reduced to its various research representations (models)—formulas, classifications, etc. One of its extreme manifestations is the *instrumentalist reduction* responsible for the above-mentioned problem of instrumentalism (see Section 3.2.1). This means that the method as such "dictates" how an *Umwelt* should be represented in cognitive situation, so an ontological comprehension of the former is reduced to just a correct application of a method [Rieppel 2007; Pavlinov 2018]. This is a characteristic feature of the entire numerical research program in systematics.

The ontic reduction of any complex biological phenomena to their constituent elements (particular *Umwelts*) means the loss of some part of biological content inherent to Nature (*Umgebung*). Therefore, this content is obviously the most comprehensive at the beginning and the least at the end of the stepwise reduction cascade. Therefore, the more reductionist a research approach is, the less the properties of objective reality (*Umgebung*) are reflected in the studied *Umwelt*, and the more the epistemic and subjective components of cognitive situation contribute to the latter. In more traditional terms, this means that the less objective (real) and the more subjective (nominal) is a conceptually delineated object (*Umwelt*) obtained by stepwise reduction. This conclusion follows from a realization that it is the subject that launches and brings to a logical end the entire reduction cascade.

In systematics, a certain "global" reduction at the level of biota is inherent in all classification approaches; at the level of the organism, it is most characteristic of taxonomic theories based on positivist philosophy (phenetics before all). The respective reduction cascade is illustrated by a procedure of "fitting" the surrounding world to the needs of the particular taxonomic theories. First, biota is "cut out" from Nature as a systemically organized and functioning aggregate of living beings. Then, one of the biota's manifestations is "cut out," so biological diversity (BD) is delineated. Further, it turns out that BD itself is complexly organized, so taxonomic diversity (TD) is "cut out" of it, with all its other manifestations (ecological, biogeographic, social, etc.) being discarded. At the next reduction step, TD undergoes decomposition by delineating its different manifestations, which are "cut out" to adjust them to the respective cognitive tasks. The latter may be defined as reconstruction of diversity of archetypes (typology) or diversity of biomorphs (biomorphics) or diversity of populations (biosystematics) or hierarchy of monophyletic groups (phylogenetics).

The phylogeny that generates the latter hierarchy can be interpreted complexly (according to Haeckel) or reduced to cladogenesis (according to Hennig). In phenetic systematics, all reasoning about structure and causes of TD is discarded, so the latter is reduced to an operationally definable set of organisms At the organismal level, reduction goes farthest in genosystematics, in which the organisms are represented by molecular sequences.

All that has just been described constitutes a taxonomically justified reduction cascade of delineating different *Umwelts* as the key constituents of the ontic component of the respective cognitive situations. Now, the sequential dynamics noticed above of the ratio of the ontic, epistemic, and subject inputs to the resulting *Umwelts*, as applied to this particular cascade, allows special attention to be drawn to the following. Indeed, it is quite obvious that no BD exists in itself out of biota as a whole; there is no phylogeny in itself out of the historical development of biota as a whole (in which there is also ecogenesis); and even more so, this is true for cladogenesis (in phylogeny, there are also semogeneses and anagenesis). At the level of the organism, neither anatomical *Bauplan* nor genome, and even more so, no molecules, exist in themselves. When considered from this standpoint, molecular phylogenetics appears to be much more "subjective" than the classical one, although "techies" are sure of the opposite; obviously, they confuse natural objectivity with an instrumental *intersubjectivity* (see [Smaling 1992] on the latter).

3.3.3　Between Realism and Nominalism

Science deals with generalizations designed to reflect somehow the phenomena that are really or at least supposedly inherent in Nature. In parametrizing natural philosophy, these are natural laws—for example, describing the motion of bodies in mechanics, the speed of chemical reactions in analytical chemistry, limiting factors in ecology, etc. In classifying natural history, such generalizations are certain natural assemblages of organisms—for example, ecosystems in ecology, natural areas in biogeography, natural taxa in systematics, etc. Each assemblage reflects a certain commonality of the properties of organisms it unites and, if naturally delineated, is ascribed a law-like status [Duhem 1954]. The main question about all such generalizations and the phenomena they pretend to reflect is the following: are they real or nominal [Grene 1987; Mamchur 2004; Devitt 2005; Taylor 2006]? In other words, do the phenomena, about which these generalizations are formulated, actually exist as objective attributes of Nature or they are just imaginary subjective products of human cognitive activity? Philosophers and theorists of science were and still are most concerned with both this question and possible answers to it, from Antiquity until recent times. This long-term concern inevitably resulted in the historical changes of the meanings of notions of reality (objectivity) and nominality (subjectivity) [Daston and Galison 2007].

Ontic realism in its "absolute" sense affirms an objective existence of all phenomena of Nature at all levels of generality: it means the endowment of an objective being (reality) of both perceived physical bodies (*naturalia*) and their intelligible assemblages and processes, to which generalizing concepts (*universalia*) correspond. The rationale for realistic ontology is an assumption that these assemblages and processes are endowed with their own essences standing for their objective existence, which allows them to be recognized in the surrounding world and makes their investigation sound. *Moderate realism* differentiates phenomena according to the degree of their reality: some are more real than others, for instance, depending on their level of generality. This ontic realism is complemented by *epistemic realism*: it means that subjective knowledge reflects more or less adequately objective reality—of course, with reservations regarding various cognitive limitations, historical conditioning, probabilistic character of knowledge, etc. It is obvious that, without such a realistic standing, natural science activity becomes groundless and aimless [Devitt 2005].

Ontic nominalism in its "absolute" version recognizes only objective existence of the directly perceived things (*naturalia*) but not their assemblages. From this point of view, any generalizations (*universalia*) are but a product of mental activity: they are convenient for describing the diversity of things, but there is nothing in the real world to whom they may correspond. The nominalistic position is one of the manifestations of reductionism: complexly arranged Nature is reduced to particular unrelated things. The version of *moderate nominalism* largely coincides with moderate realism.

In the classical scientific paradigm, only realism and nominalism are usually considered. In the non-classical one, the third way of considering the problem of reality is of greater importance: it is *conceptualism*, the core of which is defined by the above-mentioned concept of ontic relativism [Quine 1996; Dupré 1993; Mahner and Bunge 1997]. In this case, it is presumed that Nature (*Umgebung*) is complexly structured and various manifestations of its structure are individualized as cognizable objects (*Umwelts*) by means of certain concepts that "construct" different conceptual realities. The latter are objective ("natural") insofar as they can be linked with certain processes that cause the respective manifestations of the structure of Nature [Putnam 1991; Quine 1996; Mamchur 2004; Khalidi 2016]. On the other hand, since any conceptual reality is "construed" by a subject and does not exist outside a specific cognitive situation, it can be considered to be "imaginary" to an extent.

In systematics, absolute realism means acknowledging the reality of all taxa and their ranks regardless of the levels of generality. Such is the position, first of all, of the adherents of taxonomic "esotericism" of the 19th century; in modern phylogenetics, monophyletic groups are endowed with reality as elements of the phylogenetic pattern [Eldredge and Cracraft 1980; Wiley 1981; Pavlinov 2015]; this is also true for "species as individuals" [Ghiselin 1997; Pavlinov 2009b, 2013a; Richards 2010]. According to moderate realism, reality is ascribed only to the lower categories, but not to the higher; it is evident in some scholastic studies (Ray, Linnaeus) and in biosystematics; such a viewpoint was called *bionominalism* [Mahner and Bunge 1997]. In contrast, an extreme nominalist position means complete denial of the reality (objectivity) of the entire taxonomic hierarchy: it is supported by adherents of the Ladder of Nature and by positivists (such as in phenetics).

Scientific conceptualism is of exceptional importance in contemporary systematics: it shows how certain biological phenomena are introduced into its cognitive situation. Realism and nominalism dispute endlessly on this matter, while conceptualism emphasizes the contextual character of all judgments that constitute taxonomic knowledge. According to this, all issues related to the problem of taxonomic reality and the "objectivity" of the respective cognitive means representing its manifestations (classification, taxon, rank, character, etc.) can be reasonably discussed only in a certain conceptual context. For example, taxa as monophyletic groups are considered real in phylogenetics but nominal in phenetics.

3.3.4 BETWEEN MONISM AND PLURALISM

Each of this pair of regulators with opposite meaning largely determines the basic motivation of all cognitive activity. One of them focuses on the cognition of the object with the aim of explaining it with a certain unified—and therefore the only true—theory. Another admits different explanations of the object with the (approximately) same truth status according to its different conceptualizations. In science, they are usually termed scientific monism and scientific pluralism, and were mentioned in Section 3.1, where classical and non-classical science philosophies were considered. They concern both the ontic and epistemic components of the cognitive situation.

Scientific monism is common to all classical cognitive doctrines; it is formalized as the fundamental *principle of the unity of both Nature and scientific knowledge* about it. The historical roots of this cognitive paradigm were laid in ancient natural philosophy; in the post-ancient tradition it was strengthened by biblical monotheistic faith, then this monistic worldview passed from natural theology into natural science, and it is based on the following assumptions [Hempel 1965; Gaydenko 2003]. At the level of ontology, it is recognized that the surrounding world (*Umgebung*) is organized strictly deterministically and structured uniformly by the action of a single global fundamental cause (*ontic monism*).[3] Accordingly, at the level of epistemology, it is recognized that there is a single acceptable way of comprehending and describing *Umgebung*—a unified scientific Method leading to a single "final" theory expressing the overall structure of Nature (*epistemic monism*). Such a theory presumes the reduction of all manifestations of *Umgebung* to a certain basic level, at which a certain initial or final cause acts; it is the latter that constitutes an ultimate end of cognition, and an aspiration to get to this end determines the cumulative equifinal development of the entire science. According to strict monism, any discrepancies reflect the incompleteness of current scientific knowledge and therefore have a temporary transitory status. In a moderate form, monism acknowledges the admissibility of different theoretical concepts that interpret the same object in different ways, but with an undoubted dominance of one of them as "the most truthful." As different scientific philosophical doctrines imply different understandings of "truthfulness" itself, a fundamental problem arises concerning the recognition of the "by far most truthful" among them.

Scientific monism tends to generate a kind of *scientific dogmatism*, akin to religious dogmatism, and intolerance inherent in it. It is the main source of the hot

---

[3] It should be kept in mind, however, that in ancient times, two fundamentally different ontologies competed, Platonic and Aristotelean, with descending and ascending cascades of causal relationships, respectively.

conflicts between the followers of different theories and research programs, each claiming to lead to "an ultimate truth" and, accordingly, to be a "guide" for the rest of the scientific community. The ethical *principle of tolerance* [Popper 1959] was proposed as a certain protection against pressure from this kind of dogmatism, calling for a more open-minded attitude towards different points of view as a normal manifestation of scientific pluralism; but it seems to work rather poorly.

Scientific pluralism can be traced back to the ancient idea of the plurality of worlds, which was later developed by Giordano Bruno and then by G. Leibniz. In modern non-classical science, the main position of this *ontic pluralism* presumes that the surrounding world (*Umgebung*) is organized quasi-deterministically by the action of a complex set of different local causes, so it is structured by their quasi-independent actions into fragments, levels, etc., each with its own essential properties. This makes them causally irreducible to either each other or any most fundamental phenomenon. At the epistemic level, pluralism began with a denial of the self-evident and unified status of Aristotelian logic, which led to the emergence of several logical systems (see Section 3.5). Modern *epistemic pluralism* forms the basis of non-classical scientific rationality supported by evolutionary epistemology [Hull 1988; Lektorsky 2001; Gaydenko 2003; Stepin 2005; Kellert et al. 2006]. It presumes a non-monolithic character of science with respect to both general principles of scientific research and scientific knowledge inferred from them. Accordingly, in constructing some general cognitive situation of natural science, it is recognized first that there are many admissible ways of reducing overall *Umgebung* to certain different local *Umwelts*. Each of the latter forms a specific ontic component of a particular cognitive situation, to which corresponds a specific epistemic component that develops a specific methodology. Within every cognitive situation thus construed, its theoretical carcass is elaborated for studying a particular *Umwelt* within a particular research program. These particular theories about different *Umwelts* are not reducible to a unified "final" theory; instead, it is their totality that, with some approximation, makes up an overall theoretical framework for understanding *Umgebung* according to the *principle of complementarity* [Armand 2008].

It follows from the last thesis that scientific pluralism, contrary to a widespread misconception, does not imply the denial of truth: it only means that there is no "universal theory of everything" with its classical "unified truth," but there is a certain set of "local" theories with their corresponding "local truths" in total compounding a non-classical "multiple truth." An important question as to whether some of them can be considered fundamentally untruthful is largely decided based on the background knowledge. For example, in a cognitive situation shaped by the strictly materialistic world picture, any reference to the supra material causes of Nature is excluded from the very beginning as fundamentally untrue.

One of the fundamental problems raised with scientific pluralism is the *incommensurability of theories* elaborated within different conceptual spaces [Kuhn 1977; Laudan 1981; Hacking 1983]. To eliminate this undesirable effect of splitting the overall cognitive situation into quasi-isolated fragments, it is necessary to develop some conceptual translators as a tool for the mutual interpretation of such theories [Wang 2002; Kellert et al. 2006]. They constitute a part of meta-theoretical knowledge, which once again brings us to a realization of the fundamental significance of the conceptual pyramid.

Taxonomic monism dominates in classical systematics: it means that taxonomic reality should be described within a unified conceptual space by means of a single (and therefore the only) "truthful" taxonomic theory. Accordingly, a unified and therefore the only classification based on such a theory should necessarily be developed. This position is most fully manifested in the conviction that "there can be only one natural classification" [Rozov 1995: 16], which indirectly presumes there is only one possible way to cut Nature at its "joints" to uncover really "natural" kinds [Lam 2020]. *Moderate monism* recognizes that there can be many taxonomic theories, but one of them can and should claim priority status; ideologists of each particular taxonomic theory (phenetic, phylogenetic, biosystematic, etc.), for quite understandable reason, attribute this status to it.

On the contrary, *taxonomic pluralism* adopted by non-classical systematics allows for a plurality of taxonomic theories and research programs implementing them, each dealing with a particular manifestation of taxonomic reality [Pavlinov 2011a, 2018, 2020]. The basis for such pluralism at the ontological level is shaped by acknowledging the irreducible multidimensionality of taxonomic reality as its fundamental property, with its manifestations being delineated naturally by reference to the respective natural causes structuring biota. Based on their properties, particular taxonomic theories (typological, phylogenetic, etc.) develop both specific criteria of naturalness of classifications and, respectively, classifications themselves meeting these specific criteria. Another important manifestation of taxonomic pluralism presumes a possibility that different taxonomic theories may be elaborated for different groups of organisms taking into account their biological specifics, including the peculiarities of the structure of their diversity. In particular, different species concepts can and should be developed for different groups with evident biological specificity of intra- and interspecific relations [Pavlinov 2013a]. At an empirical level, taxonomic pluralism presumes a multiplicity of classifications related to the same manifestation of taxonomic reality: this position is formalized by the fundamental *principle of taxonomic uncertainty* [Zarenkov 1988; Pavlinov 2018], and one of its epistemic causes is an irreducible multiplicity of classification algorithms [Sneath and Sokal 1973].

3.4 KNOWLEDGE AS AN INFORMATION MODEL

For a mythological and intuition-oriented mind, a cognizable object and knowledge about it do not differ too much: this is quite clearly expressed by the ancient thinker Parmenides with his aphorism "a thought and what it is about" [Gaydenko and Smirnov 1989; Corazzon 2019]. In rational science, an object and knowledge about it are separated: the first belongs to the cognized reality, the second is information about it.

In order to connect rationally cognizable reality with knowledge about it, it is reasonable to consider the latter as an *information model* of the former functioning as its *cognitive model*; the latter's main task is to *represent* a studied object in the cognitive situation [Bertalanffy 1968; Wartofsky 1979; Fraassen 2008; Frigg and Nguyen 2016]. From a theoretical viewpoint, of greatest importance is a *conceptual model* in the form of theory, concept, etc., which shapes the ontic component of the cognitive situation. Such a model is both selective and reductionist, as it outlines a particular *Unwelt* by considering only those object's properties that are essential for a particular research task. One of the main requirements for a model is to be isomorphic (or similar in the sense of [Suárez 2010]) to the respective *Umwelt*; this is formalized by the *principle of model isomorphism*. In this regard, the fundamental epistemic *principle of cognizability* is refined to the *principle of representation*, according to which an object is considered cognizable and knowledge about it is considered true ("objective") to the extent that its cognitive model is isomorphic to it. So this principle is a rather powerful epistemic regulator of all the cognitive activity.

Conditions of the principle of model isomorphism bring two very serious epistemic problems into the cognitive situation. Firstly, since the model is construed within a certain conceptual space, the degree of its isomorphism can be assessed only within this space, which yields a kind of tautology. Secondly, since the model is always simpler than the simulated object, an isomorphism between them is inherently incomplete, which means an irreparable *incompleteness of knowledge* about the studied object; this is sometimes considered as one of the most fundamental epistemic principles.

Semantically, cognitive models can be explanatory or descriptive, stationary or dynamic, causal or phenomenological, etc. Semiotically, their forms are very diverse: models can be notional (explicit) or mental (implicit), verbal, mechanical, mathematical, geometric, formulary or graphic, analogous or digital, etc. A special case of such a model is a *metaphor* that represents an intuitively non-obvious object in more familiar notions, images, etc.: for example, "the Book of Nature," "Ladder of Nature," "flow of time," "spiral of development," etc. [MacCormac 1985; Gibbs 2008]. In biology, metaphors occupy a very prominent place: "evolutionary tree," "epigenetic landscape," "morphospace," "taxonomic map," "body plan," etc. [Lewontin 1964]. Their special value in cognitive situations of natural science is largely because natural phenomena cannot be unambiguously reflected in a strictly fixed system of notions (see Section 3.5). From this point of view, their definitions, insofar as they cannot be rigorous and exhaustive, are metaphoric to a degree.

In systematics, the principle of representation presumes that any particular taxonomic knowledge of whatever content and form is a cognitive model (representation) of a particular manifestation of taxonomic reality. From this standpoint, the main property of this knowledge is to be isomorphic as much as possible to this manifestation. A particular classification is a *classification model* representing the structure of some fragment of taxonomic reality, so the naturalness of classification can be refined in terms of its isomorphism. Respectively, a taxon is a model of a certain group of organisms as an element of this structure. A tree diagram in different interpretations (phylogram, cladogram) represents different manifestations of the phylogeny.

A research sample is a primary *empirical model* of taxonomic reality, with the particular forms of a depiction of specimens in it being their *descriptive models*, which can be word-based, character-based, graphic, etc. (see Section 3.7.2).

3.5 THE LOGICAL BASES

The general idea of rationality is most consistently implemented by the development of logical foundations of scientific activity, including justification of the scientific status of knowledge. According to the key *principle of logical consistency*, the latter is scientific if (other things being equal) it is based on certain logically consistent inference procedures. This general idea is captured by an iconic phrase, *the logic of scientific research* [Popper 1959; Shchedrovitsky 2004].

In natural science, the historically primary embodiment of rationalism was the *classical* (Aristotelian) logical system, which in the Antique and early post-Antique periods was considered self-evident and the only one possible [Gaydenko 2003]. Its fundamentals are based on the *principle of the excluded third*, which makes the whole classical logic *two-state*; its basic principles (axioms), as expressed in the terms of systematics, can be summarized in a simplified form as follows. The sequential division of taxa is carried out on a *single basis* (combined use of incompatible characters is prohibited), must be *exhaustive* (a taxon is divided into subtaxa without a remainder), provides *discrete* taxa (those of the same rank must be mutually exclusive), and their hierarchy is *continuous* (all designated categories are fixed without any gaps).

Significant changes in the understanding of the methods of construing logical systems began in late Medieval times and reached their apogee in the second half of the 19th and the early 20th centuries; as a result, the self-evident and exclusive status of Aristotelian logic became questioned [Berkov and Yaskevich 2001; Shuman 2001; Russell 2019]. This trend became most clear with the mathematization of logic based on set theory axiomatics: with it, *intensional* logic of classes was complemented by the *extensional* logic of sets [Pellegrin 1987]. Recognition of *deductive* and *inductive* logical systems directly affected classification procedures in systematics. Development in this direction created preconditions for the emergence of *non-classical* logic, which is now a rather ramified set of various formal systems, each with its own set of rules for inferring some true judgments from others [Mikhailov 1983; Shuman 2001]. Of fundamental importance is an understanding that the initial premises for the development of these systems are rooted in personal knowledge and cannot be considered strictly and unambiguously given [Kline 1980; Russell 2019]. This conclusion is especially impressive in the case of the foundations of mathematics, which is most usually considered a paradigmatic instance of an "exact" and "objective" science: its foundations appear to be initially "subjective" [Shapiro 1997; Perminov 2001].

Classical two-state logic, by recognizing only true and false statements, contradicts a more complex content of natural science, which deals with judgments that are most usually not reducible to the classical *"true or false"* alternative. Respectively, non-classical *multi-state* logic is more appropriate, in which judgments other than truth or falsity are presumed. Among its versions, the most popular is *three-state logic*, which considers *indefinite* judgments that are always present in any cognitive activity aimed at Nature. One of its most important versions is *probabilistic logic* associated with the calculus of probabilities of the truth of judgments. Of fundamental importance for non-classical science is *quantum logic*, which formalizes consistent procedures of uniting formally incompatible statements about the same object by their complementarity relation [Vasyukov 2005]. *Fuzzy logic* deserves special mention: in this, all judgments and their truth assessments are context-dependent and therefore cannot be strictly and universally defined [Zadeh 1992; Dompere 2009]. *One-state logic* is of particular interest, as it asserts the meaningfulness of only positive judgments as *A* and the meaninglessness of negative judgments as *non-A* [Vasiliev 1989; Bazhanov 2009]. Of importance is *modal logic*, which involves specific estimates of the truthful status (plausibility) of the scientific hypotheses. *Mereology* formalizes the whole–part relationship and serves as a basis for partonomic divisions; it became fundamentally important for the development of ideas about possible structures of diversity [Luschei 1962; Calosi and Graziani 2014]. The above-mentioned important principle of theory incompleteness is logical in its origin; it imposes certain restrictions on the possibility of the exhaustive substantiation of natural science theories, including taxonomic ones.

Scholastic systematics developed in the 16th to 18th centuries basically as an application of formal classical logic, as it was implemented in the generic–species classification scheme, to the classification of plants and animals. The emergence of post-scholastic thoughts in the 19th century diminished its significance in favor of a more substantive foundation of biological classification. However, the logician John Mill at the end of the 19th century fixed the understanding of classification as essentially a logical procedure [Mill 1882], which became widely adopted. From this standpoint, biological definitions of genus, species, character, etc. are just particular substantive interpretations of the Aristotelian logical categories. For researchers inclined to classical rationalism, such a "logical" vision of taxonomic procedures is attractive because of its rigor: unambiguous logic presumably leads to logically justified unambiguous classifications. This approach is most fully embodied by one of the versions of episto-rational systematics (see Section 5.2.2).

Consideration of biological systematics as a field of application of various logical systems yielded several important results. Thus, the theoretical biologist Joseph Woodger substantiated the need to separate notions of taxonomic category and taxon, and defined phylogenetic hierarchy as belonging to the sphere of mereology [Woodger 1937, 1952]. The last point prompted researchers to carry out a careful analysis of the differences between logical systems of classes and parts in taxonomic research (see below). Explication of basic taxonomic concepts in the terms of Cantor's set theory [Gregg 1954] yielded the so-called "Gregg's paradox": its careful analysis showed insufficiency of extensional logic for describing a biologically meaningful taxonomic hierarchy [Buck and Hull 1966; Löther 1972; Griffiths 1974a; Shatalkin 1988, 1995, 2012; Mavrodiev 2002; Pavlinov 2018].

Solutions of some important theoretical and practical issues in systematics break the requirements of classical logic in favor of ideas of different versions of the non-classical one; examples are as follows.

A requirement of an exhaustive division of a taxon into subtaxa contradicts the usual practice of identifying "taxonomic remnants," which are groups of unclear taxonomic allocations (*incertae sedis*). In classical systematics, this is considered as an indication of insufficiently developed classification. In non-classical systematics, the validity of incomplete division is justified by those "remnants" that may reflect the probabilistic nature of both taxonomic diversity and taxonomic knowledge [Zarenkov 1988; Pavlinov 2006]. The probabilistic character of taxonomic knowledge makes it more consistent with the conditions of fuzzy logic. The latter presumes a possibility of non-discrete interpretation of taxa and taxonomic categories, including partial overlapping of taxa in a classification [Tchaikovsky 1990; Gordon 1999]. Besides, fuzzy logic substantiates a context-dependent analysis of characters with conditional probabilities [Pavlinov 2018].

A requirement of the continuity of taxonomic hierarchy is one of the most significant in classical systematics: it presumes ordering of all taxa in a given classification according to a unified ranking scale. Instead, the *principle of non-strict hierarchy* is introduced to relax this requirement, and is implemented quite consistently by cladistic systematics [Pavlinov 1990, 2005, 2018].

The principle of single basis of division makes taxon definition monothetic. It was actively used by scholastic systematics, though with reservations. In post-scholastic taxonomy, it is rejected in favor of the *principle of multiple basis of division* presuming polythetic taxon definition.

The classical binary division of notions, according to which both positive *A* and negative *non-A* judgments are equally significant, is fundamental for most branches of systematics. It means that taxa can be consistently characterized by both the presence and absence of characters: a well-known example is the separation of vertebrate and invertebrate animals by Lamarck. However, in the context of one-state logic, the definition of a taxon by a "negative" character is not consistent [Mahner and Bunge 1997]; this condition is implemented by cladistics [Pavlinov 1990, 2006, 2018] (see Section 5.7.3).

Awareness of important differences between class logic and mereology led to some significant conclusions about the content of biological systematics [Woodger 1952; Mayr 1982; Mahner and Bunge 1997; Chebanov 2007]. A classification in its traditional sense involves relationships between logical classes, while mereological division presumes relationships between partitions of groups as "parts" of natural "bodies" [Tversky 1989; Mahner and Bunge 1997; Calosi and Graziani 2014]. These two logical systems outline two fundamental aspects of taxonomic reality, namely taxonomic and partonomic (= meronomic) [Lyubarsky and Pavlinov 2011; Pavlinov 2018]. In the context of this dilemma, such issues are considered as the relationship between definition *vs.* description of taxon as a class or as a quasi-individual, significance *vs.* irrelevance of similarity of organisms as a condition of their membership in a taxon, etc. [Buck and Hull 1966; Hull 1978a; Mahner 1993, Mayr and Bock 2002; Rieppel 2006].

As a consequence of the development of modern non-classical rationality, the ineffectiveness of the subjugation of natural scientific knowledge to strictly logical rules was acknowledged [Popper 1959; Hempel 1965]. Any logic is just a set of rules for ensuring logical consistency ("truth") of inferred statements with respect to certain initial ones accepted *a priori* as true within a certain logical system, but their "logical truth" says nothing about the truth of the judgments addressed to Nature being investigated [Schuman 2001]. Of importance is an awareness that different ontologies may require specific analytical tools that are the most fitting to them, including logical ones [Berkov and Yaskevich 2001; Shuman 2001], therefore "the role of logic, adequacy of logical categories to the structure of the studied reality [...] changes significantly depending on the nature of the subject areas" [Subbotin 2001: 23].

This conclusion is formalized by the above-considered principle of onto-epistemic correspondence, which implies that, within a particular cognitive situation, epistemic principles, including logical ones, should fit in some way ontic assumptions [Pavlinov 2018] (see Section 3.2.1). Theoretical "foundations of systematics lie in [the objective] ontology, not subjective epistemology," so "if logical models are to have any value, they must represent the structure of reality as closely as possible" [Griffiths 1974a: 85, 104]. This consideration leads to a fundamental problem of the justified choice of particular logical systems that are most suitable for analyses of particular manifestations (aspects, fragments, etc.) of taxonomic reality. It is obvious that such a choice in each particular case requires special analysis on substantive rather than formal grounds in the context of some biologically sound theory.

3.6   ARGUMENTATION SCHEMES

Argumentation schemes are designed to derive and substantiate knowledge about things and ideas. Their development is one of the key tasks of the branch of epistemology that underlies the methodologies of scientific research. Their content and ways of implementation depend on a general understanding of the structure of a cognitive situation, including relations between its three basic components—ontic, epistemic, and subject.

In the case of ontology, first of all, of prime importance are natural-philosophical ideas concerning the place of a phenomenon under study within the network of causal relations in Nature. At the level of epistemology, the principal issues concern interrelations between initial and derived judgments, how their plausibility (truth) is justified, and the function of background knowledge. In the case of the subject component, the role of the knowing subject (in its general understanding) in inferring scientific knowledge is relevant.

These schemes vary from the most general to quite particular. The former function as the principal methodological regulators of cognitive activity, and there are three of them—inductive, deductive, and hypothetico-deductive—which are considered in this section. Particular schemes are concrete algorithms designed for solving particular research tasks, and these will be considered in Section 3.7.2.

The inductive argumentation scheme is founded on the idea that scientific knowledge, to be objective and thus true (plausible), should be both theory- and subject-neutral, and this is achieved by following conditions. In both cases, neutrality is ensured by minimizing the metaphysical content of the cognitive situation, which strongly depends on personal preferences. The background knowledge is based on the Aristotelian ontology of the ascending cascade of causal relations in Nature: lower-order (atomic) phenomena are causes of high-order ones. The "logic of Nature" thus organized is followed by the respective inductive logic, in which high-order judgments about the properties of cognizable phenomena are inferred from the lower-order (atomic) ones inferred directly from the empirical data. Regularities in these data are supposed to reflect regularities in Nature; the latter's causal explanations (law-like models) constitute the respective highest-order generalizations.

One of the main tools for inductive inferring is the *principle of total evidence* developed by positivism [Carnap 1969], according to which the more of empirical data evidence for a hypothesis, the higher its plausibility. Hypotheses thus derived are tested by their *confirmation* by other empirical data taken as an empirical proof of the initial inferences: among competing hypotheses, the more plausible one is that confirmed by more data. An additional supporting argument is provided by *analogism* (analogical reasoning): the plausibility of a judgment about new data is supported by analogous judgments on similar data, the truth of which was proved earlier [Uemov 1974; Bartha 2013].

In systematics, inductive reasoning presumes that (a) as many characters as possible should be used without *a priori* selection of the most significant ones, and (b) classification should reflect the structure of similarities network by these characters and be construed in a "down–up" manner, i.e., from lower to higher taxonomic categories. This scheme was first mastered by natural systematics (in its narrow sense) at the end of the 18th century; in the first half of the 20th century it became a methodological fundament of classification phenetics and biosystematics with reference to positivist philosophy [Gilmour 1940; Sokal and Sneath 1963; Lines and Mertens 1970]. The principle of total evidence is discussed actively in cladistics [Eernisse and Kluge 1993; Kluge 1998; Rieppel 2004, 2005a, 2009]. Analogism is implemented by the *principle of actualism* (uniformitarianism) for reconstruction of the past [Simpson 1970]; it is important for revealing homologies [Rautian 2001; Rasnitsyn 2002]; in cladistics, it appears as the *rule of reciprocal illumination* [Hennig 1966; Wiley 1981].

In post-positivist science, strictly empirical inductively inferred scientific knowledge is acknowledged to be impossible. This is because empirical data (observations, experiments, etc.) are theory-laden: they become scientifically significant "facts" only within the context of a certain background knowledge with a richer high-order metaphysics than is supposed by positivism [Popper 1959; Carrier 1994; Quine 1996].

The deductive argumentation scheme substantiates lower-order (particular) judgments with reference to the higher-order (general) ones, with the latter constituting the respective background knowledge. Accordingly, the ontic basis of the cognitive situation includes a rich metaphysics (of Platonic kind) with a descending cascade of causal relations between natural phenomena. From such an idea of Nature, it follows a reasoning scheme based on deductive logic: comprehension of Nature should start with the highest-order judgments about the whole, which are considered true, and from which lower-order ones are inferred as detailing of the former. As the basic ontology is thought of as "absolutely" true, this scheme does not imply any influence of a subject on the content of inferred knowledge, if deductive logic is applied properly. Deductive reasoning does not presume obligatory empirical verification of its lower-order judgments: a crucial criterion of their truth is their compliance with original natural philosophy. In general, an ultimate knowledge thus inferred is theory-dependent and (supposedly) subject-neutral.

This kind of reasoning is implemented by the generic–species scheme of division of notions developed by neo-Platonists and scholasticists. According to it, classification should be built in an "up–down" manner, starting from the highest categories and sequentially dividing "genera" into "species" in a descending order. Scientific systematics began with mastering this scheme in the 16th–18th centuries. Its subsequent conceptual history was driven by developing various basic ontologies, such as the super-organismal character of Nature, Pythagorean harmony of numbers, etc. Classical typology with its key idea of general body plans is close to this, but certain inductive elements are noticeable: ideas about these plans are derived from observations of the organisms. Elements of deductive reasoning evidently occur in *a priori* weighting of characters based on their essential interpretation; such weighting is part of the scholastic and earlier post-scholastic (including natural systematics) approaches.

Deductive reasoning is severely criticized by adherents of the empirical doctrine of natural science for being strongly theory-laden and empirically non-testable. A negative attitude toward the deductive scheme became an important stimulus for early emergence and development of post-scholastic systematics.

The hypothetico-deductive argumentation scheme implies quite a rich basic ontology, in which Nature is represented as a complexly organized hierarchical system, ordered by both ascending and descending cascades of causal relations. The corresponding epistemology develops an argumentation scheme with elements of both inductive and deductive reasoning. It differs from a strictly deductive scheme by the recognition of the need for empirical justification of the plausibility of judgments about reality. It differs from a strictly inductive scheme by acknowledging the impossibility of inferring scientifically sound judgments on a purely empirical basis; they are inferred in the context of background knowledge and thus are more or less theory-dependent. With this, the subject's active occurrence in the cognitive situation is also acknowledged, which makes all judgments based on this scheme subject-dependent to an extent.

The core of this scheme may be represented as follows. An explainable empirical knowledge is considered within a background theoretical context that provides the former's general meaningful interpretation, while its particular interpretation (*explanance*) takes the form of a testable hypothesis. For example, for the elaboration of a biologically sound classification, it is necessary (or at least desirable) to realize that the procedure of classifying is not just a comparison of some specimens in a sample, but an investigation of a certain manifestation of the structure of biota. This implies a general understanding of what both biota and its general structure are, in order to avoid confusion between different aspects of diversity of orgnisms (e.g., between species and sex differences). With this, within the framework of this scheme, lower-order judgments are developed as potentially testable hypotheses about certain manifestations of Nature. Under some circumstances, such hypotheses may be inferred following an *abduction*, which judgments are a kind of "surmise" about data in the context of background knowledge [Ruzavin 2005, Aliseda 2006].

A key part of the hypothetico-deductive scheme is a special procedure of empirical testing of plausibility (truth) of hypotheses based on their *falsification* (refutation) [Popper 1959]. This is formalized by the *principle of falsification* presuming that rejections are more significant than approvals for deciding whether a tested hypothesis is plausible or not. So, the potential falsifiability of hypotheses is one of the main criteria for their scientific status. The entire scheme includes the juxtaposition of consequences derived from them as predictions (prohibitions or prescriptions) against empirical data corresponding to the events predicted by those consequences. In general, a hypothesis is considered plausible if it is not falsified by the respective test data. This criterion may be supplemented by *corroboration*, which is similar to inductive confirmation but has no decisive approving power.

An important rational basis for developing a hypothesis based on this scheme is the *principle of parsimony*, which imposes certain restrictions on background knowledge. It implies minimization of initial assumptions in that knowledge by "cutting off" non-significant (within a particular cognitive situation) ones with "Occam's razor." Accordingly, among competing hypotheses, preferred is the one that is less loaded with such assumptions and is more "parsimonious" in this sense.

The hypothetico-deductive scheme in its original version is applicable only to *strictly universal* law-like judgments (generalizations) about natural phenomena that do not have space–time boundaries [Popper 1959]. However, it is now adopted in a weaker form supported by the following ontic considerations. Synergetics claims that one of the fundamental properties of all complex systems is their irreversible non-strictly quasi-deterministic development leading to their structuring, including hierarchization [Prigogine and Stengers 1984; Barantsev 2003]. Therefore, all large-scale natural phenomena are bounded locally in time and space and by their position in the hierarchy of Nature, with each of them possessing specific emergent properties, making them "unique" to a certain degree. Accordingly, natural science actually deals with *quantitatively universal* judgments that are naturally restricted in their applicability, because the natural phenomena under study are more or less locally specific in their manifestations. From this, it follows that the consequences from scientific hypotheses about these phenomena should not necessarily be true across the entire Universe.

In addition, taking into account the probabilistic nature of scientific knowledge, it is recognized that any scientific hypotheses and their tests are probabilistic [Hempel 1966], so the scheme in question deals with *parafalsifications* [Ivin 2005]. In such a situation, if some consequences from a hypothesis appeared unconfirmed, one should not reject it as falsified and thus implausible, but rather reduce its plausibility and/or narrow the scope of its applicability [Lakatos 1978].

The main argument from an inductivist standpoint against the hypothetico-deductive scheme is that it seemingly implies a kind of *circular reasoning*, in which investigation begins with an explanatory judgment (*explanandum*) about a certain empirical observation (*explanans*), though actually the former is to be proved at the end of an investigation. However, this is not the fact: the reasoning in question is based not on a closed logical circle, but rather on a cognitive *hermeneutic circle*, which implies the need for a general understanding by an explorer of what is being explored [Gadamer 1960; Shchedrovitsky 2004; Spirova 2006].

A possibility of application of the hypothetico-deductive scheme in systematics has been actively discussed over the last several decades [Platnick and Gaffney 1977; Ruse 1979; Wiley 1981; Pesenko 1991; Kluge 1997, 2009; Hull 1999; Rieppel 2003; Rieppel et al. 2006; Stamos 2007b; Fitzhugh 2016]. The reason is quite obvious: the accomplishment of this scheme is currently considered one of the key criteria for the scientific character of knowledge in natural science. Therefore, in order to demonstrate the scientific status of systematics, it is necessary to show that the classifications it develops can be treated as potentially falsifiable hypotheses. Orthodox Popperism is considered inapplicable in systematics for the obvious reason that judgments about the groups of organisms represented by taxa cannot be considered strictly universal [Kitts 1977; Hull 1978b]. With the provisions of the principle of falsification being weakened, it seems to become more adequate to the properties of the taxonomic reality and methodology of systematics; of special importance is the conclusion that probabilistic taxonomic hypotheses cannot be ultimately refuted by any data [Rasnitsyn 2002; Pavlinov 2018]. As the principle of falsification is weakened, the principle of total evidence becomes more important [Eernisse and Kluge 1993; Kluge 1998; Rasnitsyn 2002; Rieppel 2008], which means a certain convergence of hypothetico-deductive and inductive reasoning in systematics. Besides, an application of abductive reasoning is considered a useful addition to the hypothetico-deductive one [Rieppel 2005a, 2008; Fitzhugh 2006].

Against the applicability of any version of hypothetico-deductive reasoning in systematics, a general argument is put forward, which is based on the treatment of classification as a primary form of cognitive activity. From this standpoint, classification cannot be preceded by any rich metaphysics, nor can it be treated as a hypothesis, so this kind of reasoning is irrelevant here [Gregg 1950; Gilmour and Walters 1963; Sokal and Sneath 1963; Lyubarsky 2018]. Actually, this argument is true for the initial forms of knowledge developed by folk systematics. However, it seems to be irrelevant from the conceptualist perspective that considers the advanced forms of knowledge elaborated by contemporary science.

3.7   METHODOLOGIES AND METHODS

The epistemic component of a rationally developed cognitive situation includes *methodology* as one of its most important parts. Its main task is to substantiate and develop methods as specific research tools built into the respective argumentation schemes. Since scientific research is impossible without methods and the development of methods is impossible without methodology, each research program, in a certain sense, can be thought of as a *methodological program* [Lakatos 1978]. The latter is a specifically organized general regulator of cognitive activity; it is a kind of *heuristic* that includes general methodological principles, particular methodologies, and finally concrete methods implementing them. It can be divided into two main components: *positive heuristic* indicates what is needed to be done while *negative heuristic* indicates what not to do in research so that its results can be considered scientifically consistent cognitive models of natural phenomena.

There is a lot of methodological programs: they are developed by science as a whole, by its different disciplines, and within them by particular research programs. In systematics, the research is generally based on the *classification methodology* [Rozova 1986] associated with the substantiation and development of technical means allowing for a representation of the structure of taxonomic reality by means of classification systems. The conceptual history of systematics began and proceeded with the development of this methodology and the "natural method" implementing it. Therefore, the main stages of this history were largely marked by the development of particular versions of classification methodology.

The structure of *scientific method* is composed of three main components, viz. conceptual, logical, and operational [Lukashevich 1991]. Its *conceptual component* corresponds to the ontic component of a cognitive situation; it is defined by the content of the respective conceptual space; in systematics, it depends on the typological, phylogenetic, phenetic, or other background of the respective taxonomic theories. Its *logical component* is a part of the epistemic component of a cognitive situation, according to which the content of a method is determined, first of all, by the major cognitive regulators (deductive *vs.* inductive, reductionist *vs.* holistic, etc.). In particular, the procedures of taxonomic research can be based on binary or multi-state logic; they should be substantiated taking into account the principles of selectivity, sampling, etc. Its *operational component* is "technical": it contains specific procedures of extracting and processing information about the studied object; its *instrumental component* largely depends on the technical capabilities of any research.

3.7.1 Scientific Status of Methodologies and Methods

In a rationally organized cognitive activity, the most general regulator of both methodologies and methods is the *principle of scientificity*, according to which the scientific character of knowledge is determined largely by the scientific character of the methods by which it is inferred. In this regard, the main question becomes what exactly makes a method scientific; obviously, possible answers depend on the basic onto-epistemologies. Therefore, in systematics, understanding of what the "natural method" should be changed significantly in the course of its conceptual history following changes in its onto-epistemic foundations.

In determining the scientific character of the methods under the provisions of nonclassical science, the starting point is the principle of onto-epistemic correspondence, from which the *principle of methodical correspondence* is explicated [Pavlinov 2018]. This means that the general parameters of research methods must comply with the requirements of the onto-epistemic model recognized as scientifically sound within a particular cognitive situation. To this, the *principle of method effectiveness* is added: a method should allow the research task to be solved effectively within the framework of the respective model according to certain *criteria of effectiveness*. The more a method meets these two conditions, the more reasonably (other things being equal) it can be thought of as scientific—certainly, not in general but within the framework of the given cognitive situation.

The *ontic correspondence* of the method can be justified in two ways, direct and inverse; a good example is provided by the methods of phylogenetic reconstructions [Pavlinov 1990, 2005]. The main requirement for them is the correspondence to the postulated properties of the phylogeny. The *direct justification* of the method means the derivation of the respective classification algorithm from an adopted ontic model through the latter's operationalization. The cladistic analysis was developed in this way by inferring its working principles (synapomorphy etc.) from certain postulated properties of the phylogeny. The *reverse justification* implies selection of suitable methods from a variety of existing ones based on an assessment of their correspondence to the respective ontic model. For instance, among hierarchical clustering algorithms, the most suitable for phylogenetics are those that allow, in contrast to phenetic ones, the "arrow of time" to be brought into a tree-like scheme so that it represents phylogeny.

The *epistemic correspondence* of the method is largely determined by its logical component. So, since scientific knowledge is probabilistic, the methods developed on the basis of probabilistic or fuzzy logic are preferable to the "exact" ones operating with binary logic. Proponents of the mathematization of systematics assess the scientific character of quantitative methods with reference to their mathematical validity [Jardine and Sibson 1971; Dunn and Everitt 1982]. However, emphasis on such a formal assessment of the numerical methods usually decreases the significance of their ontic correspondence and eventually leads to a reduction of the biological content of taxonomic research.

The *effectiveness* of the method means its ability to solve a specific research task with certain required precision within a reasonable time. For example, in phylogenetic systematics, the effectiveness of its methods depends on their ability to reconstruct the phylogenetic pattern. Thus, this important characteristic is context-dependent: its assessment depends on how the respective *Umwelt* is defined.

A special emphasis on analytical methods in some methodological programs gives rise to the already-mentioned *problem of instrumentalism* [Rieppel 2007; Pavlinov 2018]. This means a certain "closure" of cognitive activity to the method as such: the latter dictates how an *Umwelt* should be analyzed, so the properties of the former indirectly shape the properties of the latter; this is an unpleasant side effect of the above principle of methodical correspondence. For instance, an application of the hierarchical classification algorithm necessarily provides a hierarchically arranged classification, even if the real diversity pattern is in fact non-hierarchical; this means that the respective classification appears to be inadequate as a cognitive model of taxonomic reality.

In considering the methodological problems of systematics, the *principle of method limitation* is of particular importance: it asserts the impossibility of developing a universal method that is capable of solving all the conceivable tasks associated with the analysis of the taxonomic reality in all varieties of its manifestations. So, with

the help of each particular method, only a certain (not any) manifestation of taxonomic reality can be effectively investigated. The causes of the method limitations
are of three kinds—conceptual, logical, and operational—which correspond to the
three main components of the method structure considered above. This circumstance
encourages methodologists to develop new means for solving the particular research
tasks as the latter diversify; this results in a growing variety of both the methods
themselves, each with its own advantages and disadvantages, and the classifications
obtained on their bases. This principle is one of many manifestations of scientific
pluralism which also involves methodological programs.

3.7.2 Basic Methods

This section describes briefly the principal methods used in taxonomic research. They
are considered from a philosophical standpoint in quite a general manner (as particular methodologies), so all technical details are omitted [Pavlinov 2018].

The sampling method is used at the stage of preparing taxonomic research.
The importance of this method is determined by the epistemic *sampling principle*,
according to which taxonomic research is always carried out not on the taxonomic
reality itself (diversity as such), but on a *research sample* representing it. The latter
acts as an *operational model* of an *Umwelt* and outlines a kind of *empirical reality*: its
formation marks the final step in the cognitive reduction of the *Umgebung*. The main
elements of the sample are organisms and their characters; or, more precisely, their
descriptions (see below). In traditional "museum" systematics, the basic research
sample is equal to the world *collection pool*, i.e., a totality of the fixed and preserved
derivatives of biological organisms in museums, herbaria, etc. [Pavlinov 2016].

The sampling method is shaped by two basic principles. One is the *principle of
sample selectivity*, which explicates the principles of method correspondence: it
presumes that the sample, instead of being composed on a random basis (one of the
conditions of classification phenetics [Sneath and Sokal 1973]), should be formed
under the purposeful guidance of a conceptually defined research task. Indeed, the
requirements for samples designed for phylogenetic or phenetic, comparative morphological or experimental physiological, macro- or microsystematic research
will be different. Another is the *principle of sample correspondence*: this means
that the sample must be *representative*, i.e., it must reflect adequately the structure of an *Umwelt* under study. Sample correspondence determines the reliability of
extrapolations of the results of its study to the respective *Umwelt*.

The main qualitative characteristic of the sample is its *composition*, which depends
on the content of the sample elements. The main quantitative characteristic of the
sample is its *size*, determined by the number of constituent elements. The cumulative
quantitative characteristic of the sample is its *dimension* defined by a combination of
the amount of specimens and their characters.

The descriptive method serves as a means of recording information about sample
elements in a form that provides the technical possibility of conducting taxonomic
research. The *primary description* constitutes a basis of empirical analysis of the raw
data themselves: it fixes in some way the information about the sample elements. This

can be a verbal description of organisms or their derivatives, their analog or digital images (drawings, photos, etc.), voice recorded on a phonogram, figures of individual measurements in a table, molecular sequences, etc. The result of the primary description of a specimen/character is reasonable to consider as its individual descriptive model. It is the latter that constitutes an individual sample element, so at an operational level, the sample-as-model is represented as an aggregate of descriptions of organisms by their features. The *secondary description* is based on a certain set of primary descriptions, and provides their generalized characteristics. This may include a summation of several comparative morphological series, a table with calculated statistical characteristics, etc., as well as a description of the results of experiments or other manipulations with objects.

The comparative method serves as an analytical tool for revealing a certain structure of diversity in the research sample by comparing its elements. The main task of this method is some ordering of sample elements according to their similarities. In the *primary comparison*, the compared objects are the sample elements themselves: specimens are compared over their characters (taxonomic analysis) or characters are compared over the specimens (partonomic analysis). In the *secondary comparison*, the compared objects are the results of previously conducted analyses: these objects may represent the results of various experiments (e.g., hybridization under different conditions) or comparisons based on different methods (e.g., different clustering algorithms), or comparisons by different sets of characters (e.g., molecular and morphological).

The comparison is *contextual* and therefore *relative*: its context is set by a certain *comparison basis* (standard), with reference to which the similarity relations of the compared objects are assessed. This means that the comparison of any two isolated objects does not have a soundly interpretable meaning; the latter is provided using a certain comparison basis and depends on it. This basis can be represented by some fixed scale (for example, "more or less") or a certain third object of the same "kind" as the compared ones, including an abstract archetype [Shreyder 1983; Rautian 2001]. Thus, in general, the comparative method is carried out in the form of the *ternary relation* between three objects, with two of them being compared while the third provides a basis for comparison.

Depending on the particular procedures, the comparative method can take the following forms. The *phenetic method* is based on comparing sample elements directly on a pairwise basis within a given sample, with the latter constituting the generalized comparison basis. In the *typological method*, a certain generalized (secondary) element of the sample, usually (arche)type or *Bauplan*, serves as a comparison basis. The *comparative historical method* includes the historical time scale as a supplementary comparison basis. If the similarity is assessed quantitatively, the comparative method becomes the *mathematical* one.

The experimental method involves direct or indirect manipulations with objects, which may be specimens or aggregates thereof. This method is only applicable to fairly simple objects, so it is of quite limited use in systematics; it is most developed in biosystematics and in some numerical approaches. Experiments can be *active* or *passive*: the former are carried out under controlled and (preferably) manipulated conditions, whereas the latter are not. If real experiments are impossible, they are replaced by imaginary (mental, virtual) ones, in which the elements of a subjective or virtual reality are manipulated.

The mathematical method is based on numerical estimates and computational operations. Depending on the nature of the latter, this method can be *analytical* (based on precise calculations by formulas), *numerical* (based on calculations by sample parameters), or *graphical* (based on graphical constructions). Currently, these operations are significantly automated by the use of computers.

Standard algorithms of the mathematical method in systematics serve as auxiliary tools for solving analytical tasks. In simple cases, this method, as part of the comparative one, implements the above operations with the objects (sample elements or assemblages thereof) on a numerical basis, including: (a) quantitative assessment of similarities between them; (b) quantitative analysis of the entire structure of similarity relations; and (c) transformation of the latter structure into classification. In more complex cases, this method, as part of the experimental one, serves for conducting virtual experiments: for example, to reveal an effect of changing certain sample parameters on the results of comparative or experimental analyses. Another area of application of the mathematical method is the analytical comparison and/or generalization of different classifications, as well as the analysis of their structure (for example, their compliance with Zipf–Mandelbrot distribution).

The classification method is a kind of quintessence of research activity in systematics as a classifying discipline, implementing the above-mentioned classification methodology at an operational level. It summarizes the results obtained by other research methods outlined above and serves as a kind of "common denominator." This method is based on the logical procedure of *classifying*: it is an operation on the descriptive models of the sample elements representing the objects in some way, but not on the objects themselves.[4] Its main result is a presentation of the similarity structure of the research sample in the form of classification.

An operational basis of the classification method is the *classification algorithm*, which recognizes particular classification units (taxa or partons) of different levels of generality and unites them in a single classification system. This algorithm is applied to estimates of similarities between objects that resulted from their previous comparative or experimental analyses. It can be deductive or inductive depending on the basic argumentation scheme. In contemporary taxonomic research, it usually incorporates certain elements of the mathematical method.

---

[4]   In this, classifying differs from *sorting* in the sense of dividing a set of physical objects into subsets; the sorting may be *extensional* (based on counting the objects) or *intensional* (based on comparing the objects by their characters). A variant of the latter is the *identification* of objects, i.e., their allocation to already-recognized classification units (for example, to different species).

4 An Outline of Taxonomic Theory

Question: What is theoretical knowledge?
Answer: An intellectual improvisation on the Theme of Being.

Igor Ya. Pavlinov

No scientific discipline can function normally without a more or less developed theoretical background: accordingly, the progress of the former depends, by and large, on the completeness of the latter. Its core is shaped by a theory, which in the most general sense means a conceptual system containing generalized knowledge about the studied object (phenomenon, etc.). It is noteworthy that, although the notion of theory is one of the key and basic ideas in science, its sufficiently clear and uniform definition does not exist [Stepin 2005]; moreover, it probably cannot exist due to the qualitative heterogeneity and dynamic nature of the entire system of scientific knowledge. A uniformly construed "final theory of everything" for the whole of science is an ideal of the classical monistic paradigm, while recognition of its impossibility is an attribute of the non-classical pluralistic paradigm. The latter presumes, however, that any natural science must comply with some general universal provisions and criteria that make its theory scientifically sound. With this, it is implied that particular science branches may and should develop their particular theories taking into account the specifics of their cognitive situations and studied objects (phenomena, etc.).

These two interconnected universal principles are certainly true for the theoretical foundations of biological systematics. There are also several other major conditions to be taken into consideration when developing these foundations. The latter is shaped by the unity of onto-epistemology as part of the cognitive triangle: one does not exist without the other; they are interconnected. This circumstance is reflected by the *principle of onto-epistemic correspondence*. The entire theoretical construct of systematics is hierarchically organized in a conceptual pyramid: more particular concepts can be defined only within the framework of more general ones. Taxonomic theory (TT) is not a finalized product but a living, evolving conceptual system, with its content changing caused by the dynamics of general philosophical scientific contexts. The vector of its development is directed towards a more complete comprehension of what and how it pretends to generalize; in our case, this is the taxonomic reality (TR). It is a fuzzy construct with a rather solid core and a labile periphery.

Elements of construing theoretical foundations of biological systematics can be found in many works, starting with those of early taxonomists (Jung, Linnaeus, de Candolle, etc.) and ending with the most recent research. In them, fundamental statements are usually formalized as "canons" or "rules," "axioms" or (more often) "principles," though without clear delineation of their basic functions. The ideas developed by logicians claimed to be comprehensive [Woodger 1937; Gregg 1954; Jardine 1969; Mahner and Bunge 1997], but they look too formal for biological systematics. Theoretical developments by biologists [Simpson 1961; Sokal and Sneath 1963; Mayr 1969; Lines and Mertens 1970; Løvtrup 1973; Sneath and Sokal 1973; Wiley 1981; Pavlinov 1990; Mayden and Wiley 1993; Quicke 1993; Schuh 2000; Epshtein 2002, 2003; Rasnitsyn 2002; Wägele 2005] are substantive and therefore more attractive, but they aimed at substantiating particular views of the scope and tasks of systematics and are too fragmentary.

Thus, the current state of the theoretical foundations of systematics is still in its infancy and requires special detailed research: so far it is not so much a theory in a certain strict sense, even "immature" [Zuev 2015], as some set of its fragments. Apparently, the main cause is that the very task of developing a kind of broad biologically sound theory was not previously posed by theoreticians who cared mostly about particular classification approaches (natural systematics, phenetics, cladistics, etc.). So, what is urgently needed now is to develop, first of all, a general understanding of how these foundations should be built to make them adequate for a contemporary understanding of the cognitive situation of systematics considered in general. The state of affairs in mathematics and physics, from which examples of developed theoretical constructions are usually taken [Sneed 1979; Rybnikov 1994; Perminov 2001], shows that the search for possible solutions of this fundamental task promises to be difficult and therefore quite lengthy.

This chapter presents the author's ideas about one of the possible ways to develop theoretical foundations of systematics in the form of *quasi-axiomatics* [Pavlinov 2011a, 2018; Pavlinov and Lyubarsky 2011]. It is not deduced formally from a certain ready-to-use high-order conceptual system, but rather is a product of the author's intuitive inspiration about a desirable structure of such a theory. So, this is by no means a completed product; it is just an attempt to look more closely at some key positions and to outline a possible direction in which rationally and constructively framed theoretical foundations of systematics can be developed. With this, it is designed to develop more clearly the general context in which basic concepts and research programs of systematics will be considered in the subsequent parts of the book.

4.1 TAXONOMIC THEORY AS A QUASI-AXIOMATICS

The main task of systematics is to study taxonomic diversity; or, more formally and comprehensively, TR (see Section 4.2.1). Its basic conceptual construct is TT, whose main purpose is to shape the conceptual space of the cognitive situation in which this general task is to be solved. More deliberately, TT is about both general properties of this reality and general principles of comprehending it.

Since the area of application of TT is TR, this is the *biologically meaningful* theory. To an extent the TR is not given but construed within a certain conceptual space, TT describing it is a kind of *constructive theory*; as noted above, it is built as a *quasi-axiomatics*. In this case, "quasi" means that its basic statements are substantively interpreted from the very beginning by reference to the natural phenomenon under study, i.e., they are introduced by and large inductively. Its representation as an "axiomatics" is important, among other things (appropriate formalizations, etc.), for distinguishing two main categories of statements. Some of them refer to the ontic component of the cognitive situation and appear as the *quasi-axioms* or *presumptions* (see Section 4.1.2 on their differences), while others refer to the epistemic component of the same situation and perform a function of the *inference rules*, and can be properly called the *principles* of taxonomic research. Both categories of statements are interconnected by the principle of onto-epistemic correspondence: this means that axioms/presumptions (ontic premises) and inference rules (principles) work conjointly, therefore they should be introduced into this quasi-axiomatics in a way that makes them meaningfully compatible. For example, if an *Umwelt* is defined phylogenetically using the respective ontic premises, then the principles should specify how a classification is to be developed to reflect just the phylogenetic relations and not anything else.

4.1.1 GENERAL AND PARTICULAR TAXONOMIC THEORIES

The TT, as it is here understood, is not analogous to what the physicalist philosophy usually considers a theory. It does not explain or predict specific facts about the TR; instead, it configures conceptually the cognitive situation in which systematics operates. Its main task is to outline correctly (including biologically meaningfully) this situation in the form of interconnected quasi-axioms/presumptions and inference rules of quite general order. On this basis, TT can be defined as a set of logically consistent and substantively compatible general judgments about the structure of TR (ontology) and the principles of its study (epistemology), reflected in the system of basic notions (thesaurus) [Pavlinov 2011a, 2018].

The basic structure of TT corresponds to that of the cognitive situation within which it is developed. This means, among other things, that TT is shaped as a conceptual pyramid: its top is occupied by a *general* TT (GTT), at its middle levels are *particular* TTs (PTTs) of different levels of generality, while at the lowest levels are their various operational interpretations. Within the GTT, the TR as a whole is outlined and its main properties (structuredness, possible causality, etc.) are indicated by ontic premises, while the basic provisions of elaborating classifications are stated by epistemic principles. At lower levels, these general statements are detailed down to more specific ones shaping the contents of different PTTs; in particular, the TR is decomposed into its different manifestations. According to the *principle of constructivity*, the statements of both GTT and each PTT should not be implied but explicitly introduced. It is possible that transitions from more general to more particular statements within this pyramid can be represented as *deducible judgments* ("theorems"), and there are attempts to do so in systematics [Løevtrup 1975; Reif 2009]; this issue is not addressed here.

Thus, the GTT is not intended to solve particular classificatory problems and tasks, but provides a framework for possible ways of formulating and solving them by PTTs. This means that the GTT makes sense not in itself but as a means of developing specific PTTs dealing with various manifestations of TR. These PTTs are designed to develop their particular quasi-axiomatics as a basis for elaborating classifications representing them as adequately as possible. By this, the GTT serves as a *metatheory* outlining something like a "space of logical possibilities" for construing specific PTTs. Therefore, bearing in mind the requirements of the *principle of deducibility*, the GTT must be built in such a way as to allow specific statements of the PTTs to be defined as explications of the statements of the GTT. This means, among other things, that the content of the conceptual space defined in terms of the GTT should be sufficient to cover possible PTTs as its particular interpretations.

Different PTTs detail certain provisions of the GTT based on some additional premises shaping their respective particular cognitive (sub)situations; by this, they structure the GTT by delineating subareas in the overall conceptual space outlined by the GTT. The three-component structure of the cognitive situation presumes recognition of three main categories of the PTTs. Ontic PTTs define the studied objects as more specific *Umwelts*: these are, for example, typological, phylogenetic, biomorphic theories. Epistemic PTTs define the way these specific objects are studied: these are, for example, logical or numerical theories. In subject (subject-oriented) PTTs, the main emphasis is on the subject of taxonomic research: for example, on personal intuition (phenomenological theory, etc.).

In the category of ontic PTTs, the lower levels of the conceptual pyramid are construed for particular manifestations of the TR by fixing the respective lower-level quasi-axioms/presumptions; several groups of such theories can be distinguished here. Some of them are *aspect-based* (*aspectual*) theories: these include theories (phylogenetic, typological, etc.) that define different aspects of the TR by fixing particular relations between organisms (kinship, structural similarity, overall similarity, etc.). Complementary to this is a group of *relational* theories, which consider the properties of both these relations as such and their interconnections. A separate group includes *object-based* theories: they deal with the structural units of taxonomic and partonomic aspects of the overall TR—taxa and partons, respectively (see the next section), as well as their ranks. In each of these groups, at a far lower level, the PTTs can be divided into groups of *causal* and *phenomenological* theories: the former take into consideration the causes of TR, whereas the latter do not. Based on another criterion, the PTTs can be divided into *structural* (typology, phenetics), *functional-structural* (biomorphics), and *processual* (evolutionary) theories.

The PTTs elaborated as aspectual and epistemic are of fundamental importance in the structuring cognitive situation of systematics: they serve as conceptual bases for the respective research programs, and some are sometimes called "systematic philosophies" [Hull 1970; Mayr 1982, 1988; Minelli 1994; Ereshefsky 2001b, 2008]. Object-based and relational PTTs are not endowed with such a function, but they are no less significant: without them, systematics cannot function normally. This circumstance is reflected, for example, in the proposal to single out a separate discipline dealing with species [Skvortsov 1967; Zavadsky 1968; Dubois 2011].

The ways of identifying categories and groups of PTTs listed above are not exhaustive. These theories can be construed to be most adequate to the specifics of both different levels of TR (micro- and macrosystematics) and different groups of organisms (for example, those with or without ontogeny, with or without biparental reproduction).

4.1.2 BASIC QUASI-AXIOMS AND PRINCIPLES

The ontic part of the GTT includes quasi-axioms and presumptions describing the TR and answering the question *"what?"*—what exactly does systematics, in its quite general understanding, investigate? In brief, the answers to this general question define both the objects and their interrelations that systematics explores.

Quasi-axioms and presumptions differ in their conditionality [Rasnitsyn 1996, 2002; Pavlinov 2011a, 2018]. The former are accepted in an "absolute" form; their truth is not questioned and is not verified within particular taxonomic research. The latter are accepted in a softer form; their truth status is accepted not "absolutely" but probabilistically, so their estimates can vary (increase or decrease) as the result of a particular study. It is clear that either the axiomatic or presumptive status of statements about ontology depends on certain background theories. For example, in systematics as natural science, the objective existence of the diversity of organisms is taken for granted as the fundamental quasi-axiom; otherwise all explorations addressed to the TR become meaningless. In contrast, an assumption of evolution as a cause of the diversity of organisms is in general a presumption depending on particular world pictures. With this, it is a quasi-axiom in phylogenetics, while it is excluded from the ontic background in phenetics. A partonomic structuredness of the organisms is expressed by the empirically substantiated quasi-axiom, while particular judgments about the homologies of their parts are always presumptions.

From the very beginning, something like the *quasi-axiom of existence* is introduced in the GTT. This states that biological organisms with their properties really exist as part of Nature—more precisely, as the elements of the TR. The quasi-axioms/ presumptions of *systemity* and *common cause* assert a non-random feature of the TR structure that is caused by an action of certain causes (initial, acting, formal, etc.). According to the *quasi-axiom of relations*, there are some non-random *affinity relations* between the diversity patterns of organisms with their properties, which makes sense of representing the TR structure by certain classifications.

The *quasi-axiom of aspectedness* formalizes the possibility of fixing various aspects of the TR. The most basic is recognition of two fundamental "orthogonal" aspects designated as *taxonomic* and *partonomic* ones, corresponding to the diversity of organisms by their features and diversity of features (partons) over all organisms, respectively [Meyen 1977, 1978; Winston et al. 1987; Tversky 1989; Lyubarsky 1996; Górska 2002; Keet and Arale 2008]. The interrelations of these basic aspects are fixed by the *quasi-axiom of taxonomic–partonomic complementarity*, which means that the taxonomic and partonomic "subrealities" are reflected onto each other in a certain regular way. The following, within each "subreality," specific forms of relations of their elements (organisms and their features) are introduced as specifications of the general quasi-axiom of relations. Some of them are taxonomic (similarity, kinship, etc.), others partonomic (similarity, homology, etc.), a set of lower-level quasi-axioms specifies interconnections between different categories of relations (say, between similarity and kinship, between similarity and homology).

The most general inference rules (principles) seem to be as follows. First, the correct "translation" of statements (quasi-axioms and principles) of the GTT into those of different PTTs is provided by the *principle of interpretability*. According to the *principle of classifiability*, the diversity of organisms and their features can be more or less adequately reflected by some classifications; it is an explication of the above-mentioned general *principle of cognizability*. The principle of *taxon–character correspondence* asserts a certain non-random interrelation between the sets of taxa and characters attributed to them; it follows from the above quasi-axiom of complementarity and ensures the possibility of recognizing non-randomly arranged and characterized classification units. The *principle of taxonomic uncertainty* fixes the impossibility of developing a single PTT and some "omnispective" classification based on it, which would reflect exhaustively the TR in its entirety.

There are several important principles of more practical meaning that regulate the proper arrangement of taxonomic research. According to the *principle of taxonomic unity*, taxa are delineated in such a way as to reflect the certain unity (affinity) of organisms by specified features and/or relationships. The *principle of criterial uniformity* asserts that classification should be elaborated based on some unified system of criteria of recognizing and ranking taxa, choice of characters, etc. It is not always possible to follow this principle for various reasons, so an alternative *principle of criterial non-uniformity* is introduced to soften the requirements of uniformity. The *principle of representativeness* asserts that, for a classification to be adequate to a certain manifestation of the TR, the research sample must be sufficiently representative. The *principle of character inequality* states that, in order to obtain a sought classification, it is necessary to select ("weight") characters according to specified criteria.

Along with these basic principles, the GTT develops some lower-order principles connecting the general ones with certain operational rules. Examples are particular specifications of the principle of criterial uniformity for particular classification tasks. Thus, considered together with the *principle of ranking*, it presumes that, in classification with a ranked hierarchy, the latter should be fixed based on a single ranking scale. The *principle (or rule) of progression* states that, within an inclusive taxon, its subtaxa of the same rank should be placed in a sequence reflecting some chosen gradient (evolutionary advancement, etc.).

The subject component of the cognitive situation is not considered in the standard axiomatic systems; however, recognition of the three-component structure of this situation obliges this. Perhaps, in order to reflect it more adequately in a quasi-axiomatics, the latter should include certain general categories regulating the very cognitive activity; an example is the *axiology* developed within the framework of modal logic [Miroshnikov 2010; Ivin 2016].

As follows from the above, the PTTs are such lower-order quasi-axiomatic systems, in which the main provisions (quasi-axioms, presumptions, principles, etc.) of the GTT are filled with different particular contents provided by different interpretations of the original statements. In developing the PTTs focused on ontology, general axioms of existence and relations are first specified by fixing certain aspects of the TR (phylogenetic, typological, biomorphological, etc.). These interpretations define particular *Umwelts*, for which the respective aspect-based PTTs are elaborated. If the PTTs are evolutionarily interpreted, their emphases can be on adaptatogenesis or cladogenesis, and parallelisms can be emphasized or ignored. With this, particular relationships between organisms can be taken into account—only similarity or only kinship, or some combination thereof. Based on these ontic assumptions, certain rules for homologization of features, selection of characters, etc. are developed. Relational and object-based PTTs specify the provisions of the GTT for developing particular quasi-axiomatics dealing with the properties of specific relations (similarity, homology, kinship) and the structural units of the TR (species, characters, etc.). In epistemically focused PTTs, particular ways of studying the TR are substantiated: for instance, the principles of taxonomic research can be deduced from certain logical procedures or subordinated to the conditions of certain quantitative methods. Finally, a version of subject-oriented PTT may be based on the predominantly pragmatic principles implementing the above-mentioned axiological consideration.

Another way of the PTTs fragmenting is presumed by the specifics of particular groups of organisms with respect to the structure of their diversity, some basic features of their natural history, etc. An example is the multiplicity of species concepts reflecting, at least in part, the diversity of natural history of certain groups of animals, plants, prokaryotes, etc. (see Section 6.7). Another example is the so-called "horto-taxonomy" of cultivated plants [Datta 2020].

4.2 DEFINING BASIC NOTIONS: TWO CASE STUDIES

Statements (quasi-axioms) that shape the ontic basis of a cognitive situation are introduced in the form of definitions of the respective notions. Following the conditions of certain general epistemic regulators, such as the *principle of constructivity*, they must be sufficiently complete in their content and sufficiently strict. However, compliance with these conditions is hindered by some serious limitations (see Section 3.2.2). Below we consider how these problems can be solved in the case of two concepts, which are among the most fundamental for systematics: TR and the classification system.

4.2.1 TAXONOMIC REALITY

In considering the ontic foundations of systematics, one of the most important and problematic issues is the correct definition of the TR as a subject area of this discipline [Zuev 2002; Pavlinov 2010b, 2011a, 2018; Pavlinov and Lyubarsky 2011]. It is problematic because TR, as a kind of conceptual reality (*Umwelt*), can be correctly defined intensionally (in a logical sense) only in a certain theoretical context. This means that understanding of what is or is not the TR changes with the development of the entire theoretical content of systematics.

The source of contemporary ideas about the subject area of systematics is the classical natural-philosophical idea of the System of Nature, in which places of organisms and their groups are determined by their intrinsic features, primarily essential. The latter are contrasted with the extrinsic properties, featured by their mainly ecological and spatial relations. Focusing on these features and relations distinguishes different biological classifying disciplines developed during the 19th century: it separated systematics, which pays attention basically to intrinsic features, from biogeography and biocenology analyzing extrinsic features and relations. This general understanding of the specificity of systematics was inherited from the essentialist scholastics, first of all, by typology and classical phylogenetics, as well as by phenetics and cladistics, though they reject an essentialist interpretation of both the taxa and their characters. However, in biomorphics and biosystematics, which are also parts of systematics, some extrinsic features are just as important as intrinsic ones, so these "essentialist" boundaries are blurred. They practically disappear with the extremely broad defining subject area of systematics as the "overall" diversity of organisms, which should be classified by "all and any characters" [Simpson 1961; Sokal and Sneath 1963; Rollins 1965]: the cause for such an erosion is that "all and any characters" may include both intrinsic and extrinsic features of organisms.

In order to get rid of an extra historical "load" in considering this issue and to place it in the proper conceptual framework, the following general scheme seems to be correct [Pavlinov 2011a, 2018]. It is based on the conceptual pyramid, according to which any concept can be correctly defined only in the context of a certain more general concept. In the case under consideration, this is the concept of the structured biota: the latter can be conditionally thought of as the *Umgebung*, in which various manifestations (aspects, levels, fragments, etc.) are recognized as specific *Umwelts* based on the specific features and relations of organisms and their aggregations. One of these *Umwelts* is the sought TR, so its intensional definition should refer to its specific features distinguishing it from other manifestations of the general structure of biota.

It is reasonable to base consideration of this issue on the following general assumptions (quasi-axioms), which intuitively seem quite reasonable. Firstly, biota is defined as a self-developing non-equilibrium system that is being structured as it develops (evolve) and functions. Secondly, an assumption is required about the non-random nature of the general structure of biota determined by the actions and interactions of various natural causes. Thirdly, it is assumed that reference to these causes allows, on a fairly natural basis, distinction between different manifestations of this general structure, one of which is TR. Further, it can be additionally acknowledged that these manifestations of biodiversity do not exist on their own as independent natural phenomena; rather, they are *epiphenomena* which reflect various manifestations of the general structure of biota [Pavlinov 2007b; Loreau 2010]. Finally, one should clearly realize that these epiphenomena are individualized as cognizable entities within the framework of a certain cognitive situation by a knowing subject by means of certain concepts. On this basis, framework conditions can be set for the correct definitions of the TR, the GTT, and its various manifestations and studied by specific PTTs.

Different manifestations of biotic structure can be fixed (individualized) at a conceptual level in two ways, either as *objects* or *aspects*. In the first case, particular natural *phenomena* can be thought of as natural entities existing outside of and besides the knowing subject; they are recognized by the latter by their specific emergent properties and relations. Examples are cells, organisms, populations, species, ecosystems, levels of hierarchical organization of biota, as well as (with some reservations) monophyletic groups. In the second case, it is not the phenomena that are cognized as such, but rather certain epiphenomena, which are aspects of the biotic structure fixed by the knowing subject based on the respective cognitive intentions. So, these individuated aspects by no means occur in the structure of biota as something subject-independent; their presence in the cognitive situation as cognizable entities depends largely on the subject's cognitive activity. However, reference to the supposed natural causes and processes structuring biota can be taken for a justification of the naturalness of its aspects being recognized in this way. Examples are the taxonomic, biogegraphic, ecological, etc. aspects (forms) of the diversity of organisms, of which the first one is the aspectually defined TR. The basic aspects of the latter are the taxonomic (in its narrow sense) and partonomic ones; the phylogenetic and typological aspects of TR can be further individualized; cladogenetic and anagenetic aspects of phylogeny are also examples of this kind. It follows from this consideration that both TR itself and manifestations of its aspectual structure cannot be specified in a single trivial way: their recognition is subject-motivated, their definitions are fuzzy.

To sum up the consideration described above, it is to be emphasized that manifestations of the biotic structure, one of which is the TR, are basically aspectual: they are "given" not by themselves but in the way they are considered by particular subjects in particular cognitive situations. Indeed, all disciplines studying biotic structure actually deal with the same totality of organisms, as well as with their properties and relationships between them. Different aspects of this structure are individualized by fixing those particular properties and/or relations considered significant for the analyses of just these and no other aspects. This point is clearly exemplified by various aspectual representations of a single organism in behavioral, ecological, physiological, anatomical, etc. studies. On a similar aspectual basis, a fundamental distinction is made between the phylogenetic, ecological, and ecomorphological aspects of the biotic structure [Pavlinov 2007d]. This means, by the way, that the particular aspect-based manifestations of this structure, in contrast to object-based ones, are completely internested.

As stated above, in the analysis of the basic structure of TR, it is of fundamental importance to distinguish between its two most general "orthogonal" aspects, usually designated as *taxonomic* and *partonomic* (meronomic) [Panova and Shreyder 1975; Meyen 1977; Tversky 1989; Pavlinov 2011a, 2018]. This distinction between the diversity of organisms themselves and their features, respectively, constitutes one of the very first cognitive acts aimed at TR [Jardine 1967; Meyen 1978; Lyubarsky 1996; Pavlinov and Lyubarsky 2011; Pavlinov 2018]. Accordingly, they outline taxonomic and partonomic "subrealities," for which specific quasi-axiomatic systems are construed and specific classifications are elaborated. Within the framework of formalizations accepted in numerical systematics, they correspond to the Q- and R-modes of the analysis of phenetic hyperspace [Sneath and Sokal 1973].

Such an aspectual givenness is evident in the case of the *Umwelts* outlined by the particular TTs with reference to certain causal relations presumably structuring biota in specific ways. By this, particular aspects of TR are individualized as epiphenomena that are studied and classified as something actually inherent to wild-life. These aspects, constituting the ontic component of the respective PTTs, can be characterized as follows: the typological aspect is formed mainly by structural causes, the phylogenetic aspect is formed mainly by historical causes, the biomorphological aspect is formed by a combination of structural and functional (ecological) causes, etc. Operationally, they "diverge" at the level of choice of the characters, which are supposed to "mark" aspects of interest.

4.2.2 CLASSIFICATION SYSTEM

The main "form of being" of knowledge in systematics is *classification*, which serves as a cognitive (informational) model of a certain manifestation of TR. In one of the approaches considering the ways of representation of the diversity of Nature, it is aptly designated as the *classification system* as opposed to the *parametric system* [Subbotin 2001]. The first reflects the relationships between different objects characterized by the same or different features and represents the categorical structure of these relations by recognizing assemblages of objects (taxa, biogeographical units, syntaxa, etc.) based on their specific features. The second reflects the relationship between different features that characterize the same or different objects and represents quantitative relationships between these features by compact formulas (the van't Hoff equation in chemistry, the Schrödinger equation in physics, the equation of allometric growth in biology, etc.).

A number of rather formalized definitions of the classification system (= classification in its general sense) was proposed that might be suitable for direct interpretation in terms of systematics [Woodger 1937; Gregg 1954; Jardine 1969; Voronin 1985]. Below is the exposition of a revised version developed earlier by the author [Pavlinov 2011a, 2018; Pavlinov and Lyubarsky 2011] based on the following notions and definitions.

The classification system **CS** reflects the general structure of the TR. The latter is decomposed into two basic aspects, taxonomic and partonomic, interrelated in a certain way. The taxonomic system **TS** is developed for the first aspect, and the partonomic system **PS** is developed for the second aspect. The general classification system is a union of these two particular classification systems.

The classification system **CS** is an ordering of the classification units C_E by establishing certain relations R_{CE} between them, so it is the *relational* system. In the simplest case it can be defined as an ordered pair $\{C_E, R_{CE}\}$. If taxonomic system **TS** is developed, its classification units are taxa **T**; the relations between them R_T are defined as similarity, kinship, rank, etc.; taxa are characterized by taxonomic characters C_T, with the respective relations R_C between them being defined (intercorrelations, weights, ranks, etc.); relations R_{TC} between taxa and characters are established by *taxon–character correspondence* (see Section 6.2). If partonomic system **PS** is developed, its classification units are partons **P** with their respective relations R_P (homology, rank, etc.) they are assigned to certain taxa P_T, establishing *taxon–parton correspondence* R_{TM} between them. Partons assigned to taxa P_T are operationally representable as taxonomic characters C_T, which makes them mutually equivalent: $\mathbf{P_T} \leftrightarrow \mathbf{C_T}$; on this basis, equivalences are established between the corresponding relations: $\mathbf{R_P} \leftrightarrow \mathbf{R_C}$ and $\mathbf{R_{TP}} \leftrightarrow \mathbf{R_{TC}}$. Together, these equivalences can be viewed as a functor connecting systems **TS** and **PS** into a single system **CS** [Shreyder and Sharov 1982]. All these parameters remain formal if they are not interpreted in a meaningful way. For such an interpretation, one more parameter is introduced—basic background theory **BT**, in which context other parameters acquire a certain meaningful interpretation.

Accordingly, the following general definition of the classification system **CS** looks as follows:

$$\mathbf{CS} \supset \{\mathbf{TS}\} \cup \{\mathbf{PS}\}$$
$$\mathbf{TS} \supset \mathbf{BT}\{\mathbf{T}, \mathbf{C_T}, \mathbf{R_T}, \mathbf{R_{TC}}\}$$
$$\mathbf{PS} \supset \mathbf{BT}\{\mathbf{P}, \mathbf{P_T}, \mathbf{R_P}, \mathbf{R_{TP}}\}$$

Position of the parameter **BT** out of the brackets means that it is not an obligatory part of the definition of either interpretations of **CS**. Omitting it makes the latter meaningfully undefined and therefore theory-neutral in any respect. If included in the definition of **CS**, it means the latter is meaningfully interpreted and thus theory-dependent; this is true for all parameters of the above definitions—taxa, partons, characters, and relationships between them. An integrative effect provided by the parameter **BT** is formalized by the above-mentioned *principle of criterial uniformity*: all parameters enclosed in the brackets {...} must have the same interpretation in all fragments and at all levels of the respective classification system. Thus introduced, the parameter **BT** fulfills two important functions. First, it serves as the basis for developing both the definition and criteria of naturalness of the classification system within the framework of the respective PTT. Second, it defines an *ordering factor* (for example, kinship relation), according to which the general structure of the classification system is shaped.

Taking into account the structure of the cognitive situation, two respective basic components can be distinguished in the parameter **BT**, viz. ontic $\mathbf{BT_O}$ and epistemic $\mathbf{BT_E}$, according to which two variants of meaningful interpretation of the classification system can be defined. If a PTT is ontology-based, this parameter is defined as $\mathbf{BT_O}$: it makes the classification system biologically sound, and its concrete values indicate the properties of TR considered significant according to this PTT. If a PTT is epistemology-based, this parameter is defined as $\mathbf{BT_E}$: it outlines the main properties of the method (in the general sense) that generates the respective classification system, which remains with no biologically sound content. Thus, the first option means the dependence of the content and structure of the system **CS** on a particular substantive background knowledge, whereas the second option means its dependence on a particular classification algorithm.

5 Major Research Programs in Systematics

> Whenever a theory appears to you as the only possible one, take this as a sign that you have neither understood the theory nor the problem which it was intended to solve.
>
> K. Popper

Every scientific discipline differentiates as it develops: this is an obvious consequence (and evidence) of its normal functioning as a conceptual system [Hull 1988]. This differentiation is due to the impossibility of reducing knowledge about some complexly organized phenomenon to a certain unified cognitive model (theory). It gives rise to the emergence and coexistence of different research programs, each corresponding to a certain understanding of both a certain natural phenomenon or epiphenomenon (an *Umwelt*) and the way of its cognition. Each such understanding is formalized as a meaningful theory developed by a certain scientific community, school, etc. [Lakatos 1978; Stepin 2005]. The stability of a research program, and with it a particular cognitive situation shaped by it, is ensured by the invariability of the respective scientific problem being resolved, which forms its stable "core," while its "periphery" can change according to changes in the general philosophical scientific context of systematics. Exhaustion of cognitive potential of a certain research program leads to a scientific revolution, which completes the development of the given program and stimulates the emergence of new ones, which means a certain change in the content and structure of the overall cognitive situation.

The latter, in its general scope embracing all the biological systematics, is developed under the integrating effect of the idea of the Natural System; different understandings of both the System itself and the methods of its exploration take a form of general *cognitive programs* and particular *research programs* detailing them. Among cognitive programs in systematics, the most pronounced are *rational* and *empirical*; in Chapter 2 they are called "methodist" and "collectionist," respectively. The first is associated with the theoretical understanding of what the Natural System is and what the ways of knowing it are; it encourages the development of theoretical (onto-epistemic) and methodological foundations of systematics. The second is associated with elaborating the practical classifications; its metaphysical component is minimized, while its methodological component is either absent or sufficiently developed depending on the particular program.

Development of general ideas about the structure of the cognitive situation, in which biological systematics operates, is one of the key tasks of the general taxonomic theory (GTT); an important part of this task is outlining both the partial taxonomic theories (PTTs) and the research programs implementing them (see Section 4.1.1). At present, there is no clear understanding of the general grounds on which these programs might and should be distinguished, and, accordingly, what the programs themselves are, how they relate to each other, etc. The reason is that there is no sufficiently developed GTT (see Chapter 4).

The first attempts to recognize explicitly and to discuss the research programs in systematics, as well as to evaluate their scientific status, were undertaken in the 1960s–1970s. There appeared to be acknowledgment of only three principal "systematic philosophies" most vividly discussed at that time, namely, phenetics, cladistics, and evolutionary taxonomy [Hull 1970, 1988, 2001; Mayr 1982, 1988; Pesenko 1989; Minelli 1994; Schuh 2000; Ereshefsky 2001b, 2008; Rasnitsyn 2002]. However, such a "three philosophies" viewpoint did not take into consideration or drastically reduce the significance of other "philosophies" that were not so actively discussed then— typology, biosystematics, ecomorphological approach, etc. Therefore, such an oversimplification provided a very distorted representation of the theoretical foundations of biological systematics, including the diversity of the research programs operating in it, their historical and philosophical roots, their mutual interactions and influences, and their contributions toward the development of systematics. In general, multiplicity of research programs in systematics is caused by great complexity of biological diversity which can be considered from very different standpoints.

In systematics, the formation of research programs, if the latter are not interpreted too strictly, is evident from the earliest stages of its conceptual history. Their fates are different: some function for a long time, while others disappear rather quickly (on a historical time scale). These fates depend, by and large, on the extent the particular research programs are in demand by the scientific community, which in its turn is determined by general philosophical scientific contexts that change over time. Therefore, a program gains recognition if it appears "in its place in due time," and it actively works within the respective context until the latter "shrinks." Another cause for the loss of popularity of a program is its inability to set and solve properly the research tasks demanded at a certain stage of conceptual history of systematics. In general, multiplicity of research programs in systematics is caused by great complexity of biological diversity which can be considered from very different standpoints.

In the latter, the first was the scholastic program based on an essentialist understanding of organisms and the deductive genus–species classification scheme. Within its framework, particular subprograms were formed, associated with different ways of understanding the essence (ontic or epistemic, division of botanists into "fructists" and "corollists"). At the end of the 18th century, the anti-scholastic revolution took place, caused by a change in the understanding of both the Natural System and the natural method; it led to the appearance of several research programs based on different onto-epistemologies, viz. typological, organismic, numerological, natural (in its narrow botanical sense). Some of them differentiated into more particular subprograms: in the typology, there were three of them (stationary, dynamic, epigenetic); natural systematics also produced three subprograms defined by the natural methods of Adanson, Jussieu, and Candolle. These two main programs were developing quite actively until the second half of the 19th century when the growing popularity of transformist natural philosophy caused the evolutionary revolution and the emergence of the research program of the evolutionarily interpreted systematics with two subprograms, namely classification Darwinism and systematic phylogeny (now called phylogenetic systematics). The natural systematics of botanists, after some confrontation with phylogenetics, nearly merged with it; typology faded into the background of the cognitive situation of systematics, while organismism and numerology almost ceased to exist.

In the first half of the 20th century, systematics underwent a positivist revolution caused by the growing popularity of the positivist cognitive program in natural science (mainly in the version of physicalism). Several research programs appeared that implemented the physicalist idea in different ways: these were biosystematics (as a successor of classification Darwinism), phenetic and numerical programs, as well as several versions of rational systematics (onto-rational and episto-rational programs). With this, the concepts having appeared in the 19th century (typology, phylogenetics) nearly lost their recognized status. In the second half of the 20th century, a post-positivist revolution caused a revival of interest in phylogenetic systematics and led to the active development of several subprograms within its framework. Evolutionary taxonomy and cladistics were initially differentiated, then the "new phylogenetics" became most popular by combining cladistic and numerical ideas superimposed on molecular data. In its shadow, evolutionary ontogenetic systematics is currently arising, and may become a new promising research program.

Particular research programs are "brought to life" by *taxonomic schools* appearing as part of the subjective component of the cognitive situation; each such school is not so much a scientific but rather a social phenomenon [Mikulinsky et al. 1977; Hull 1988; Mishler 1991; Rozova 2014]. For this reason, in the formation and maintenance of many schools, an important role is played not only by its attractive (demanded) conceptual framework but largely by a personality factor. The latter means that an authoritative scientific leader usually acts as an organizing force by promoting a particular way of posing and solving a certain research problem. In systematics, famous examples are C. Linnaeus in the 18th century, A.-P. de Candolle and G. Cuvier at the beginning of the 19th century; C. Darwin and E. Haeckel in its second half; and E. Mayr, G. Simpson, and W. Hennig in the second half of the 20th century.

A specific social aspect of the activity of the taxonomic schools is manifested in their competitive relations [Hull 1988]. The reason is that their leaders, committed to a monistic cognitive idea, claim their exclusive role in the "right" understanding of how to interpret and solve the key problems of biological systematics. Therefore, the value judgments of those leaders about the scientific status of their respective schools most usually turn into self-praising, very reminiscent of the meaning of Lenin's thesis (once famous among Soviet people) that "the [...] doctrine is omnipotent because it is true." Lenin himself mentioned Marx's doctrine; if you replace it with those of Linnaeus, Darwin, Hennig, or somebody else, it does not make much difference.

An important part of the subject component of the contemporary cognitive situation of systematics is orientation of the organizational support of the taxonomic research towards a "mainstream." It brings a certain "mass kitsch" phenomenon to the formation of a social status of the respective research programs. Currently, technological approaches gained an over-popularity and became a kind of "scientific fashion" by making it possible, in the shortest time and with minimal intellectual effort, to obtain required results that fit well into the "mainstream" [Hołyński 2005]. For the normal development of systematics as a scientific discipline, this yields a serious problem of significant disparities in the ratio of theoretical and practical research: the former are pushed to the periphery of the cognitive situation, so the latter remain without proper theoretical basis and comprehending.

This chapter describes the main research programs of biological systematics whose development shaped the content of the latter's conceptual history throughout the 20th century. The order of their presentation in subsequent sections is set basically by the gradient of biological meaningfulness of the corresponding taxonomic theories. Accordingly, the chapter begins with the simplest classification phenetics, continues with numerical programs, the next is typology with several programs close to it, and, lastly, evolutionarily interpreted theories and programs are considered. These programs are characterized according to a more or less standard scheme to make them mutually comparable. The scheme is based on consideration of the following key features of the programs: what their historical roots are; how their ontic bases are formed; how the respective classifications are understood substantively; how taxonomic unity, the choice of characters, delineation of taxa and their ranks, etc. are interpreted; what are the area of their application; and what are their advances (what new aspects the approach provides for the development of systematics) and shortages (the limitations of the approach).

5.1 THE PHENETIC PROGRAM

This program, based on the "phenetic idea" [Ereshefsky 2001b] and formed in the first half to middle of the 20th century, rationally develops and formalizes the basic empirical tradition in the modern conceptual framework. Its purposeful formation began with the works of the "anti-scholastics" of the second half of the 18th century. The combinatorial approach of M. Adanson is usually mentioned in this connection, to the extent that the founders of the phenetic systematics call it "Adansonian" [Sneath 1958; Sokal and Sneath 1963; Burtt 1966]. However, this epithet is not correct [Winsor 2004; Pavlinov 2018], as the phenetic idea was most fully expressed at that time by J. Blumenbach (see Section 2.4.2).

In the 20th century, the phenetic program, in its rather strict sense, is most consistently developed by the *phenetic theory*; the taxonomic school developing it is called *phenetic systematics*, and sometimes also *classification phenetics* or *phenonics*. The very term "phenetics" was introduced into the scientific vocabulary to designate an ordering of organisms by their overall similarity [Cain and Harrison 1958]. In one of its early versions, this program was defined as systematics of phenotypes, as opposed to that of genotypes [Sokal and Sneath 1963], but later the phenetic idea was also extended to genetic data [Sneath 1995]. Its general provisions are summarized in the form of a system of so-called "Adansonian axioms" [Sokal and Sneath 1963; Sneath and Sokal 1973].

A key role in the formation of the phenetic program was played by a direct appeal of its ideologists to the philosophy of positivism [Gilmour 1940; Sokal and Sneath 1963], which characterizes it as the most reductionist in all points. In its cognitive situation, any substantive background knowledge is minimized as metaphysical and therefore "unscientific." Accordingly, an *Umwelt* is defined simply as a set of physically observable and measurable bodies—organisms and their features. One of the principal conditions for the phenetic systematic to be scientifically consistent (from the positivism viewpoint) includes a claim of operational definitions of its basic concepts, viz. characters, similarity, taxon, classification, etc.; i.e., the phenetic program is designed for developing *operational systematics* [Sokal and Sneath 1963; Sneath and Sokal 1973]. This condition presumes maximal automation of the entire classification activity claimed to minimize the contribution of a knowing subject to it. All this, from the positivist standpoint, presumes the "objective" status of both the phenetic procedures and the classifications resulting from them.

Respectively, the traditional general understanding of the naturalness of classification is replaced, with reference to the philosophy of J. Mill, by its prognosticity (predictivity), according to which the *prognosticity criterion* is introduced to assess the consistency ("naturalness") of phenetic classification. From this standpoint, the best (most natural) classification is the one with the greatest predictive capacity [Gilmour 1940]; it is sometimes called "Gilmour-natural." Such a classification is developed as purely empirical (in a philosophical sense): it "must provide nothing more than a summary of observed facts" [Colless 1967]. At the operational level, the predictive value of classification is replaced by its *informativeness*, which is defined as the amount of information about the diversity of organisms it stores. According to this criterion, such classification, to be maximally informative, should encompass information on the greatest possible number of characters. It is additionally assumed that various classifications, if based on large enough subsets of different characters, should converge asymptotically to become eventually indistinguishable (the *principle of convergence* of mathematics); this supposes the potential attainability of a single most informative classification as an ideal of phenetic systematics [Sneath and Sokal 1973]. The most informative classification that can be used to solve many different research and/or practical tasks is called the *general purpose classification* [Gilmour 1940, 1961]; it performs a reference function and therefore is also called the *general reference classification* [Gilmour and Walters 1963; Sokal and Sneath 1963; Sneath and Sokal 1973]. Along with it, different "locally useful" *special-purpose classifications* can and should be developed for particular different tasks; there may be a lot of them, and they can be very different in their content. So, the cognitive position of classification phenetics presumes a kind of moderate pluralism.

An elementary unit of phenetic analysis is an organism ("entity") as an aggregate of particular modalities of characters by which it is described. The characters are interpreted as "observed properties of [...] entities, without any reference to inferences" [Colless 1967]. Such interpretation is inconsistent because characters cannot be distinguished without prior homologization of the organismal features (structures, etc.), which is inconceivable without the substantive "inferences." A character is considered a "logical" construct which is used as a kind of "unit of information" in comparing different "entities" [Gilmour 1940; Sokal and Sneath 1963; Colless 1967]. This is formalized by the *concept of unit character*, which has a double meaning: (a) it is not logically divided into characters of less informational capacity;

and (b) correlations between characters are supposed to be minimal, which makes them independent both meaningfully and logically. To be free from any "inferences," the sample of characters for the phenetic analysis should be composed randomly with respect to a certain prior substantive motivation. When developing classifications, preliminary "weighting" (assessment of significance) of characters is minimized since it always relies on certain metaphysical considerations (background knowledge); this, generally speaking, means an *equivalent weighting*.

The substantive basis for distinguishing unit characters is *operational homology*, the concept of which is developed with reference to the *theory of correspondences* [Woodger 1952]. It is carried out based on the similarity of features (attributes, properties, etc.), which are considered "operationally homologous [if] they are very much alike in general and in particulars" [Sneath and Sokal 1973]. This definition of homology makes the latter quantitative: the features can be of various degree of homology measured by a proportion of coinciding unit characters in their total number used in the analysis. It is assumed that judgments about operational homologies of features "need no inferences about [their] previous homologies" [Sneath and Sokal 1973], but this is incorrect. In fact, such features are distinguished, prior to their description by unit characters, based on just the "previous homologies," which are not discussed in the framework of particular phenetic research because these homologies seem obvious from a common-sense standpoint, to which Sneath and Sokal refer.

One of the key suggestions of phenetic systematics to operate with as many equivalent characters as possible yields a serious methodological problem. Contemporary systematics can operate with an infinite number of various kinds of characters taken from quite different sources (parts of organisms). To avoid the problem of "bad infinity," the *principle of non-specificity* of characters is introduced: characters taken from different parts of the organism, if they are numerous enough, presumably provide "approximately the same" classifications (the above-mentioned principle of convergence). This justifies the possibility of operating with a not too large sample of characters representing randomly an infinitely large general population thereof [Sokal and Sneath 1963]. This idea can be considered as an extension of de Candolle's principle of character interchangeability (see Section 2.4.2); however, the very idea of random character sampling is inconsistent from the conceptualist standpoint (see Section 3.7).

A kind of "quintessence" of phenetic systematics is the concept of *overall similarity* as a summation of the unit similarities/differences between organisms by all available unweighted unit characters. It is equivalent to the *phenetic relation*, the totality of which shapes the *phenetic pattern* as a special formalization for a more "neutral" concept of phenetic diversity [Cain and Harrison, 1958; Gilmour 1961; Sokal and Sneath 1963; Sokal and Camin 1965; Williams and Dale 1965]. This concept was repeatedly criticized from different positions. From the basic premises of phenetics, the main drawback is a non-objectivity of similarity: the latter is "given" not by itself, as something directly observable, but in assessments that depend on background knowledge and particular assessment methods [Tversky 1977; Pavlinov 2005, 2018; Pavlinov and Lyubarsky 2011; Bartlett 2015] (see Section 6.4). Therefore, the similarity can be assessed in different ways, so it turns out to be inexpressible "in general," which makes this concept non-operational [Shatalkin 1990a; Morreau 2010; Pavlinov 2018]. To get around this problem, the concept in question is reduced to a more operational *concept of relative similarity* as a "sample estimate" of the desired overall similarity.

Operational representations of the phenetic pattern are of two kinds; both are actively used in numerical taxonomy. One is the multidimensional *phenetic hyperspace*, which is an abstract model based on an analogy with physical space. The axes of this hyperspace are the characters used for comparing organisms; the scales of these axes are determined by the respective character states; the organisms compared are represented as the points in hyperspace; their relative position in the latter is determined by the character states attributed to them. Another representation of the phenetic pattern is a hierarchical classification tree called *phenogram*.

The only basis for grouping organisms to produce their *phenetic classification* is overall (or rather relative) phenetic similarity. Accordingly, the general principle of taxonomic unity turns into the *principle of phenetic unity* (homogeneity). In terms of classification phenetics, the structure of taxonomic diversity is represented operationally as the filling of phenetic hyperspace [Sneath 1961; Sneath and Sokal 1973]. The classification procedure is strictly inductive. It begins with combining elements of the research sample into the most homogeneous *operational taxonomic units* (OTUs) considered as the objects of application of classification algorithms. The latters' principal task is to distinguish groups of OTUs of different levels of generality, so as to minimize differences within groups and maximize differences between groups. These groups constitute the taxa of phenetic classification, which are generally termed *phenons*; their particular interpretations are the *smallest inclusive* or the *smallest recognizable* units [Sarkar and Margules 2001; Zachos 2016]. They are defined *polithetically*: this means that each phenon is not necessarily characterized by all characters in the analysis. It is assumed that the closest phenons of the same level of generality may overlap if the available characters do not allow for making them discrete [Michener 1962]; such a possibility is profoundly justified by the fuzzy set theory [Bezdek 1974].

If the phenetic classification is built as hierarchical, its hierarchy is described quantitatively based on a ratio of differences within and between phenons. For the two phenons in the same classification to be assigned the same level of generality (rank), they must be characterized by a similar ratio of these differences. This general idea is formalized as *the principle of uniformity of the levels of differentiation of taxa* [Starobogatov 1989]. Such quantitative hierarchization of the phenetic classification makes it possible to abandon traditional ranked hierarchy with its standard terminology [Michener and Sokal 1957; Sokal and Sneath 1963].

The development of phenetic classifications based on overall similarity by a large number of unit characters yields the possibility of active use of quantitative methods, and they are actually used. Therefore, phenetic and numerical research programs are often not distinguished, as shown by the term "numerical phenetics." However, their identification is incorrect: phenetic theory deals with what is researched (ontology), whereas numerical theory deals with how it is researched (epistemology). In fact, the classifications, phenetic in their "spirit," can be developed without employing numerical procedures: this is evidenced by the practice of an enormous army of empirical systematists, as well as by the calls of some traditionalists to develop "omnispective" classifications.

Considering the results of the development of the phenetic research program in systematics from a general philosophical scientific perspective, it should be noted, first of all, that both its basic features and limitations are rooted in the positivist philosophy of science underlying it. Contrary to its initial assumptions, phenetic theory is not free from certain substantive theorizing and therefore is not strictly empirical: from a conceptualist standpoint, phenetic classifications, if actually independent of any meaningful theory, become meaningless [Rosenberg 1985; Pavlinov and Lyubarsky 2011; Pavlinov 2018]. Well-known objections to the theory-free interpretation of random research sampling and assumed to be "as if objective" overall similarity are considered elsewhere in this book (see Sections 3.7 and 6.3). Therefore, if the rise of the phenetic idea was due to an enthusiastic acquisition of positivist philosophy by systematics, then debunking of the claims of this philosophy to an "ultimate" philosophical scientific consistency caused the growing influence of post-positivist philosophy of science, leading to a decline in interest in this research program.

At the same time, some important provisions of the phenetic theory found a fairly wide application in those research programs that do not identify themselves with it. Thus, in phylogenomics, the idea of reduction of the organism to a set of automatically recognized uncorrelated unit characters is brought to perfection: nucleotide bases in DNA or RNA molecules are interpreted in this way. In cladistics, a clear "phenetic spirit" is introduced by the principle of total evidence put forward in the early 20th century by positivist methodology; it is also present implicitly in phylogenomics in the form of whole-genome analyses [Savva et al. 2003].

5.2 RATIONAL SYSTEMATICS

Rationality constitutes the basis of science, distinguishing it from other forms of cognitive activity. In its most general and rather ordinary sense, it means that scientific knowledge should be based on arguments of reason instead of faith or feelings. According to this, such an argumentation scheme becomes of paramount importance, in which some judgments are derived from others according to certain rationally fixed rules. This scheme presumes that (a) if initial judgments are true and (b) the inference rules are consistent (that is, they are also "true" in a logical sense), then (c) inferred judgments are also true, so (d) the knowledge containing them is rational and "scientific." With respect to systematics, the rationalist biologist Alexander Lyubishchev formulated this condition as follows: "a rational system should be understood as such, all elements of which are inferred based on some general principles, a certain theory" [Lyubishchev 1975: 164]. On this basis, the *rational research program* in its general form is developed in systematics. It is worth recalling that the notion of rational classification was introduced at the beginning of the 19th century by A.-P. de Candolle, though its content was understood somewhat differently [de Candolle 1819] (see Section 2.4.2).

Any rationally arranged taxonomic conception is evidently theory-dependent. Initial judgments, from which rational considerations begin, are of two kinds depending on how the "certain theory" mentioned above is interpreted. Some such theories are *ontic* and consider the very subject of research, while others are *epistemic* and consider the argumentation schemes. Thus, there actually are two research programs of the rational kind, *onto-rational* and *episto-rational*.

5.2.1 THE ONTO-RATIONAL PROGRAM

This program has deep natural-philosophical roots; generally speaking, it embodies the general idea that any classification is biologically sound only if it is elaborated within a substantive context set by some general biological theory. Therefore, any theory and research program is rational in this general sense, as far as it (a) is developed within the cognitive situation that includes a certain natural-philosophical concept (organismic, numerological, phylogenetic, typological, etc.) and (b) infers its key classification principles (say, the principle of taxonomic unity, character selection, etc.) from this concept.

The key idea of *onto-rational systematics*, in its rather strict sense, was formulated by the biologist and natural philosopher Hans Driesch. In his opinion, an ordered diversity of biological objects should be generalized in a form of laws similar to those in chemistry about chemical elements or in geometry about geometric figures [Driesch 1908; Naef 1917]. In the earliest versions of this kind of rationality, general structural plans of organisms were derived from the basic types of symmetry, so the respective onto-rational classifications appeared formally similar to the classifications of crystals or minerals [de Candolle 1819; DeCandolle and Sprengel 1821; Haeckel 1917].

The greatest advantage of such onto-rational systems of living beings is that they become a powerful heuristic for explaining known and predicting yet unknown biological forms. Therefore, systematics aimed at developing such rational classifications can be considered *nomothetic*, i.e., revealing the general laws of the diversity of organisms [Lyubishchev 1969, 1972; Meyen 1978]. The nomological essentialism, with its concept of the *nomological space of states* [Mahner and Bunge 1997], can be considered its general foundation. Similarly, it can be thought of as *causal*, explaining the ordered diversity of organisms by referring to the natural causes responsible for it [Rosenberg 1985; Subbotin 2001; Pavlinov 2011b, 2018]. According to one of its interpretations, "the question of rational systematics is a particular instance of the more general question of the 'logic of morphology' " [Webster and Goodwin 1996: 9]. Such systematics is aimed at developing classifications of not only real but also conceivable organismal forms, which is one of the ideals of the morphology of rational kind [Driesch 1908; Lyubishchev 1968, 1972; Beklemishev 1994; Lyubarsky 1996; Webster and Goodwin 1996; McGhee 1999]. Thus, such systematics can be rightly called "imaginary," linking its development with an "imaginary biology," so called because of its analogy with Lobachevsky's "imaginary geometry" [Morgun 2006].

Most devoted proponents of this taxonomic theory rate onto-rational classifications higher than the natural ones understood in the Linnaean sense: the former are nomothetic (law-like), whereas the latter are ideographic (descriptive). According to Driesch, a particular rational classification, logically speaking, can be thought of as a "species" relative to a "genus," which is some law of nature that lies behind a certain aspect of the ordered diversity of biological forms. In contrast, a "Linnaean" natural classification has the meaning of a catalog, therefore, in it, "some particular taxonomic explanations [...] cannot be inferred from others" [Zarenkov 1976: 26]. According to other authors, the rational system as an expression of a certain natural law of transformation of biological forms is equal to the natural system as a law-like generalization, so they are synonymous [Rozova 1986; Zabrodin 1989; Subbotin 2001].

The formalization of the concept of Driesch–Lyubishchev provides *systematics of natural kinds*, if the latter are understood as classes of objects having particular essential properties in common [Mahner and Bunge 1997]. Closely related to this is interpretation of the natural kind as an aggregate of objects tied by a certain law of mutual transformation, such as Goethe's metamorphosis [Webster 1993; Webster and Goodwin 1996; Zakharov 2005]. These authors emphasize that natural kinds are theoretical constructs, whose recognition depends on a chosen aspect of considering the diversity of forms (*ontological relativism*; see [Quine 1996]). Such essential properties or transformation laws of various levels of generality shape a natural hierarchy of organismal diversity. Each natural kind of a given level of generality outlines a certain group of organisms, thus yielding a hierarchically organized rational taxonomic system, which may be considered natural. From the ontological relativism perspective, position of onto-rational systematics is evidently pluralistic.

One of the versions of onto-rational systematics is aimed at developing *periodic systems* of organisms, for which Mendeleev's periodic system of chemical elements is usually taken as a model [Driesch 1908; Lyubishchev 1923, 1972; Tchaikovsky 1990; Popov 2002, 2008]. This taxonomic theory is based on the assumption that there are certain natural laws of diversity of biological forms that are manifested in periodic changes of "secondary" features of organisms arranged according to the unidirectional gradient of a certain "primary" feature. A legacy of the idea of the Ladder of Perfection is clearly visible; the difference is that, in the latter, the entire diversity is linearly ordered along a predominating gradient, whereas it is folded into a solenoid in a periodic system making periodic changes of the "secondary" features more evident. The main problem here is that the biological forms are much more complex as compared to the chemical elements, therefore it is difficult to recognize in them a single primary feature, around which orderliness the entire periodic system could be consistently built. The complexity of organisms is usually considered as such a "primary" feature, but it does not lend itself to a strict and general enough definition that would allow the development of a unified scale of complexity or progressiveness [Salthe 1993; Adami 2002; Giampietro 2002].

The newest *epigenetic* version of onto-rational systematics is designed to uncover static regularities of biological forms generated by dynamic regularities of ontogenetic patterns [Ho 1990, 1992, 1998; Ho and Saunders 1993]. As the biologist Mae-Wan Ho, its most active proponent, asserted, ontogenetic "development is by and large an orderly and hierarchical process which in turn imposes a hierarchical structure on the adult organization of living things" [Ho and Saunders 1993: 290]. According to her, the mechanisms that generate an ordered diversity of biological forms "are atemporal and universal," so referring to them provides a kind of universal classification, in which each taxon is "a class of individuals sharing a common developmental process" [Ho 1992: 199]. Such a class is the above-mentioned natural kind, so it is reasonable to consider Ho's taxonomic theory a variant of systematics of natural kinds. On the other hand, it is close to ontogenetic systematics in its basic premises (see Section 5.4.2). Apparently, this supposed universality may only be true for multicellular organisms with rather advanced ontogenetic development.

5.2.2 THE EPISTO-RATIONAL PROGRAM

This program implies subordination of all classification procedures to certain formally built universal inference rules (algorithms, etc.), i.e., to a unified Method in its classical understanding dating back to Antiquity. In this case, Lyubishchev's "a certain theory," mentioned at the beginning of this chapter, takes the form of a unified, formally construed logical scheme, from which a universal classification algorithm is to be inferred according to certain general inference rules. It was this rational idea that stimulated the development of early systematics; its first version was the logical genus–species scheme of scholastics, then it was embodied by the method of natural systematics (in its narrow botanical sense). In the 20th century, the formal deductive scheme was revived to shape the *episto-rational program* of contemporary systematics.

Epistemic rationality attracts those taxonomists who strive to deontologize and depersonalize completely classification activity and, unlike empiricists, rely on "objective" formal methods that are free from both speculative metaphysics and personal imagination and therefore are themselves capable of giving "objective" classifications. However, in contrast to initial assumptions, belief in a "true" method makes it a kind of manifestation of quite religious consciousness that has a deep metaphysical or even natural-philosophical background [Pavlinov 2011b, 2018; Pavlinov and Lyubarsky 2011]. This belief goes back to the awareness of the ancient natural philosophers of a certain isomorphism between the movements ("dialectics") of both cognizable Nature and cognizing reason, with the respective Method as a product and a form of the latter's activity [Akhutin 1988]. From this viewpoint, the "true" method is the one whose structure (logic) is isomorphic to the structure of the cognizable reality. If we discard such a rationale and assume the method is purely formal, that is, not correlated some way with Nature, the fundamental question remains unanswered: to what extent are certain fundamental properties of classification—say, its hierarchical structure—a result of the method applied (say, the hierarchical genus–species scheme) or do they reflect the essential features of Nature (whether as a hierarchy of either super-organisms or phylogenetic pattern)?

In contemporary episto-rational systematics, this general idea is embodied in specific PTTs by two main approaches [Pavlinov and Lyubarsky 2011; Pavlinov 2018, 2019, 2020]. One is based on logical schemes, whereas the other relies on numerical methodology; accordingly, two more particular research programs are distinguished, logical and numerical.

"Logical" systematics is based on the conviction that its particular classification theory should be inferred from a certain general theory of classification, which, in turn, is but a kind of "general logic" applicable to any object regardless of its nature [Thompson 1952; Lyubishchev 1972; Zarenkov 1988]. Currently, this idea is developed in quite a general form by *classiology*, aimed at elaborating a general theory of formal (logical) classification [Pokrovsky 2014]. Strangely enough, the adherents of this idea, when appealing to some "general logic," do not take into account that the latter does not seemingly exist: as was noted in Section 3.5, there are several formal logical systems, therefore there may be just as many formal classification theories that might implement their logical reasonings. This is tantamount to an acknowledgment that a single universal theory of "classiology" is hardly possible, and is doomed to be pluralistic due to multiplicity of logical systems.

This general research program also includes attempts to formalize the foundations of biological systematics in a language borrowed from formal axiomatic systems like the set theory [Woodger 1937; Gregg 1954; Voronin 1985; Baldwin 1987; Shelah 1990; Mahner and Bunge 1997]. Methods of foundation of numerical taxonomy also contain obvious elements of axiomatization [Sokal and Sneath 1963; Jardine and Sibson 1971; Sneath and Sokal 1973; Gordon 1999].

The numerical version of the episto-rational program occupies a particularly noticeable place in contemporary systematics, so it is considered as a separate research program in the next section.

5.3 THE NUMERICAL PROGRAM

The numerical idea ("mathematism"), which constitutes the core of the *numerical research program* in systematics, as was just noted, is a version of the episto-rational idea and program. It took shape in the 20th century and is usually associated with the positivist philosophy of science. However, its foundation has a deep historical and natural-philosophical background, dating back to early Antiquity and developing throughout subsequent history. At its beginning, there was the Pythagorean aphorism "everything is Number," which affirms the subordination of Nature to the laws of harmony of numbers (on this basis, a numerological program was formed in systematics at the beginning of the 19th century, see Section 2.4.4). Then Galilei, in the 17th century, relying on the "book" metaphor of Aurelius Augustine, announced that "The Book of Nature is written in the language of mathematics." Following this, I. Kant at the end of the 18th century, in his work with a very characteristic title, *Metaphysical Foundations of Natural Science*, expressed one of the key ideas of the physicalist conception of modern natural science: "in any special doctrine of nature there can be only as much proper science as there is *mathematics* therein" (cited after [Kant 2009: 235; italics in the original]).

In early systematics, the first attempt to describe organisms "in the language of mathematics" was made by J. Jung in the 17th century [Jung (1662) 1747]. The next to express one of the main points of the numerical program was probably

H. Strickland in the middle of the 19th century: based on the "taxonomic map" meta-phor (see Section 2.4.1), he compared the similarity between groups of organisms to the distance between territories on a geographical map [Strickland 1841]. At the beginning of the 20th century, the development of quantitative methods of biometrics made a certain contribution to the initial development of the numerical program: they were used to solve elementary tasks of the pairwise comparisons of populations and to assess their similarities/differences numerically [Fisher 1925; Simpson and Roe 1939]. Simultaneously with this, the zoologist Evgeny Smirnov expressed the key idea of the program in question as follows: it is necessary "to establish those rules and laws that determine the mutual arrangement of the phenomena under study. Expression of these regularities in the form of mathematical formulas is the highest goal towards which the taxonomist strives" [Smirnov 1923: 359]. Smirnov called this systematics "exact" [Smirnov 1924, 1969]; somewhat later, its supporters gave it the epithet "numerical" and finally, quite bluntly, "mathematical" [Jardine and Sibson 1971; Dunn and Everitt 1982]. Assimilation of the positivist philosophy of science by systematics played a key role in this movement: it stimulated the development of a phenetic idea and, with it, an idea of possibility and even necessity for quantita-tive estimation of the similarity by multiplicity of characters between organisms and taxa [Cain and Harrison 1958; Cain 1959a; Sneath 1961; Sokal and Sneath 1963]. A purposeful development of the quantitative methods of elaborating classifications began in the late 1950s and early 1960s [Michener and Sokal 1957; Sokal and Sneath 1963; Williams and Dale 1965]. This marked an actual rise of numerical systematics, which is sometimes called *taxometry* or *taxonometry* [Rogers 1963; Williams and Dale 1965; Abbot et al. 1985; Jensen 2009].

5.3.1 MAJOR FEATURES

The ideological "core" of the numerical research program can be represented by two main positions. Firstly, the relationships (similarity, affinity, etc.) between organisms and their aggregates can and should be evaluated (measured) quantitatively. Secondly, the structure of relations measured somehow can and should be translated into a clas-sification based on quantitative methods. All this taken together makes it possible to present the classification procedure in an algorithmic form, which makes it strictly analytical and "transparent" for verification, and reproducible. According to the idea going back to Kant (see above) and strengthened by positivist philosophy, this allowed the numerical program to claim to be the only one deserving the status of scientific in biological systematics.

It should be emphasized that, within the framework of the numerical program, the object of taxonomic research itself is not considered, but rather the research method in its specific "numerical" meaning. This means that numerical systematics, like the entire episto-rational systematics, does not have a subject area of its own: it belongs to the epistemic component of the cognitive situation. However, unlike self-sufficient "logical" systematics, it functions as a kind of "supplement" to ontic-based research programs (phylogenetic, biomorphic, phenetic, etc.). In this capacity, if it touches upon certain issues concerning the object itself, it is only in such a way as to "adapt" it to the needs of the numerical program to make it suitable for the application of quantitative methods. Due to this, numerical systematics, together with classification phenetics, promotes the creation of rather a specific *Umwelt*, in which there is no living nature at all, but only various kinds of abstractions—formalized representations of both organisms and their characters, and formalized assessments of the relations (affinity). Accordingly, a specific language for describing such *Umwelt* is developed: for example, there are no populations or species in it but the above-mentioned OTUs, to which various technical means of description and comparison are applied.

One of the most important abstractions of this kind, which became very widespread both in systematics and outside it (for example, in ecology of communities, biogeography), is the geometric interpretation of similarity relations (see next section for details). It is based on the phenetic idea idea of *phenetic hyperspace* with its axes being the characters and with OTUs distributed in it. "Numerists" successfully applied Euclid's theorem of the right-angled triangle to this hyperspace and thus obtained a simple way to calculate the pairwise *phenetic distances* between these OTUs as analogs of real geographic distances. These numerically expressed distances in total compose the *distance matrix*, to which quantitative methods are applied to transform the initial phenetic hyperspace filled with OTUs into some other distribution models, one of which is the desired classification.

As in scholastic systematics several hundred years ago, the main emphasis of the numerical program is transferred to the classification method as such, so both its main problem and task become a justification of this method in a specific "numerical" manner. Following rigorous understanding of the "systematic philosophy" under consideration, it is argued that numerical methods should be deduced from certain well-formulated mathematical theories [Jardine and Sibson 1971; Dunn and Everitt 1982; Semple and Steel 2003]. Respectively, consistency of a method thus substantiated yields consistency of a classification elaborated with it. However, many of the numerical methods, borrowed from biometrics or developed by numerists, though intuitively understandable and therefore quite popular, turn out to be without a serious mathematical background and are sometimes criticized for this reason [Williams and Dale 1965].

Several methodologies are distinguished within numerical systematics, and they differ in some key assumptions. According to the content of the background knowledge, *numerical phenetics* and *numerical phyletics* are separated: no causes of taxonomic diversity are considered in the former, which makes it theory-neutral from an ontic perspective, while phylogeny is considered as such a cause in the latter, which makes its ontology burdened with metaphysics. At the level of analysis of characters, the proper numerical taxonomy [Sneath and Sokal 1963; Sokal and Sneath 1963], taximetry [Abbot et al. 1985], and taxonomic analysis [Smirnov 1969] can be distinguished: in the first, characters are generally introduced with equal weights, whereas the other two presume their differential weighting; Smirnov's analysis includes elements of typology.

Algorithms and particular methods of numerical systematics are quite numerous, and they differ regarding certain principles of transition from raw data to the classifications. This variety is unavoidable, as it is largely due to the above-noted arbitrary nature and therefore potential multiplicity of the initial axiomatic systems used for substantiation of particular methods. Each of them may be good in itself within the framework of its respective axiomatics, but they are fundamentally irreducible to each other or to some general "supermethod" to the extent that they are based on different axiomatic systems. Therefore, generally speaking, there is no single numerical classification method that is equally applicable in all taxonomic research [Sneath 1995].

The areas of correct application of the numerical methods in systematics depend on two main factors: (a) to what extent the features of the organisms can be formalized by unit characters and (b) to what extent these characters are comparable (homologous) in different organisms. For obvious reasons, the effectiveness of the methods decreases with increasing complexity of organisms and the degree of differences between them. Therefore, the problems and technical difficulties caused by these factors are most relevant in the case of complex morphological macrostructures and are minimal when working with biochemical (molecular) characters. For the same reason, it is easier to compare species of the same genus than different orders of the same class. Both the universality and simplicity of molecular structures make numerical methods applicable almost equally effectively at all taxonomic levels up to the highest; due to this, the contemporary reconstruction of the global "Tree of Life" by means of numerical phyletics becomes possible [Cracraft and Donoghue 2004].

Many numerical methods allow study of the structure of both taxonomic and partonomic aspects of taxonomic reality; that is, to compare both taxa (organisms) and their characters on the same methodological basis. The areas of their application can be represented as a result of the decomposition of general phenetic hyperspace into the *I*- (organisms) and *A*- (characters) subspaces [Williams and Dale 1965]; they correspond to taxonomic and partonomic aspects of the taxonomic reality (see Section 4.2.1), respectively. Technical means for them are denoted as *Q*- and *R*-analyses [Sneath and Sokal 1973], respectively. *Q*-analysis involves a comparison of the OTUs by their characters and estimation of similarity relationships between them. *R*-analysis examines interrelations between characters as a degree of coincidence of their distributions over the set of OTUs.

With some reservations, it is permissible to refer to the tasks of numerical systematics not only elaboration of the classifications but also exploration of their structure characterized by certain quantitative parameters. These tasks belong to the field of *comparative systematics*. Among them, of great interest is analysis of the already-mentioned Zipf–Mandelbrot rank distribution, which describes an inverse relationship between the number and size of taxa [Fairthorne 1969; Orlov 1976]. Comparing different classifications based on quantitative methods is another important task [Rohlf and Sokal 1981]. Besides, it is possible to study numerically the distribution of characters on the classification trees.

Apparently, the further development of the numerical program in systematics will be associated with a more active development of the sufficiently flexible probabilistic approaches (such as Bayesian), including those based on fuzzy logic [Amo et al. 1999; Scherer 2012]. This will make numerical methods more suitable for the research logic of biologists.

5.3.2 TWO BASIC VERSIONS

As noted above, numerical systematics became divided very soon after its formation into two directions differently implementing its research program. Although some technical developers of the latter do not see an essential difference between them, it is very significant: it is determined not so much by their methodological apparatus as by contents of their *Umwelts*. They "serve" different ontic-based research programs, and their substantive difference is unambiguously reflected in their respective names: one of them is *numerical phenetics*, while the other is *numerical phyletics*.

Numerical phenetics provides the technical means for implementation of phenetic and partly biosystematic research programs. In this case, in the development and application of quantitative methods, ontic considerations are almost disregarded. The basis of the classification procedure is quantitative assessment of similarity as such, i.e., not interpreted from any substantive standpoint. The unit characters are entered into similarity analysis with equal "weights," with the main task being to differentiate phenons. For this, both ordination and clustering methods are considered equally suitable. Hierarchical clustering provides a tree-like *phenogram* rooted at the level of the least similarity between phenons. The result is a strictly *phenetic classification* performing mainly the function of a reference system.

The most popular version of numerical phenetics in its fairly general interpretation is that of Sokal and Sneath; the taximetrics of Abbott et al. is also fairly widely use, while the taxonomic analysis of Smirnov is almost forgotten (see references to them above). Some approaches of biosystematics [Solbrig 1970; Stace 1989] and early numerical phyletics [Cavalli-Sforza and Edwards 1964] also operate with the phenetic pattern, so methodologically they are focused on the phenetic idea.

An important part of the methodology of numerical phenetics is the above-mentioned geometric interpretation of similarity relations. The latter is a function of the distance between OTUs in phenetic hyperspace: the less the similarity, the greater the distance. This function is defined by a system of axioms, according to which distances can be metric, ultra- or pseudometrics; a function with metric properties defines the *Euclidean distance* [Sokal and Sneath 1963; Jardine and Sibson 1971; Sneath and Sokal 1973; Dunn and Everitt 1982].

The quantitative measures of phenetic similarity are numerous; they generally give non-coinciding estimates, and there is no reason to suppose relations between different categories of coefficients are described by a monotonic function [Sneath and Sokal 1973; Sneath 1995]. Some have significantly different meaningful interpretations: for example, in the case of measurable traits, distance coefficients estimate differences in size, whereas correlation coefficients estimate differences in proportions [Sneath and Sokal 1973]. An important property of most of the coefficients is that they presume similarity and difference to be symmetric, linked by a simple linear transformation [Williams and Dale 1965; Gordon 1999; Dunn and Everitt 1982].

The main task of numerical phenetic analyses is to distinguish phenons by minimizing differences within and maximizing differences between them. An assessment of the consistency of the results depends on how they are represented and particular research tasks. For instance, in the case of cluster analysis, consistency of results is estimated by *cophenetic correlation*, defined as correspondence between the structure of similarity relations between OTUs in the original distance matrix and in the resulting phenogram [Sneath and Sokal 1973].

The hierarchical phenogram obtained as a result of cluster analysis, in which phenons of different levels of generality are distinguished, can be quite simply (if no additional criteria are specified) transformed into a hierarchical classification by the interpretation of phenons as taxa. The hierarchization scale is determined by a range of values of the distance coefficients employed; since this range is continuous, the discrete ranks (if any) are fixed arbitrarily [Sokal and Sneath 1963; Sneath and Sokal 1973]. With this, since the phenogram is an ultrametric tree, all taxa of the same generality level are assigned the same rank (as in cladistics). The overall structure of both the phenogram and the respective classification depends on the similarity measures and clustering methods employed and, therefore, may be different for the same raw data. All this makes the results of numerical phenetics analyses strictly comparable only if the standard (comparable) conditions for their conducting are observed, including use of the same characters, the same similarity measures, and the same algorithms for distinguishing phenons.

Numerical phyletics develops quantitative methods to implement the phylogenetic research program in a specific way. Its designation, consonant with numerical phenetics, emphasizes both difference and significant similarity between these two branches of numerical systematics in their underlying assumptions, procedures, and results [Eades 1970; Pesenko 1991; Pavlinov 2015, 2018]. Methods of numerical phyletics are designed to reconstruct phylogenetic relations. These reconstructions are based on a prior *evolutionary scenario*, by which the background phylogenetic model is operationalized, and which indicates probable changes of the characters in the presumed evolution of the studied group. Similarity is considered as an indicator of kinship relations; accordingly, characters are "weighed" to select the most reliable indicators of kinship; so, formally this is a branch of taxometrics (see previous section). The main operational task is to distinguish *phylons* (monophyletic groups) that unite OTUs most closely in the phylogenetic sense. The only appropriate way of representing results is a phylogenetically interpreted dendrogram (phylogram, cladogram, molecular phylogenetic tree) rooted at the level of the least closeness among phylons. The main result is a phylogenetic classification reflecting the hierarchy of phylons of different levels of generality recognized in the dendrogram. All these operations are carried out through the analysis of similarity relations, with phylogenetic relations appearing as a result of their speculative interpretation at the final stage of the numerical analysis based on the background phylogenetic model.

The first studies within this subprogram [Cavalli-Sforza and Edwards 1964; Camin and Sokal 1965; Fitch and Margoliash 1967] were basically phenetic in their methodology. The truly phylogenetic reconstructions based on the quantitative methods began to develop in the 1970s in a combination of numerical technology and cladistic methodology. Such development of the cladistic version of phylogenetics was predetermined by and large in that it presupposed the possibility of using quantitative methods of estimating similarity relations [Hennig 1966; Farris et al. 1970]. This possibility is presumed by the *principle of summation of synapomorphies*, according to which the more synapomorphies characterize a group, the more consistently (all things being equal) it may be considered monophyletic [Hennig 1966; Farris et al. 1970; Estabrook 1972; Wiley 1981; Pavlinov 1989, 1990, 2005]. Numerical phyletics received a powerful stimulus for its development after molecular genetic data became available [Swofford et al. 1996; Nei and Kumar 2000; Felsenstein 2004].

To emphasize the highly technical nature of the approaches that make up numerical phyletics, it is sometimes referred to as *computational phylogenetics* [Warnow 2017] or, since its most popular methods are probabilistic, as *statistical phylogenetics* [Felsenstein 2004]. As far as these methods can be technically represented as algorithms for "translating" information in raw data into tree-like phylogenetic schemes, computational phylogenetics can be thought of as a section of *phyloinformatics* that includes all information resources and methods of their processing related somehow to phylogenetic reconstructions [Page 2005].

A "canalized" development of this research program in a purely technical way simplified significantly the ontic component of the cognitive situation, in which numerical phyletic tasks are solved. The main prerequisite for this became the *principle of parsimony*, which determined a reductionist nature of the entire "new phylogenetics." In this case, it is based on the assumption that the process of evolution is fundamentally "parsimonious" so that the diversity of characters for a given group is achieved by a minimum number of evolutionary steps, which presumes a minimum number of parallelisms and reversions. In particular studies, this leads to minimizing prior speculations about the evolution of the studied group—about its directions, its adaptive meaning, the evolutionary significance of organismal features, etc. In an extreme case, all such speculations are discarded, which is tantamount to the model of random evolution of "unweighted" uncorrelated characters. With reference to the above principle, one of the leaders of numerical phylogenomics, Joe Felsenstein, suggested replacing the hypothetical-deductive model of phylogenetic reconstructions with a statistical model [Felsenstein 1988]. All these considerations, together with the emphasis on the method as such, deprive numerical phyletics of its evolutionary ontic foundation and focus it on instrumental problems, just as in the case of numerical phenetics [Pavlinov 1990, 2007b, c, 2015, 2018; Faith 2006; Rieppel 2007].

Numerical phyletics includes a fairly diverse set of tree-inferring algorithms; they are somewhat conventionally divided into cladistic (usually for working with morphological data) and molecular phylogenetic (phylogenomic). The main goal of all of them is to obtain a *phylogenetic tree* as a representation of the structure of phylogenetic relationships.[1] It differs "technically" from a phenogram by its rooting, which corresponds to an initial event in the evolution of the studied group, and thus a proposal to use the phenogram as an estimate (in a statistical sense) of phylogeny [Colless 1970; McNeill 1982] is incorrect. This tree can be either a *cladogram* or *phylogram*: formally speaking, the former is an ultrametric and the latter is a metric tree.

Taxonomic interpretation of the results of numerical phyletic analysis, just as in phylogenetics in general, presumes the transformation of the branches in

---

[1]   In contemporary literature, this tree is usually, although incorrectly, called "phylogeny"; such a pure technical designation is nonsense, as phylogeny is the historical development of organisms and not a formal tree-like scheme.

a tree into the taxa in a classification. A specific "numerical" accent to this is added by taking into consideration some quantitative aspects of tree topology. In particular, a quantitatively measured "amount of evolution" may serve as the basis for ranking respective taxa, which means an evolutionary interpretation of phenetic distance. In genosystematics, the latter can be interpreted, with reference to the molecular clock hypothesis, as an indicator of the "time of evolution" to build a unified global ranking scale [Avise and Johns 1999; Avise and Mitchell 2007; Avise and Liu 2011]. However, uneven "running" of this clock [Ayala 1999; Schwartz and Maresca 2007] casts doubt on the meaningfulness of calibrating a unified ranking scale in this way.

5.3.3   BASIC CONTROVERSIES

The characteristic properties of methods of numerical systematics are considered by its supporters as advantages, while its opponents consider them shortages; the sources of the controversy of these standpoints are briefly discussed below.

The formalized character of its apparatus, which makes it possible to algorithmize the classification procedure, is thought of (or at least presented) by its supporters as the *objectivity* of both this apparatus and the results obtained with it. In this case, objectivity is understood as an alternative to *subjectivity*, i.e., as independence from (or minimal dependence on) a subjective factor, including subjective speculations about the background metaphysics. This general declaration is incorrect because of the complete irrelevance of the philosophical category of objectivity to this issue. To begin with, axiomatic foundations of highly formalized methods are arbitrary to a greater or lesser degree and therefore not "objective" [Kline 1980; Perminov 2001]. Then, a strong formalization of the quantitative methods leads to their "emancipation" from cognized reality, which seems true for the results obtained with them, as well. In this regard, it is quite meaningless to apply the category of "objectivity," in its real-istic meaning, to the assessment of the results of numerical systematics, i.e., to speak of their "truth" as a correspondence to what is implied to be "a matter of fact" [Ruse 1973; Rasnitsyn 2002; Pavlinov 2018]. This is just what Albert Einstein asserted, with all his professional comprehension of the subject of mathematical physics: "to the extent that the proposals of mathematics relate to reality, they are not reliable; insofar as they are reliable, they are irrelevant to the reality" (cited after [Vollmer 1975: 28]). Therefore, from a philosophical perspective, as far as the methods and results of numerical systematics are concerned, it is more correct to speak not about their objectivity, but about their *intersubjectivity* in its rather technical (methodo-logical) meaning [Smaling 1992], i.e., about agreement in the results obtained by different researchers applying similarly organized methods to similarly organized raw data.

The thesis about the *empirical* nature of numerical systematics is also hardly true. Philosophically, the emphasis on methods makes the entire numerical approach *rational* rather than empirical. Numerically based research involves analysis not of the natural objects themselves, but their cognitive models in form of a combination of formalized variables (characters). In such an operation of ideation, the natural object disappears, and with it disappears empiricism with its focus on an object as such.

Indeed, a "sample centroid" in numerical systematics, just as an "ideal blackbody" in physics, is but a kind of semblance of a Platonic *eidos*; more precisely, not an *eidos* itself, but rather its specific numerical representations.

The formalization and algorithmization of numerical methods are firmly associated with their *accuracy* and *repeatability*; here the problems are as follows. The accuracy of each such method is not universal and not "absolute," but local and "relative": it is specified only for certain formalizations that serve as a rationale just for this particular method, and may not be so for any other formalizations [Williams and Dale 1965; Shatalkin 1983]. The same is true for repeatability, which is associated with the above-mentioned intersubjectivity: it is fulfilled only under certain standard conditions of solving standard tasks. Thus, both these "advantages" relate to the method as a "thing in itself," and not to the substantive content of the research tasks, with respect to which at least "accuracy" turns out to be a fake.

The inevitable variety of numerical methods gives rise to a serious methodological problem of the reasonable choice of a particular method as a means of solving a particular research task. An important part of this problem is the requirement to define the basis for this choice; it is the same as a choice among the logical grounds considered above (see Section 3.5). The central point here is the definition of criteria for assessing the *consistency* (effectiveness) of numerical methods. Proponents of mathematical systematics customarily validate a method by reference to a well-founded mathematical theory underlying it; as was emphasized above, such a rationale makes each method consistent only with respect to its own initial formalizations. However, from a biological (substantive) perspective, this is not enough: "as applied to the natural sciences, any mathematical method makes sense not in itself, but in connection with the purpose for which it is used" [Shatalkin 1983: 52]. As far as the meaningfulness of the entire taxonomic research is determined within the context of some basic substantive theory, the consistency of the method should be assessed within the context of this theory; this is the main provision of the epistemic principle of methodical correspondence (see Section 3.7).

Since different methods lead to different classifications, this generates specific *taxonomic uncertainty*: the cumulative result of applying different "exact" methods to the same raw data turns out to be very fuzzy in admitting different particular taxonomic solutions [Sneath and Sokal 1973; Pavlinov and Lyubarskiy 2012]. All this is far removed from the unambiguity that is anticipated by the ordinary users of numerical methods attracted by the slogans of the latters' proponents.

The orientation of numerical systematics towards the method as such, and thus towards de-ontologization of the taxonomic research, plunges the latter into the already-mentioned fundamental epistemic problem of instrumentalism [Rieppel 2007; Pavlinov 2018] (see Section 3.2.1). The latter means that the quality of classifications is determined not through their compliance with the structure of taxonomic diversity, which they are designed to represent as their cognitive models (the condition of realism), but through the formally substantiated quality of the methods themselves (the condition of instrumentalism). With this, an accentuation on the method as such results in an "inverse correspondence": it appears that it is the method as such that dictates how an *Umwelt* should be analyzed, so the properties of the former indirectly shape the properties of the latter.

Leaving aside an "anti-mathematism" inherent in intuitionist researchers and their objections, the following important point should be emphasized. Biological systematics solves substantial tasks; the numerical program develops formal methods for solving these tasks. This means that the numerical theory as such does not have an independent meaning for systematics: certain formal results that it produces, like any other formalisms, are subject to biological interpretation, so it is the latter that should serve as the basis for meaningful taxonomic conclusions [Moss and Hendrickson 1973; Pavlinov 2018]. It is noteworthy that J. Gilmour, one of the first and leading ideologists of the positivist treatment of systematics, pointed out that fascination with quantitative methods may bring the illusion that taxonomic research reaches a conclusion with the results of their application; in fact, it merely begins with them [Gilmour 1961].

Generally speaking, the modern mathematization of biological systematics is much more than just its "digitization" limited to the employment of certain quantitative methods in solving certain classification tasks. As a matter of fact, it plays a very significant role as an epistemic regulator of many aspects of research activity in this discipline; in fact, it leads to the latter's formalization, operationalization, de-ontologization, de-subjectivation, etc. [Hagen 2003; Sterner 2014]. Considered from this perspective, numerical systematics provides no fewer problems than it pretends to solve, and these problems turn out to be quite fundamental. The current controversy around the numerical research program in systematics is caused by a shift in the scientific paradigm: the post-positivist conception of systematics significantly differs from the positivist one in understanding the structure of cognitive situation, including substantiation of both taxonomic knowledge and its scientific status. Those simple solutions offered by numerical approaches seem hardly adequate to the complexity of the entire problem of comprehending and describing taxonomic reality. Accordingly, at present, the most urgent task is to realize this inadequacy in order to assess properly both the results obtained and possible prospects of the development of this research program of systematics as *the* biological discipline.

5.4 THE TYPOLOGICAL PROGRAM

The typological way of perceiving Nature is one of the basics, and it underlies many aspects of cognitive activity. It begins with thinking of Nature in *gestalts*, i.e., discrete integral mental images expressing the essential features of its manifestations. The result of such an intuitive "qualitative" perception is transferred to Nature itself, which is becoming thought of as consisting of certain (quasi)discrete blocks. This worldview was supplemented with an idea of prototype (archetype), which arose in early Antiquity as an ideal form ("matrix") giving rise to real forms; it had been a predecessor of Plato's *eide*. The general idea of prototype was then built into the natural-philosophical concept of the Ladder of Nature, which had a significant impact on the formation of the worldview of many naturalists and taxonomists of the 18th–19th centuries. Such a vision of Nature is a very characteristic feature of *typological thinking* [Lewens 2009a; Witteveen 2018].

It is customary to derive typological views in systematics from essentialist ones, but this is hardly true. As was noted in Section 2.4.3, the Aristotelian understanding of essence (ousia), which formed the basis of essentialism, is functional; it appeared just in this capacity in the works of many scholastic systematists (Cesalpino and others), as well as of some early post-scholastics (Jussieu, Strickland, and others). In contrast, the basis of typology was formed by anatomist zoologists in the late 18th and early 19th centuries based on the concept of proto- or archetype in its structural understanding.

Starting from the second half of the 19th century and especially in the 20th century, typological ideas in systematics appeared in a "shadow" of evolutionary ones and were almost completely rejected by positivist ideas. To a certain extent, this situation was aggravated in the most recent taxonomic research as it used the features (biochemical, molecular genetic, etc.) not amenable to a trivial typological interpretation [Shatalkin 2002]. At the same time, the typological views that emerged at the turn of the 18th–19th centuries were in demand in the second half of the 20th century and supplemented by others combining classical typology with other concepts, primarily evolutionary.

5.4.1 Major Features

The classical typology was initially developed in three basic versions, viz. stationary, dynamic, and epigenetic (see Section 2.4.3). In the *stationary typology*, the organism is characterized through its general structural plan (*Bauplan*), determined by the ratio of its constituent parts; organismal diversity is formed through the details of this general plan (Vic d'Azir, Cuvier, Saint-Hilaire). Accordingly, the typological unity of taxa is thought of as the unity of the general plans of the organisms belonging to them. The *dynamic typology* considers the general structure of organisms as a result of mutual transformations (metamorphoses) of the constituent parts of an imaginary archetype (Goethe). These transformations are not "physical," but rather imaginary, i.e., "logical" in a sense. In this case, taxonomic unity is defined as the unity of transformations of the parts of organisms allocated to the respective taxa. Mutual interpretability of stationary and dynamic concepts is provided by the possibility of representing the Goethean archetype as a whole of mutually transforming parts of generalized body plans [Naef 1931; Meyer-Abich 1949; Levit and Meister 2006; Riegner 2013]. In the *epigenetic typology* (Baer), the structural plan is regarded as a specific type (trajectory) of development in ontogenesis and eventually in phylogenesis [Kaspar 1977; Lenoir 1988; Amundson 2005]. The reference to the structural plan connects epigenetic typology with stationary typology, whereas treating the type as a process connects it with dynamic typology.

Within the framework of the cognitive situation of systematics, all three interrelated versions (as well as modern ones; see Section 5.4.2) can be combined into a single *research program of classification typology*. It is based on the idea of the diversity of organisms as a hierarchically ordered diversity of (arche)types, from most general to most particular. Based on this understanding of the System of Nature, a general idea of natural classification is developed as the one that most fully expresses the (arche)typal hierarchical structure of the living matter (see Section 6.5 for details). Respectively, the taxa are distinguished following the *principle of unity of type*: each taxon is defined through an (arche)type of the respective level of generality that is most fully implemented in the organisms allocated to this taxon. For this, the characters used to distinguish taxa are selected ("weighted") in a special way: the most significant are those that most fully characterize the respective (arche)types. With this, the characters are ranked according to the levels of generality of the (arche) types described by them: dominant characters define taxa of higher ranks, whereas subordinate ones do so with taxa of lower ranks.

Thus, in the elaboration of typological classifications, an analysis of (arche)typal diversity is primary relative to the analysis of taxa; this is formalized as a version of taxon–character correspondence, in which a character "precedes" a taxon (see Section 6.2). This means that, in taxonomic research, a hierarchy of (arche)types is first revealed, then the characters specific to them are recognized, and finally, on the basis of characters that are properly selected and subordinated, typological taxa are distinguished. According to the *principle of rank coordination* and *rule of unified levels*, the characters describing the (arche)types of the same level of generality diagnose taxa of the same rank [Starobogatov 1989; Vasilieva 1989, 1992; Lyubarsky 1996]. The last author considers such an operation as a specific objectification of taxonomic ranks [Lyubarsky 2018], but this conclusion is hardly true: according to him, the hierarchical structure of the general archetype is revealed relativistically, that is, depending on the point of view from which a researcher considers respective anatomical structures [Lyubarsky 1996].

Analysis of (arche)types and their characters, including their ranking, is carried out based on comparative anatomical studies of particular organisms and the identification of general and special features in their structural or dynamic plans. Such studies require a special "biological way of thought" [Beckner 1959]; first of all, a deep comprehension of organisms as structured anatomical wholes and therefore the capability of perceiving each structured whole as "coomon in particulars" and "constant in transient" [Lennox 1980; Lyubarsky 1996]. This means a significant influence of the subject component on the route and results of typological research: this is because typology deals mainly with macromorphological (anatomical) structures, and distinguishing and ordering them cannot be rigorously formalized [Lyubarsky 1996, 2018]. However, at present, there is an attempt to extend the general provisions of classical typology to molecular structures, biochemical and physiological processes, and behavioral stereotypes: this is due to the need to elaborate standard homology criteria for such "non-classical" attributes, which requires their typological comprehension (see Section 6.6).

The typological program most fully implements its general ideas at macrosystematic levels, where (arche)typical differences are most evident. At the generic and more so at species levels, its capabilities are much less since the differences between organisms at these levels usually do not affect their structural plans. And yet, typological consideration occurs indirectly in analyses of the taxonomic reality at its lower levels. This is because all comparisons of organisms are based on analyses of their characters, and the latter are individualized based on the homology of organismal features, which presumes the typological reconstructions of relations between these features [Shatalkin 2002, 2012].

Throughout almost the entire 20th century, the typological program in systematics was sharply criticized by adherents of those programs that are based, in one form or another, on the positivist philosophy of science (phenetics, biosystematics, "new phylogenetics"). This attitude was provoked by those typologists who explicitly designated their approach as "idealistic" [Naef 1919] and, in an extreme form, associated it with Platonism [Troll 1951]. Positivists criticized typology also for an unobservability of the (arche)types, judgments of which were based on background metaphysical knowledge; with this, they also addressed the same criticism to the evolutionary interpreted systematics. As was emphasized above, the latter objection is considered largely untenable within the framework of post-positivist philosophy of science (see Section 3.1): scientific empirical knowledge is impossible without certain theoretical considerations, and the criterion of observability does not have an absolute regulatory power. On the part of evolutionists, the main criticism is addressed to the "stationary" nature of the typological world picture; however, some recent typological versions combined with the evolutionary idea (see the next section) largely refute this criticism.

The main argument in favor of the consistency of the classification typological program of systematics is provided by reference to a causal model of biological diversity, according to which the latter is structured by different categories of causes [Pavlinov 2018]. The initial causes are responsible for the evolutionary aspect of biodiversity, which is manifested in genealogical relationships among organisms studied by phylogenetics. Material causes are responsible for the structural aspect of biodiversity, which is manifested in the qualitative structural relations of organisms studied by typology. To the extent that both these categories of causes and, due to this, their effects do not coincide, the phylogenetically and typologically defined groups of organisms are different. But this hardly means that some of them are "natural" while others are not: after all, such an assessment depends on a particular concept of naturalness. As a matter of fact, a denial of scientific consistency and biological significance of studying structural disparity of organisms yields a significant simplification of the metaphysical model of biota by excluding the structural pattern of organismal diversity. Accordingly, this means a similar simplification of the entire cognitive situation developed by biological systematics.

5.4.2 CONTEMPORARY DEVELOPMENTS

As noted above, typological ideas in systematics throughout almost the entire 20th century developed under tough pressure from evolutionarily and phenetically minded theoreticians, who unanimously declared typology idealistic and therefore "scientifically obsolete." The perception of typology as marginal is evident from the contents of manuals, textbooks, and theoretical reviews, in which it is only mentioned with negative connotations [Simpson 1961; Hull 1965, 1970; Mayr 1969, 1988; Quicke 1993; Ereshefsky 2001b].

And yet, despite this external pressure, typology develops quite actively and incrementally. On the one hand, this is facilitated by the typologists themselves, in every possible way expanding the spheres of application of the typological cognitive program, for example, by identifying typological and comparative methods [Rautian 2001] or by referring the study of any ordered disparity of organisms to typology [Tchaikovsky 1990; Lyubarsky 1996; Chebanov 2016]. On the other hand, attitudes towards typology are gradually changing in connection with a consideration of the structural aspects of developmental and evolutionary biology in a new way (more precisely, in "a new old way") [Hall 1996; Amundson 1998; Shatalkin 2002, 2003, 2012; Walsh 2006; Witteveen 2018]. This general trend is indirectly supported by a renaissance in interest in essentialism [Ellis 2001; Rieppel 2010b; LaPorte 2017].

At a theoretical level, the main trends in the recent development of classification typology fit into the same basic concepts that emerged at the beginning of the 19th century. In addition, new "hybrid" concepts appeared, taking into account the latest advances in biology in general and systematics in particular provided by certain evolutionary, empirical, and some other ideas.

Thus, stationary typology is evidently manifested in biomorphics [Pavlinov 2010a, 2018]. Modern constructive morphology developing the ideas of dynamic typology [Schmidt-Kittler and Vogel 1991; Beklemishev 1994; Lyubarsky 1996] serves as the basis for onto-rational systematics (see Section 5.2.1), which was declared to be a "new typology" [Lyubarsky 1996].

A combination of typological and evolutionary views in a unified taxonomic theory yields *evolutionary typology*. It is quite popular in the humanities under different terms (diachronic, phylogenetic, comparative-historical) as opposed to structural (synchronous, morphological) typology [Vinogradov 1982; Filatov et al. 2007; Yudakin 2007]. In biological systematics, a taxonomic theory of similar meaning goes back to classical (Haeckelian) phylogenetics, which is sometimes called "evolutionary typology" [Di Gregorio 2008]. An instance of "semi-formal" combination of typological and evolutionary concepts is provided by a phylogenetic substantiation of the Cuvierian principle of taxonomic ranking of characters [Starobogatov 1989; Vasil'eva 1992, 2007]. At the beginning of the 20th century, the zoologist typologist Adolph Naef defined his vision of typology as "developing a phylogenetic-morphological approach for [comprehending] the types" [Naef 1913: 75]. Reconstructions of ancestral forms that played an important role in phylogenetic research are based on typological analyses of recent and fossil organisms [Naef 1931; Remane 1956; Tatarinov 1977; Meyen 1984]. The concept of *phylotype* is close to such an interpretation of an ancestor: it is a historically emerged structural plan that characterizes a certain taxon [Sander 1983; Slack et al. 1993; Hall 1996; Richardson et al. 1998].

The epigenetic typology of von Baer is integrated by several modern concepts that can be combined under the common name of *ontogenetic systematics*. For the first time, the latter was denoted by the zoologist Grace Orton as "developmental systematics," according to which the "judgment in taxonomy [...] should be based on a unified summary of the data obtainable from the entire developmental pattern rather than on any single stage" [Orton 1955: 76]. The authors of one of the versions of onto-rational systematics believe that the natural classification should be based on the dynamics of ontogenetic processes of transformations of biological forms [Ho 1990, 1992, 1998; Ho and Saunders 1993]. The leaders of structural cladistics advocate this idea, proceeding from the assumption that the only reliable empirical source for judgments about evolutionary transformations of anatomical structures is their ontogenetic development, so all taxonomic judgments should be based on the latter [Nelson 1978; Patterson 1983]. As a particular taxonomic theory, ontogenetic systematics was denoted most recently by Alexander Martynov [Martynov 2011, 2012; Pavlinov 2013c].

The combination of ideas of phylogenetic and ontogenetic systematics makes this concept part of the evolutionary ontogenetic research program that now emerges based on the ideas of "evo-devo" [Minelli 2015; Pavlinov 2018, 2019] (see Section 5.8).

The main content of ontogenetic systematics is determined by a specific inter-pretation of taxonomic unity, according to which a taxon is defined as "a class of individuals sharing a common developmental system" [Ho 1992: 199]. The results of studying the diversity of ontogenetic patterns can be presented in the form of a "taxonomic map" of possible transformations of forms [Ho 1990, 1992]; another way of representing it is the "morphogenetic tree" [Rieppel 1990], which M.-V. Ho called "Baerean" [Ho 1992]. Both "transformation map" and "morphogenetic tree" can be translated into a classification following the same general rules as in other sections of typology or phylogenetics.

One of the directions of the contemporary development of the typological idea can be called *empirical typology*: it tries to substantiate the connection of this idea with the key requirements of empiricism by minimizing prior judgments about typicality. It dates back to the first half of the 19th century (A. Quetelet); the use of statistical methods to determine the type [Smirnov 1925] made it possible to denote this version as *statistical neotypology* [Sokal 1962; Sokal and Sneath 1963] and to consider typological almost all numerical systematics based on comparing samples by their calculated centroids [Remane 1956; Simpson 1961]. Another option is the inherently phenetic interpretation (and representation) of the type as a condensation of points in multidimensional hyperspace [Sokal 1962; Read 1974; Wagner and Stadler 2003].

5.5 THE BIOMORPHIC PROGRAM

Until recently, ecological and phylogenetic aspects were distinguished in the basic structure of biota [Schulze and Mooney 1994; Faith 2003]. The first is studied by ecology, and is out of the sphere of interest of systematics; the second is explored by phylogenetics, which shapes the phylogenetic theory and research program of systematics. This two-component concept significantly impoverishes the overall con-cept of biological diversity, omitting its other significant manifestations. One of them is structural disparity studied by onto-rational and typological taxonomic theories. Another is the eco- or biomorphological aspect of biodiversity, in which the ecological (position in the structure of ecosystems) and the associated morpho-physiological specificity of organisms is manifested [Pavlinov 2007d]. This last aspect is explored by *biomorphics*, which is a PTT developed by the *biomorphic research program* in systematics [Pavlinov 2010a, 2018, 2020]. Its main task is to analyze the diversity of morpho-ecological adaptations of living beings, so it is one of the manifestations of the general trend of the contemporary "biologization" of systematics.

The idea of developing the natural classification of the "basic forms of life" instead of artificial diagnostic keys was declared by Alexander von Humboldt at the beginning of the 19th century [Humboldt 1806]. This might have become an important conception in early post-scholastic systematics, but Humboldt's idea left no evident traces because of the prevalence of other taxonomic theories that are more consistent with the previous essentialist doctrine. Sufficiently developed classifications of life forms began to appear in first half of the 20th century (i.e., [Warming 1908; Kashkarov 1938]). By its end, movement in this direction led to the emergence of a fairly developed approach, denoted as *ecomorphology*, with its taxonomic application being called *ecomorphological systematics* [Aleyev 1986; Leontiev and Akulov 2004]; the above-mentioned term "biomorphics" was introduced in consonance with phylogenetics and phenetics [Pavlinov 2010a].

Ecomorphology has two essentially different meanings, viz. taxonomic and partonomic. In the first case, the diversity of organisms is studied according to their ecomorphological similarity [Aleyev 1986; Mirabdullaev 1997; Leontiev and Akulov 2004; Pavlinov 2010a]. In the second case, the goal of ecomorphology is defined as "analyses of the adaptiveness of morphological features and all dependent, correlated topics such as the comparisons of adaptations in different organisms" [Bock 1994: 407]. Some ecomorphologists characterize this general approach so vaguely (i.e., [Schoute 1949; Winkler 1988]) that it fits equally both of these general meanings. Such a confusion justifies the proposal to denote this taxonomic conception as biomorphics to separate it from the partonomic conception, i.e., ecomorphology proper.

Biomorphics considers the diversity of organisms in connection with their adaptations, manifested in ecomorphological or (more broadly) ecomorphophysiological features. Biomorphs are distinguished based on particular combinations of these features, with the assumption that these combinations are causally (adaptively) conditioned and, in this sense, "natural" [Serebryakov 1962; Aleyev 1986]. Thus defined biomorphs "are units [...] of certain communities and ecosystems, and not at all of the species, families, or other taxonomic groups" [Kirpotin 2005: 246], so this research program of systematics partly intersects with biocenology.

An elementary unit of biomorphological diversity is not an organism as such, but its ontogenetic phase that is specific in its ecomorphological features [Aleyev 1986; Kirpotin 2005]. This means that, if the differences between the ontogenetic phases of the same organism are significant (for example, larval and post-larval stages in many animals), they belong to different biomorphs. By this, the biomorph is likened to the semaphoront of cladistics and differs drastically from the ontogenetic circle of ontogenetic systematics (see Sections 5.7.3 and 5.4.2 on these terms). The definition of biomorph does not include the indication of a particular place and time of occurrence of organisms; when separating biomorphs, the general adaptive complexes of features are considered without distinguishing between homologies and analogies. Such an interpretation of a biomorph as a classification unit (taxon) makes the research program under consideration a very specific branch of biological systematics. In general, from an ontic standpoint, the biomorph, as a unit of the structural functional organization of biota, can be reasonably considered as the natural kind.

The classification system representing the diversity of biomorphs (or ecomorphs) was proposed to be named *ecomorphema* in its consonance with *phylema* representing the phylogenetic pattern [Kusakin 1995; Mirabdullaev 1997; Leontiev and Akulov 2004]. The ecomorphema possesses certain features of an onto-rational system: it is deduced from the causality responsible for the patterning biomorphological aspect of the taxonomic reality. The hierarchical classification of biomorphs is built mainly deductively: first of all, a certain common basis of division of all living organisms is fixed, such as the type of their metabolism, then the classification is built "from top to bottom" strictly following the chosen basis [Aleyev 1986; Leontiev and Akulov 2004]. In many proposed versions, biomorphs are ranked and denoted in a way similar to the "Linnaean" system (phyla, classes, orders, etc.).

The methodology of biomorphics as a classifying discipline includes differential *a priori* weighting and ranking (subordination) of features according to their role in the respective adaptations and the latter's levels of generality. According to Yuriy Aleyev, "the hierarchy of ecomorphological adaptations at the organismal level determines the hierarchy of ecomorphs [and] the more universal a system of ecomorphological adaptations is, the higher the rank of the respective ecomorph" [Aleyev 1986: 194, 195]. Thus, its methodological foundation is quite similar to the natural method developed in the late 18th to early 19th centuries by Jussieu and Cuvier. In particular, in biomorphics, just as in typology, taxon–character correspondence is adopted in a version of "precedence" of character to taxon. Besides, biomorphic classification theory includes a noticeable element of Aristotelian ousiology, as the key characters of biomorphs are defined through their performance of certain adaptive functions.

Adherents of biomorphics hold a moderately or fully realistic interpretation of biomorphs of all levels of generality; their position with respect to biomorphological classifications can be monistic or pluralistic. In the case of monism, it is argued that "there is only one unified system of ecomorphs [which] is as real as the [phylo]genetic system of organisms" [Aleyev 1986: 195]. According to the pluralistic position, different systems of biomorphs can be developed with reference to the particular ecological factors structuring biota [Du Rietz 1931; Remane 1943], so interrelations between particular organisms can be different in different ecomorphemes. The latter viewpoint allows emphasis of an aspectual nature of the classification of biomorphs.

It is suggested that biomorphs as real natural units can be distinguished only within particular groups ranked not higher than family [Lyubarsky 1992] or class [Krivolutsky 1971]. However, many authors consider the main traditionally recognized megataxa of organisms, such as animals, plants, fungi, lichens, and certain groups of unicellulates, as biomorphs (life forms); in fact, they are components of the structure of biota and in this sense are quite real as the higher-order natural kinds. In lower organisms (prokaryotes, etc.), in which phylogenetic relationships are very fuzzy [Doolittle 1999], biomorphological classifications are more relevant [Lyubarsky 1992; Zavarzin 1995]. In higher organisms, biomorphological and phylogenetic classifications can be clearly distinguished: they are complementary, reflecting different aspects of the entire taxonomic diversity [Mirabdullaev 1997; Pavlinov 2007d].

5.6 THE BIOSYSTEMATIC PROGRAM

The term biosystematics has two meanings. On the one hand, it is often used to refer to the entire biological systematics—it is simply a contraction of these two words. On the other hand, this is the designation of one of the branches of systematics associated with the study of predominantly intraspecific diversity. In this book, this term is used in the second sense.

The *biosystematic research program* thus understood is one of the sections of evolutionary interpreted systematics; it is a continuation of classification Darwinism, which was formed in the second half of the 19th century (see Section 2.5.1). Its emergence became one of the responses of systematics to the challenges of the "new biology" of the early 20th century [Dean 1979; Mayr 1982, 1988; Pavlinov and Lyubarsky 2011; Pavlinov 2018]. From the very beginning, it was contrasted with classical, or "orthodox" systematics, which was declared "morally obsolete" [Turrill 1940; Gilmour and Turrill 1941; Heslop-Harrison 1960]. They were divided as follows: classical *alpha*-systematics is engaged in the primary inventory of fauna and flora, whereas *beta*-systematics deals with the in-depth study of their constituent populations [Turrill 1938]; in addition, Ernst Mayr distinguished a detailed study of intraspecific differentiation as *gamma*-systematics [Mayr 1942, 1969]. In biosystematics, just like in phenetics, the classical idea of the Natural System lost its relevance; instead, an *omnispective* (all-embracing) classification summarizing all categories of data was declared as the ideal [Turrill 1938, 1940; Blackwelder and Boyden 1952; Stuessy 2008]. An achievement of such classification was designated as *omega*-systematics by William Turrill [Turrill 1938, 1940].

One of the main tasks of biosystematics, as a specific research program, was declared to be the comprehension of "evolution in action" [Huxley 1940b; Mayr 1942]; that is, the study of the processes of adaptation and divergence of species and intraspecific forms, instead of just their recognition, diagnosis, and naming [Heslop-Harrison 1960; Stebbins 1970; Dean 1979]. In an extreme case, it was proposed that biosystematics should be viewed as a branch of evolutionary biology [Ferris 1928], adherents of which "were synthesizing genetics, ecology, and taxonomy to answer evolutionary questions" [Kleinman 2009: 75]. Because of this, it played an important role in the formation of the evolutionary concept then called "The Modern Synthesis," uniting Darwin's concept of natural selection, population biology, and genetics [Mayr 1988].

This section of systematics appeared to be "lucky," in a specific way, with the number of terms denoting it. It was probably the first to have been called *evolutionary* systematics [Hall and Clements 1923]; to emphasize its "Darwinian roots," it was called *Darwinian* systematics [Mayr and Bock 2002]. In order to indicate its main emphasis on processes occurring in nature, and not on museum collections, it was called *bio*systematics [Camp and Gilly 1943; Camp 1951], or "*the new* systematics" to demarcate it from the "old" one [Turrill 1938, 1940; Huxley 1940a, 1940b; Mayr

1942]. The terms *micro*systematics and *population* systematics indicated the level of taxonomic differentiation it dealt with [Mayr 1942, 1969]. It was denoted as *differential* systematics [Vavilov 1931; Ipatiev 1971] to emphasize its attention to the differentiation of populations, while its designation as *synthetic* systematics [Turrill 1940] indicated its aim to grasp all features of organisms.

The specific tasks posed and solved by biosystematics resulted in special attention being paid to the structure and evolution of populations and their local aggregates, recognized as the only real and natural units deserving consideration, so the species category was initially excluded from its objects [Mayr 1942; Dean 1979]. The respective conceptual frame was shaped by treating these units as primarily *genetic ecological* as opposed to the traditional *systematic* ones [Turesson 1922; Thorp 1940; Rozanova 1946; Heslop-Harrison 1960]. Göte Turesson [Turesson 1922] termed *ecospecies* the unit comparable in rank to the "Linnean" species (*linneon* of early geneticists), and its subdivisions were denoted *genotypes* and *ecotypes* depending on the main presumable causes of their differentiation. With this, it was emphasized that the intraspecific units of biosystematics should not necessarily coincide with the varieties and subspecies of "orthodox" systematics, since they were distinguished on different grounds [Valentine and Löve 1958]. The structure of these units became very detailed with a complex hierarchical subordination; at the beginning of the 20th century, about a dozen of the units were recognized, and soon their number reached several dozens [Du Rietz 1930; Gregor 1942; Camp and Gilly 1943; Sylvester-Bradley 1952]. By the end of the 20th century, almost a hundred were accumulated, but no more than a dozen remained in common use [Stace 1989]. In botany, these units are mainly associated with local forms [Styles 1987], while zoologists usually pay more attention to a large-scale picture of intraspecific differentiation [Heslop-Harrison 1960; Mayr 1969].

> In addition to the classification units, it was proposed to record continuous trends of geographic and ecotypic variation as *clines* [Huxley 1939]. This concept enjoyed some popularity as an alternative to subspecies, so it was proposed to abolish traditional intraspecific systematics operating with discrete units [Arnoldi 1939; Kiriakoff 1947; Wilson and Brown 1953; Edwards 1954; Terentiev 1957; Endler 1977; Rossolimo 1979]. From the taxonomic viewpoint, it was criticized as the clines described changes in individual characters instead of differentiation of the populations [Mayr 1942, 1969; Heslop-Harrison 1960], so it belonged more to partonomy than to taxonomy [Pavlinov 2018].

Increased attention to natural populations led to a negative attitude towards museum collections and herbaria [Hall and Clements 1923; Ferris 1928; Turrill 1940; Quicke 1993; Hedberg 1997; Feliner and Fernandez 2000]. As the zoologist George Myers wrote, "classification is [...] primarily a method of expression of what we know about nature and relationships of living populations," therefore what a taxonomist "is dealing with is not a set of dead specimens, but living, changing populations out of doors" [Myers 1952: 107–108].

Those approaches of biosystematics that are focused on the comparative analysis of intraspecific units employ all available categories of characters, which presumably allows discovery of most completely evolutionary affinities of natural populations and construction of really evolutionary classifications [Hall and Clements 1923; Takhtajan 1970; Quicke 1993]. Thus, this approach realizes in part the phenetic idea: as a matter of fact, if evolutionary affinity is measured by the similarity of genotypes, which is not a given in itself, then it is necessary to assess an overall phenotypic similarity as its indirect indicator [Mayr 1942; Sokal and Sneath 1963]. Therefore, from the very beginning biosystematic research was based on methods of biometrics, albeit quite simple [Quicke 1993].

Within the framework of biosystematics, a special research subprogram was developed, namely *experimental systematics* [Rozanova 1946; Myers 1952; Heslop-Harrison 1960; Vasilchenko 1960; Stace 1989; Quicke 1993]; it is sometimes considered a separate program [Stuessy 2008]. It is based on a physicalist idea that all judgments about evolutionary relations of natural populations should be subjected to experimental verification. Accordingly, two main approaches were developed to realize this idea [Quicke 1993]. One examines the effect of external conditions on the stability of features of organisms, for which they either are transferred from one biotope to another or, taken from different natural biotopes, grown under the same laboratory conditions. Another involves the analysis of the genetic compatibility of organisms using two main methods: (a) immunodistant analysis provides an assessment of compatibility at the tissue level (*immunosystematics*), or (b) hybridological analysis investigates pre- and post-copulation isolating mechanisms: this is considered important in the biological species concept. The use of experiments drew criticisms from those "orthodox" who, with reference to the classical empirical tradition, argue that systematics is a science of observations and comparisons [Bremekamp 1931].

Biosystematic research was and remains the most popular in botany [Lines and Mertens 1970; Solbrig 1970; Takhtajan 1970; Stace 1989; Hoch and Stephenson 1995; Feliner and Fernandez 2000; Kleinman 2009]. One of its leaders, Armen Takhtajan, defined its modern contents as "a branch of botany that studies the taxonomic and population structure of species, its morphological, geographical, ecological and genetic differentiation, origin and evolution" [Takhtajan 1970: 331]. In zoology, biosystematics (more usually, "new systematics") was initially promoted most actively by Ernst Mayr, but then interest in it almost disappeared. In this regard, it is curious that, at the beginning of his scientific career, Mayr denied the importance of the species category for "the new systematics" [Mayr 1942], while later he defined systematics as the "science of species" [Mayr 1969].

One of the newest variants of biosystematics is *integrative taxonomy*, which is being developed in opposition to the so-called "turbo-taxonomy"[2] dealing with rapid descriptions of species based on molecular genetic data [Dayrat 2005; Padial et al. 2009, 2010; Schlick-Steiner et al. 2010; Riedel et al. 2013; Goulding and Dayrat 2016]. At the beginning of the 20th century, supporters of biosystematics insisted that the descriptions of species should not be limited to their diagnoses inferred from museum specimens; so at the beginning of the 21st century, developers of integrative taxonomy similarly protest against reducing systematics of species to their

---

[2] This term was coined recently to denote "speedy" description of new species for the needs of a more complete biodiversity inventory in "reasonable time" [Krell 2004; Riedel et al. 2013].

recognition based on molecular data only. "Integrativists" rightly believe that this is but an initial stage of the study of species diversity, the main part of which is exploration of all aspects of species biology. Emphasizing the "multi-character" nature of this approach, microbiologists call it *polyphase taxonomy* [Vandamme et al. 1996; Gillis et al. 2001].

5.7 THE PHYLOGENETIC PROGRAM

Biological systematics got in touch with the evolutionary idea for the first time at the turn of the 18th to 19th centuries (Lamarck), though this attempt was unsuccessful. More active and productive mastering of this idea occurred in the second half of the 19th century (Darwin, Haeckel), and resulted in the emergence of *evolutionary interpreted systematics*. In the 20th century, the latter took a leading position; its claim to leadership was "morally" supported by the maxim that "nothing in biology makes sense except in the light of evolution" [Dobzhansky 1973]. In complete agreement with this are assertions that systematics not relying on an evolutionary idea lacks its organizing principle [Huxley 1940b], and a taxonomist rejecting this idea is like a chemist rejecting the idea of periodic law [Bonde 1976].

The development of evolutionary interpreted systematics proceeded from the very beginning in two main versions (see Section 2.5.1). One was classification Darwinism with the emphasis placed on intraspecific diversity, and this gave rise to contemporary biosystematics, considered in the previous section. Another was the systematic phylogeny of Haeckel associated with the historical interpretation of macrotaxa; this is now known as *phylogenetic systematics* (= phylosystematics according to [Reif 2009], or phyletics according to [Stuessy 2009]).[3] The latter is developed in the 20th century by the *phylogenetic research program*, currently presented in two main versions, viz. *evolutionary taxonomy* and *cladistic systematics*, each with several modifications The main difference between them is in their basic ontology: in the former, evolution is interpreted as a complexly organized process of gaining morphobiological adaptations, in which both clado- and anagenetic components are equally significant; in the latter, phylogenesis is reduced to cladogenesis, and its other manifestations are discarded.

5.7.1 MAJOR FEATURES

Building the cognitive situation of phylogenetic systematics presumes recognizing that (a) evolution is an important cause of the diversity of organisms, therefore, (b) the study of this diversity is at least incomplete and at most incorrect without taking into account its evolutionary cause, and (c) the same is true for the taxonomic system, which cannot be considered natural if it is not based on the phylogenetic background. This means that the entire quasi-axiomatics of phylogenetic taxonomic theory, including its quasi-axioms (ontology) and inference rules (epistemology),

---

[3] At present, phylogenetic systematics is most usually seen as synonymous with cladistics, which is incorrect: actually, the content of the former is much wider than that of the latter. Its original (Haeckelian) broad treatment, rather than the most recent (Hennigian) one, is followed here.

should be initially evolutionarily interpreted. Thus, this theory is construed as partly onto-rational and fully causal: its taxonomic judgments are inferred from prior evolutionary models, including indication of causes of the diversity of organisms.

The basic ontology of this theory is decomposed into two main components: the *processual* one corresponds to the process of evolution (phylogeny), and the *structural* one corresponds to the diversity of organisms generated by this process, usually called the *phylogenetic pattern* [Eldredge and Cracraft 1980; Wiley 1981; Pavlinov 2005, 2018]. Although the latter is supposed to be a consequence of the process of phylogeny, they are not the same. Systematics, as a classifying discipline, studies structural components, while the processual one, constituting part of its background knowledge, is studied by phylogenetics. Therefore, the current popular identity of phylogenetics and phylogenetic systematics is no more correct than the statement of the most enthusiastic taxonomists of the early 20th century that systematics deals with the process of evolution.

The processual part of basic ontology has the form of an *evolutionary model* as a set of statements about the properties of the evolutionary process that assumed most significance within the respective evolutionary concept and a taxonomic theory based on it. In the phylogenetic program, the evolutionary model describes phylogenesis in its general (Haeckelian) understanding. With this, two basic aspects of phylogenesis are distinguished, viz. the phylogenesis of organisms and the phylogenesis of their features [Zimmermann 1931, 1934]. Currently they are designated as *cladogenesis* and *semogenesis*, respectively; this is formalized by the *quasi-axiom of decomposition* [Pavlinov 1990, 2005, 2018]. These two aspects are interconnected by Darwin's formula "descent with modification": each evolutionary event involves the nascence of both a new species and a new feature, and this is formalized by the *quasi-axiom of isomorphism of cladogenesis and semogenesis* [Pavlinov 1990, 2005]. In addition, two other complementary aspects are distinguished in this process, viz. *cladogenesis* (lineage branching) and *anagenesis* (evolution within a single lineage) [Huxley 1958]. Description of specific evolutionary trends of a particular group of organisms and their features is referred to as an *evolutionary scenario*.

The structural part of the basic ontology is inferred from the processual one, and the properties of the structure of diversity are correspondigly determined. It consists of groups, their features, and the relationships between them; it is generated by a single evolutionary process, which allows one to ascribe, by the respective set of quasi-axioms, to the structure of diversity the same general properties that are ascribed to the underlying evolutionary process. Thus, the former corresponds to a certain aspect of the entire taxonomic reality fixed by the phylogenetic PTT; it is decomposed into clado- and semogenetic aspects as particular manifestations of the taxonomic and partonomic aspects of this reality, with the taxonomic aspect being referred to as the above phylogenetic pattern. Besides, clado- and anagenetic aspects are distinguished in the latter to characterize the results of the respective aspects of evolution.

From a taxonomic perspective, the main characteristic of a group as an element of the phylogenetic pattern is its *phylogenetic unity*. The latter is determined by interrelations between group members. They are decomposed into two main components, viz. *kinship* and *similarity* generated by the processes of clado- and semogenesis, respectively. The general structure of kinship relations, with reference to divergent evolution, is described by the *quasi-axiom of kinship irreversible decrease* [Starobogatov 1989; Pavlinov 1990, 2005, 2018]: the earlier two groups diverged, the less their kinship. The general structure of similarity also possesses this property, but is not as clearly expressed: the earlier two groups diverged, the less their similarity, which can "locally" increase due to parallelisms and convergences. Based on the above quasi-axiom of isomorphism, a certain coherence of these two basic components of relations is postulated, and this is formalized by the *quasi-axiom of similarity–kinship correspondence*. It is most important for the entire evolutionary interpreted systematics [Pavlinov 2005, 2018], and is sometimes referred to as the *principle of inherited similarity* [Rasnitsyn 2002]. In the most general case, this axiom is expressed by the formula "the closer the kinship of organisms is, the more similar they are," justifying the possibility of judging kinship by similarity. Because, in the course of evolution, similarity decreases not so monotonously as kinship, this correspondence is not strict, this is fixed by the *quasi-axiom of similarity–kinship (phylogenetic) uncertainty*: not every similarity indicates kinship [Pavlinov 1990, 2005, 2007c, 2011a, 2018; Huelsenbeck et al. 2000; Rangel et al. 2015]. This uncertainty is aggravated by different interpretations of both kinship and similarity and the quasi-axiom connecting them in different taxonomic theories: for example, kinship relation may include only a strictly understood monophyly (holophyly) or also paraphyly (see Sections 5.7.2 and 5.7.3).

In construing epistemology of phylogenetic taxonomic theory, an evolutionary interpretation of the general principle of naturalness is of key importance: it provides the *principle of phylogenetic naturalness* going back to Darwin and Haeckel. According to this, taxonomic system is natural to the extent that it reflects phylogenetic pattern; this provision is formalized by the *principle of phylo-taxonomic correspondence*. Compliance with its conditions is limited by specific *uncertainty relation* between the respective structures of phylogenetic pattern and taxonomic system reflecting it [Starobogatov 1989; Pavlinov 1990, 2005, 2018]: the more detailed the former is due to comprehensive reconstruction of the continuous process of evolution, the less adequately it can be represented by a classification.

The phylo-taxonomic correspondence can be adopted in several particular interpretations. Its *strong* form means that the taxonomic system should *completely correspond* (ideally, one to one) to the reconstructed phylogenetic pattern: this position is adopted by cladistics. Its *weak* form means that the taxonomic system must be *consistent with* (not contradict to) the respective pattern: this is the position of evolutionary taxonomy. Because of the uncertainty relation, the first interpretation is like an unattainable ideal, while the second one looks more constructive.

The main task of the epistemology of phylogenetic systematics is the development of methodological principles that ensure implementation of the principle of phylo-taxonomic correspondence. Two basic problems are met here: (a) kinship as such is not observable, so it is reconstructed based on similarity as its indirect evidence, and (b) congruence of similarity and kinship relations is non-strict (the similarity–kinship (phylogenetic) uncertainty described above). Requested maximization of phylo-taxonomic correspondence is provided by weighting the characters and/or similarity that facilitates recognition of phylogenetically most natural groups. These criteria are inferred from background evolutionary models: for example, when evolution is interpreted as adaptatiogenesis, features characterizing adaptive specificity of the groups are of the greatest importance; if evolution is reduced to cladogenesis, those features most reliably indicating kinship relations of groups are most suitable. In addition, principles of *special similarity* and *overall similarity* are introduced; in cladistics, they take the form of the principles of synapomorphy and total evidence, respectively (see Section 5.7.3).

When reconstructing phylogenetic pattern and translating it into classification, the *problem of NP-completeness* is highly relevant [Garey and Johnson 1979; Semple and Steel 2003; Felsenstein 2004; Pavlinov 2005]. In phylogenetic systematics, this problem means the following: the more complex a background evolutionary model is, the less attainable a phylogenetic classification that adequately represents the respective phylogenetic pattern due to both decrease in the reliability of particular phylogenetic reconstructions and increase in uncertainty relations. Therefore, a simplified phylogenetic model in cladistics is preferable to a more complex one in evolutionary taxonomy.

The phylogenetic tree serves as the standard form of representation of the structure of both evolutionary process and phylogenetic pattern. Its main properties are defined by branching sequence, length, and direction of its branch: the former reflects the temporal sequence of cladogenetic events, while the other two correspond to anagenetic events in the reconstructed history of the group studied. In the phylogenetic pattern, the temporal sequence defines inclusive hierarchy of groups, while to others define their dissimilarities. This tree can represent a particular history in a more or less reduced form depending on the information it contains. The most content-rich is the *phylogram* representing directions and stages of evolution of the group and its subgroups, their evolutionary advancement and diversity, with all this superimposed on the geological time scale. In contrast, the *cladogram* shows only a sequence of cladogenetic events (splitting of phyletic lineages). These differences in tree constructions are reflected in the contents and structure of the phylogenetic classifications inferred from them.

The phylogenetically natural classification appears as a result of interpretation of the original phylogenetic tree in terms of systematics. This procedure can be imagined, figuratively speaking, as a "sawing" of the tree into various fragments, which are interpreted as taxa of different levels of generality. Two "sawing" techniques are employed, *vertical* and *horizontal*, depending on the adopted ratio of clado- and anagenetic components. In the first case, attention is focused on the cladogenesis and particular branches are cut off to individualize the *clades*. In the second case, attention is focused on anagenesis: the tree and its large branches are "sawed" across their "axes," so the *grades* are individualized. The resulting phylogenetic classifications can be of two extreme types, *cladistic* and *gradistic*, with either a "vertical" or "horizontal" component dominating in them, respectively.

The inclusive hierarchy of phylogenetic classification can be determined either *absolutely* or *relatively*. The first case depends on a certain external unified ranking scale, e.g., chronological [Hennig 1950; Avise and Johns 1999; Avise and Mitchell 2007; Avise and Liu 2011]. In the second case, it is determined entirely by the branching sequence of the phylogenetic tree, so the ranks of different taxa in a classification can be defined only with respect to each other. Strictly speaking, the relatively defined ranks are not comparable among different classifications, while the use of the same ranking scale with uniformly ordered and denoted ranks makes the respective hierarchy close to absolute.

For the elaborating phylogenetic classification, several general rules determining the relative position of taxa in it are followed:

- the *rule of phylogenetic proximity*: the closer two branches of the phylogenetic tree are, the closer the corresponding taxa are in the classification
- the *rule of ranking*: the higher the order of the branch of the phylogenetic tree is, the higher the rank of the corresponding taxon
- the *rule of progression*: the sequence of taxa of each rank corresponds to the gradient of evolutionary advancement, with the ascending regression (from less to more advanced organisms) being conditionally adopted; this rule echoes the natural-philosophical Ladder of Perfection.

Objections to the key ideas of phylogenetic systematics can be summarized as follows.

The empirical argument presumes that no sufficiently speculative background knowledge should precede an empirical one; this argument is put forward by empiricists of various kinds and typologists. From a post-positivist perspective, this argument is hardly correct: empirical knowledge without background theoretical (speculative) context is scientifically untenable. An objection by typologists looks pretty anecdotal: typological knowledge is no less speculative than phylogenetic, as the archetypes are not observable but rather inferred based on a certain natural-philosophical worldview.

The logical argument, partly relating to the previous one, proceeds from the idea that classification is an *explanandum* (explainable), while the evolutionary model is its *explanans* (explanation). Any argumentation scheme, in which *explanans* precedes *explanandum*, involves circular reasoning, thus making it inconsistent. This criticism is incorrect from the point of view of the hypothetical-deductive argumentation scheme: the latter involves not a logical closed but a cognitive *hermeneutic* circle (see Section 3.6 for details). This means that a general evolutionary model serves as a conceptual framework for inferring the particular phylogenetic hypotheses to explain the data observed.

The substantive argument is put forward by adherents of onto-rational taxonomic theory, who believe that the historical causes are not the most important ones among those that shape the overall structure of the diversity of organisms; no less (or even more) important are the structural causes. A solution to this contradiction seems to be provided by taxonomic pluralism: this presumes that different taxonomic theories and their respective classifications can and should be developed based on different causal ontological models outlining local manifestations of the same taxonomic reality [Pavlinov 2018, 2020].

One of the problems phylogenetic systematics faces, also used in its criticism, is the variety of evolutionary concepts that result in a variety of particular taxonomic theories and their respective classifications. From a positivist perspective, this is considered a disadvantage, since it yields the need to change particular classifications following a change of particular evolutionary concepts [Sokal and Sneath 1963]. From a post-positivist perspective, this is a normal manifestation of the paradigmatic nature of the developing natural science; after all, in systematics, such changes are scarcely fundamentally different from those in physics or chemistry.

Since the middle of the 20th century, the phylogenetic research program has functioned in three main versions, viz.: classcal phylogenetics, evolutionary taxonomy, and cladistics; the former was characterized briefly in Section 2.5, the latter two are considered in this section. They differ basically in their interpretation of the phylogenetic naturalness of classifications, including (a) the ratio of the components of the evolutionary process taken into consideration, and (b) phylo-taxonomic correspondence.

5.7.2 EVOLUTIONARY TAXONOMY

Evolutionary taxonomy in its key positions is quite close to Haeckelian classical phylogenetics; it was thus baptized by its prime theoretician, the zoologist George Simpson in his book *Principles of Animal Taxonomy* [Simpson 1961] to distinguish it from evolutionary systematics based on classification Darwinism. It differs from cladistics mainly (a) in a richer evolutionary model, including both clado- and anagenetic components and reference to evolutionary mechanisms, and (b) in a weaker phylo-taxonomic correspondence [Simpson 1961; Ashlock 1971; Bock 1974, 1977; Hecht and Edwards 1977; Shatalkin 1991; Szalay and Bock 1991; Pavlinov and Lyubarsky 2011; Pavlinov 2018, 2020]. It differs from the Haeckelian version by the great attention that is paid to the interpretation of evolution as adaptatiogenesis, with the *concept of the adaptive zone* playing an important role [Simpson 1961; Van Valen 1971]. The latter is a complex of major environmental conditions that determines the general type of adaptation and morphobiological specificity of a group formed by similar reactions of closely related organisms with similar biological organization to similar environmental factors. According to this evolutionary model, "most higher taxa involve some basic adaptation that [...] usually occurs when there is a shift from one fairly distinct adaptive zone to another," with such shift involving "parallel trends [...] going on throughout a higher taxon" [Simpson 1961: 222, 224].

The phylogenetic tree is adopted in its classical interpretation with a rather complex configuration: it indicates sequences of splitting the groups of organisms (lineages), their time of existence (appearance and extinction), and dynamics of diversity, as well as successive stages of parallel evolutionary transformations within the dominating trends. When transforming such a tree into a classification, it can be "cut" into fragments both vertically and horizontally to reflect most adequately the morphobiological specifics of the respective groups. Because of the designation of this approach as "evolutionary," the classifications developed within its framework are also usually called not phylogenetic but rather *evolutionary*.

The principle of monophyly is adopted in a weak form: it is the so-called *broad monophyly*, which is defined as follows: "monophyly is the derivation of a taxon through one or more lineages [...] from one immediately ancestral of the same or lower rank" [Simpson 1961: 124]; this allows consideration of the paraphyletic groups formed as a result of parallel evolution as "broadly monophyletic." According to such an understanding of monophyly, the relation between similarity and kinship is equally broadly understood: this position is reflected by the concept of *basic similarity*, including (a) similarity of the ancestor with its descendants and (b) similarity of the latter due to their parallel evolution [Bigelow 1958]. As a decisive argument in favor of this position, a kind of *non-symmetrical* character of the evolutionary process is referred to: a group does not lose its morphobiological specificity and integrity after giving birth to one or several descendant branches with their specific adaptations [Simpson 1961; Mayr 1969, 1988; Bock 1974, 1977; Szalay and Bock 1991; Brummitt 1996; Rieppel 2005b; Hörandl 2006; Pavlinov 2018]. As a result, the problem of opposing the taxa interpreted as grades or clades [Huxley 1958] is mostly irrelevant: they are equally natural from an adaptationist standpoint.

When distinguishing macrotaxa, the principle of monophyly is supplemented by the *principle of decisive gap*: the amount of differences between groups acquired in the course of their divergent evolution serves as a measure of their taxonomic distance [Mayr 1942, 1969; Simpson 1961; Ashlock 1971; Stuessy 1997; Rasnitsyn 2002]. The extinction of intermediate forms is also considered as one of the arguments for the "naturalness" of such gaps separating extant groups [Simpson 1961; Mayr 1969; Løvtrup 1975; Wiley 1981; Kemp 2016]. G. Simpson provided an adaptive-evolutionary interpretation to this (formally phenetic) principle: "the eventual rank of the taxon [...] is usually proportional to the degree of distinction of the zone entered, hence the amount of basic divergence involved" [Simpson 1961: 222–223].

Acknowledging the complexity of the evolutionary process and the impossibility of presenting its results unambiguously by evolutionary classification leads to a weak formulation of the phylo-taxonomic correspondence. According to Ernst Mar, "the classification does not and cannot express phylogeny; it is merely based on phylogeny" [Mayr 1942: 277]. G. Simpson expressed this idea in an even "softer" form: "it is preferable to consider evolutionary classification not as expressing of phylogeny, not even as based on it [...], but as *consistent* with it" [Simpson 1961: 113; italics in the origin].

An emphasis on steady evolutionary trends implies that the taxa are characterized by both conservative and innovative features, including those acquired as a result of the parallel evolution of organisms belonging to them. The character selection is based on their differential weighting, with the following main criteria taken into account:

- *functional* and *adaptive meaning*: the greater the significance of a feature, the more reliably its diversity can be interpreted functionally and understood as a result of adaptatiogenesis
- *orderliness of evlolution*: the greater the significance of a feature, the higher the probability of its unidirectional changes; a part of this orderliness is also evolutionary parallelism as evidence of monophyly (proposed by Darwin)
- *paleontological data*: the significance of a feature is greater if it is available on paleontological material, since it increases the reliability of judgments about probable ancestral forms.

In connection with such a "soft" general position, evolutionary taxonomy in its original version lacks strict algorithms for deriving phylogenetic schemes and constructing corresponding classifications. An important part of its subsequent development became an articulate algorithmization of both phylogenetic and classification procedures based on a combination of classical and cladistic phylogenetics and phenetics using quantitative methods. In the concept developed by the zoologist Alexander Rasnitsyn, the central point is an understanding of the taxon as a *monophyletic continuum*: it "should be phenetically homogeneous within it and heterogeneous outside it, and at the same time be monophyletic" [Rasnitsyn 1983: 28]; this variant is termed *phyletics* (in consonance with phenetics) or *phylistics* (in consonance with cladistics) [Rasnitsyn 1983, 1996, 2002]. Another approach with a rather advanced numerical methodology was developed by the botanist Tod Stuessy, who called it *explicit phyletics* [Stuessy 1997, 2009].

The broad biological contents of evolutionary taxonomy gave pheneticists and cladists reason to accuse it of eclecticism or syncretism [Sneath and Sokal 1973; Farris 1983]. Its supporters, however, believe that their classifications are biologically more meaningful in comparison with phenetic and cladistic ones, since they encompass more parameters of the diversity of organisms [Simpson 1961; Mayr 1965, 1969; Bock 1974, 1977; Rasnitsyn 1983, 2002; Stuessy 1997]. Emphasizing this, Roman Hołyński called evolutionary taxonomy *synthetic* [Hołyński 2005] (not in the sense of [Turrill 1940]).

5.7.3 CLADISTIC SYSTEMATICS

Cladistic systematics (cladistics, cladogenetics) is a version of phylogenetic systematics, in which a one-to-one correspondence between phylogenetic pattern and phylogenetic classification is the guiding idea, and genealogy is the key parameter underlying the classification. To implement this idea, phylogenesis is reduced to cladogenesis as the only parameter that can be considered universal and capable of being reflected by a classification with minimal distortions. Such a reductionism means that claiming this approach most fully embodies the fundamental aims of evolutionary interpreted systematics [de Queiroz 1988; de Queiroz and Gauthier 1990, 1992] is not correct.

The cladistic concept originated in the depths of the German school of morphologists and phylogeneticists, in which Haeckelian systematic phylogeny emerged in the second half of the 19th century [Hamilton 2014; Rieppel 2016]. The botanist Walter Zimmermann was the first to formulate its main provisions and formalize some important concepts [Zimmermann 1931, 1934]. His approach was then further developed by the zoologist Willie Hennig: the first complete version of Hennig's phylogenetics, published in 1950 in German [Hennig 1950], like Zimmermann's papers, was left almost unnoticed; the beginning of the "cladistic revolutions" was laid by the publication of its revised edition in English [Hennig 1966]. Zimmermann and Hennig identified their simplified phylogenetic concept with the whole of phylogenetics, and with this, phylogenetics with phylogenetic systematics. Both identities are hardly correct because phylogeny is a multifaceted natural process, of which cladogenesis is but a part, and phylogenetics studying it is not identical to systematics construing classifications (see above). Therefore, the designation of Hennigean phylogenetics, with its focus on cladogenesis, as *cladistics* [Mayr 1965] seems more than justified. The latter's main principles are presented in a number of monographs [Nelson and Platnick 1981; Wiley 1981; Ax 1987; Pavlinov 1990, 2005; Wägele 2005]; besides, many most recent textbooks on systematics deal basically with its cladistic version.

The conceptual space of cladistic systematics is built on a fairly rational basis using some elements of axiomatization. The respective formalisms were present first in the works of W. Zimmerman and used quite actively subsequently by some more recent authors [Løvtrup 1975; Gaffney 1979; Wiley 1981; Pavlinov 1990, 1998, 2005, 2018; Reif 2009]. For this reason, cladistics is distinguished by a fairly well-developed specific procedure of taxonomic research enriched with specific terminology.

The general position of cladistics with respect to the *Umwelt* it constructs and investigates is declared to be realistic: according to Hennig, all monophyletic groups (in their cladistic interpretation) "from species to higher category rank, have individuality and reality" [Hennig 1966: 81]. However, it was emphasized in Chapter 3 of this book that, the more reductionist an *Umwelt* is, the less of an objective reality of the entire *Umgebung* it embraces. This is obviously true in the case of cladistics: its evolutionary model is much more reductionist and therefore is less "realistic" in comparison with the classical phylogenetic one [Pavlinov 2005, 2018]. Be that as it may, cladists believe that their approach "provides all parts of the field studied by biological systematics with a common theoretical foundation" [Hennig 1965: 101] and therefore it is the phylogenetic (actually, cladistic) classification that has the status of "the universal reference system of biology" [Hennig 1966; Wiley 1981; Williams and Ebach 2008].

The cladistic evolutionary model implies a predominantly divergent evolution, in which the proportion of parallelisms and reversions is assumed to be minimal. According to the *principle of dichotomy*, cladogenesis is a dichotomous branching process; this is strengthened by the (rather artificial) supposition that, in each evolutionary event, the ancestral species "dies out," giving rise to two descendant species [Hennig 1966]. The monophyletic group is specified as including the ancestral species and *all* its descendants, with its ancestor being by obligation a sole species and not any supra-species group [Hennig 1950, 1966; Wiley 1981; Ax 1987; de Queiroz 1992; de Queiroz and Gauthier 1992]. Such a group is denoted as *holophyletic*, and the kinship relation that unites its members is denoted as *holophyly* [Ashlock 1971; Mayr and Ashlock 1991; Mayr and Bock 2002; Envall 2008]. A group that includes *not all* descendants of the ancestral species is denoted as *paraphyletic*; if a group is derived from several ancestral species that are not closely related, it is *polyphyletic*. To emphasize these differences, the holophyletic group can be designated as *inclusive*, while the paraphyletic group is *non-inclusive* [Ebach et al. 2006]; according to another terminological variant, they are *monocladistic* and *paracladistic* groups, respectively [Podani 2010]. Connecting their recognition with the topology of the phylogenetic tree, it was proposed to designate holo- and paraphyletic groups as *convex*, and polyphyletic as *concave* [Meacham and Duncan 1987]. The term *metaphyly* was proposed for the situation where it is impossible to distinguish clearly between holo- and paraphyly [Kluge 1989].

The holophyly, or *cladistic relationship*, is outlined as follows:

taxa (or organs) B and C are closer to each other than to A [if] the common ancestor of B and C (X2) is later than the common ancestor (X1) of all three taxa or organs [...] the temporal relationship between the ancestors of X1 and X2 is the only direct assessment of phylogenetic relationships.

[Zimmermann 1931: 989–990]

Thus, this relation is set *relatively*: two groups are related by closer kinship than each of them is with some third group, if the closest ancestor of the first two groups is not also the ancestor of this third group. The latter can be either a supposed ancestor of the first two groups or some (non-ancestral) group "external" to them. Thus, the concept of *outgroup* is introduced, and substantiated by the non-operationality of the concept of ancestor [Engelmann and Wiley 1977; Nelson and Platnick 1981; Watrous and Wheeler 1981; Wiley 1981; Pavlinov 1990, 2005; Nixon and Carpenter 1993; Reif 2005a]. Accordingly, holophyly is defined strictly and operationally as follows: two groups constitute a holophyletic group if it is shown that they are close with respect to a certain outgroup; the former are usually called *sister groups*. This definition leads to exclusion of the ancestor–descendant relationship from the definition of monophyly and makes, paradoxically enough, the entire cladistics non-historical, since it implies neither ancestor nor origin, and therefore neither does it imply history [Reif 2003].

The concept of holophyletic group (= clade), central for cladistics, contains serious shortcomings, making it rather problematic [Pavlinov 2005, 2007c, 2018]. Theoretical ones are caused by the so-called *Platnick's paradox*: a monotypic genus cannot be considered holophyletic, since (by definition) it includes only one species, which is obviously a descendant, which leaves the latter without any ancestor membership in the respective genus and thus makes the latter paraphyletic by definition [Nelson and Platnick 1981; Wägele 2005]. This shows that the strict theoretical definition of monophyly in cladistics turns out to be logically contradictory [Ashlock 1984; Pavlinov 1990, 2005, 2007c; Reif 2005b; Envall 2008; Kwok and Bing 2011; Aubert 2015], and this is an example of the inverse relationship between the rigor and meaningfulness of the concept (see Section 3.5 on the latter). Among the "semi-theoretical" shortcomings, a clear dependence of the interpretation of monophyly on the scale at which a particular phylogeny is considered is evident. From this it follows that determining the holophyletic status of a high-rank group (say, a class) with reference to a single ancestral species is non-constructive and therefore not very meaningful [Pavlinov 1990, 2005, 2007c; Gordon 1999]. Finally, at an empirical level, any cladistically defined taxon in practice is always paraphyletic, since it is theoretically impossible to encompass all its representatives, both extant and even more so extinct ones, because of the limited nature of any empirical knowledge. However, this difficulty is easily removed by constructive interpretation of monophyly: when defining a holophyletic group, only those members should be presumed that are *known at the time of its study* [Pavlinov 2005, 2007c].

The procedure of reconstructing cladogenesis, including analysis of characters and similarities, is generally referred to as *cladistic analysis*. It is quite highly formalized, especially in numerical phyletics, in which many procedural elements are borrowed from phenetics. Substantially, it corresponds to the development of a *cladistic hypothesis* at two basic levels: lower-level hypotheses include *cladistic characters* as judgments about particular semogeneses, whereas the higher-level one is the final hypothesis about the relationship between holophyletic groups [Neff 1986], and operationally it is a *multisemogenetic hypothesis* as a generalization over the entire set of cladistic characters [Pavlinov 2005, 2007c]. Besides, it is reasonable to identify third-level hypotheses in the form of posterior evolutionary scenarios [Eldredge and Cracraft 1980; Pavlinov 2005]: they can be involved in the iterative procedure of the sequential weighting of characters, in which their weights can change during the study.

An elementary operational unit of comparison is the character-bearing *semaphoront*, defined as "the individual during a certain, however brief, period of time" [Hennig 1966: 6]. This reduction radically distinguishes cladistic phylogenetics from the classical one, in which an organism is considered as a developing whole, which allows inclusion of embryological data in phylogenetic reconstructions. The basic units of cladistic analysis are *terminal groups* corresponding to the vertex nodes of the cladogram; groups corresponding to its internal nodes and (potentially) interpretable as ancestors are not considered.

The cladistic character is interpreted operationally as a *transformation series* presuming to reflect a historical sequence of transformations of a certain feature of organisms [Estabrook 1972; Mickevich 1982; Pavlinov 1990, 2005; Pogue and Mickevich 1990; Grant and Kluge 2004; Harris and Mishler 2009]. Its modalities are classified into two main categories: *plesiomorphy* denotes an initial state, while *apomorphy* denotes a derived state of the respective feature; the characters themselves are often thus denoted. This distinction sets the *polarity* of the cladistic character, with several empirical rules being used to determine it [Hennig 1966; Hecht and Edwards 1977; Wiley 1981; Pavlinov 1990, 2005]; some of them can be traced back to the works of early post-scholastic systematics. For example, the *criterion of commonality* corresponds to the principle of constancy by Jussieu and Cuvier, whereas the *rule of reciprocal illumination* is a variant of the principle of congruence by Adanson. Cladistic character is treated basically as an indirect indicator of cladistic events and the respective holophyletic groups; its classification significance is determined by its contribution to the final cladistic hypothesis. According to the *principle of polarity*, its significance is inversely proportional to the probability of its evolutionary reversions and parallelism. According to the *principle of congruence*, the significance of a set of characters is greater the more consistently they indicate the same sequence of cladstic events. According to Darwin's principle, the cladistic significance of characters is inversely proportional to their adaptive significance. In all other respects, cladistic characters are assigned the (nearly) equal weights of indicators of holophyletic groups.

From the above definition of cladistic relationship it follows that, to identify a holophyletic group (clade), it is necessary to use only that similarity that indicates the position of its members relative to their closest ancestor or an outgroup. To fulfill this condition, the concept of special similarity is introduced based on the key *principle of synapomorphy*. For this, two components are distinguished in overall similarity: *synapomorphy* is similarity in apomorphic states and *symplesiomorphy* is similarity in plesiomorphic states. For the recognition of holophyletic groups, only synapomorphies are significant; symplesiomorphies are not taken into account. With this, any differences between groups are also discarded when inferring the overall structure of cladistic relationships, which is formalized by the *principle of irrelevance of differences* (also goes back to Darwin); the latter makes cladistically treated similarity and difference asymmetrical, in contrast to phenetic. Synapomorphy is interpreted in two ways: *true synapomorphy* refers to the closest ancestor, while *underlying synapomorphy* means similarity due to parallel evolution within a monophyletic group [Saether 1979, 1982; Pavlinov 1990, 2005]. An advanced character state unique to a monotypic group is interpreted as the latter's *autapomorphy*.

It is possible to consider the principle of synapomorphy as a special kind of *similaritiy weighting* [Pavlinov 1990, 2005, 2018]. It is based on one-state logic (see Section 3.5): only positive judgments (the presence of synapomorphies) are significant, while negative ones (the absence of synapomorphies) are insignificant for delineating holophyletic groups.

The principle of synapomorphy implies a quantitative assessment of the validity of holophyletic groups, which is formalized by the *principle of summation of synapomorphies*: the more apomorphies characterize a group, the more reasonably it can be treated as holophyletic [Farris 1983, 1986; Pavlinov 1990, 1998, 2005]. From this follows the relevance of the *principle of total evidence*, which presumes the inclusion of as many characters as possible in the analysis [Eernisse and Kluge 1993; Kluge 1998; Rieppel 2004, 2005a, 2009]; in phylogenomics, this condition is implemented by whole-genome analysis [Savva et al. 2003].

In cladistic analysis, special attention is paid to the *principle of parsimony* [Farris 1983, 1986, 2008; Kluge 1984; Pesenko 1989; Pavlinov 1990, 2005, 2018], which is sometimes emphasized by the definition of cladistics with direct reference to this principle [Kitching et al. 1998]. The latter is built into either the ontological background (evolutionary parsimony) in the form of the *concept of minimal evolution* [Camin and Sokal 1965] or the algorithms of cladistic analysis to minimize the amount of parallelisms and inversions in the resulting cladistic hypothesis (methodological parsimony) [Farris 1983].[4] In structural cladistics, the principle of parsimony serves as the basis for excluding any evolutionary model from the initial conditions of the elaboration of cladistic classifications [Nelson 1978, 1979; Platnick 1979; Nelson and Platnick 1981; Patterson 1983] (see below).

The structure of cladistic relations is represented by a stylized form of the phylogenetic tree called *cladogram*; in terms of similarity, it reflects the hierarchy of synapomorphies, i.e., it is *synapomorphogram* [Sneath 1963; Pavlinov 1990, 2005; Williams and Ebach 2008]. With reference to Hennig's evolutionary model, it is

---

[4] It must be emphasized that methodological parsimony presumes implicitly evolutionary parsimony, because of the instrumentalism effect: no other evolutionary scenario can be inferred from the "most parsimonious" cladistic hypothesis than the one according to which evolution itself is parsimonious [Pavlinov 2005, 2007c].

assumed that each node of the cladogram must be dichotomous, in which case it is considered *fully resolved*; this serves as one of the criteria for its optimality [Nelson 1979; Nelson and Platnick 1981]. The base of the cladogram corresponds to an initial event in the phylogeny of the group studied and defines indirectly the entire hierarchy of its holophyletic subgroups. Each of the latter can be defined in one of the following ways: *node-based* by reference to a cladogram node, *stem-based* by reference to its internode, and *apomorphy-based* by reference to the synapomorphies [de Queiroz and Gauthier 1990, 1992; Nixon and Carpenter 2000; Sereno 2005].

One of the important properties of the synapomorphogram is a decrease in the number of synapomorphies from top to bottom. Since holophyletic groups are determined through synapomorphies only, this generates a specific *cladistic uncertainty* that increases in the same direction. This means that judgments about cladistic relationships between taxa are least reliable at the base of the cladogram [Pavlinov 1990, 2005, 2018]. This regularity is responsible for frequent controversies in treating cladistic relations among basal groups.

The basis for the development of a cladistically natural classification is the interpretation of a cladogram in taxonomic terms, i.e., translation of its hierarchy into a taxonomic one by representing clades as taxa [Hennig 1950, 1966; Nelson and Platnick 1981; Wiley 1981; Williams and Ebach 2008]. To emphasize the specificity of such a taxonomic construction, in which only holophyletic (synapomorphic) groups are present, it was proposed that it should be called not a classification, but a *cladification* [Mayr and Bock 2002; Hörandl 2010]. In the latter, the inclusive hierarchy of taxa is determined entirely by the sequence of the nodes in the respective cladogram: this means that the vertical component is dominant and the horizontal component is minimized. This condition is formalized by the *principle of equality of ranks of sister groups*: all clades converging to the same node of the cladogram are represented by the taxa of the same rank in the respective classification. The horizontal component is implicitly introduced by an analogy of the classical *progression rule*; in cladistics, a progression is defined by an increasing number of synapomorphies characterizing the respective groups [Wiley 1981; Pavlinov 2005]. The correspondence between the hierarchical structure of cladogram and cladistic classification may be *strong* or *weak*: in the first case, all sister groups identified in the cladogram must be reflected in the cladification; in the second, this is not necessary but the prohibition of paraphyletic taxa remains crucial.

The strong correspondence can make the cladification hierarchy very detailed and requiring the use of many additional taxonomic categories, therefore the *principle of non-strict hierarchy* is frequently followed in two versions. According to one of them, *plesions* as taxa with an unfixed rank are used, especially for fossil groups [Wiley 1981; Schoch 1986; Pavlinov 1990, 2005]; in another variant, taxa of different ranks can be assigned to the same level of generality [McKenna and Bell 1997]. Interpretation of the hierarchy resulting from an asymmetric cladogram also poses a special problem: the ranks of taxa in different parts of the same cladification turn out to be unable to be compared if they are separated by a different number of nodes [Pavlinov 2005]. To avoid all problems with ranked hierarchies in cladifications, it was suggested that they should be abandoned and instead the rankless hierarchy should be used [Løvtrup 1977; de Queiroz and Gauthier 1992; Ereshefsky 1997, 2001a; de Queiroz and Cantino 2001; Reif 2003; Mishler 2009; Zachos 2011]; by this, cladistic hierarchy is formally similar to that of the scholastic genus-specific scheme (see Section 2.2.2).

The taxa in cladifications can be assigned special designations reflecting their phylogenetic status; such cladifications are called *annotated* [Wiley 1979, 1981]. The entire study group is denoted as *pantaxone*; *crown groups* and *stem groups* are distinguished within it [Ax 1987; Shatalkin 1988; Forey et al. 1992; Sereno 2005]; it is proposed that the taxa corresponding to the stem groups should be called *adokimic*, i.e., "untrue" [Boger 1989], or *parataxa* [Meier and Richter 1992], or *plesions* (mentioned above).The groups of uncertain phylogenetic status are termed *metataxa*, or they may be *mixotaxa* or *ambitaxa* depending on their position in the cladogram [Gauthier et al. 1988; Archibald 1994].

As was noted in Section 2.7, development of the cladistic research program, especially coupled with molecular phylogenetics, because of its reductionist character, led to a significant "de-biologization" of biological systematics. Consequently, ideas of possible perspectives of the latter's post-cladistic development are beginning to be discussed [Wheeler 2008; Williams and Knapp 2010; Zander 2013; Pavlinov 2019, 2020]. However, this program still holds a leading position in systematics, judging by both regularly published "primers" exposing its methods and the aggressive attitude of many journals and their peer reviewers against other approaches.

The post-Hennigian development of cladistics led to its differentiation into several subprograms [Charig 1982; Hull 1988; Pavlinov 1990, 2005, 2018; Ebach et al. 2008; Williams and Ebach 2008; Pavlinov and Lyubarsky 2011; Quinn 2017]. They are united by the use of a few emblematic terms: cladogram, holophyly, synapomorphy, etc. With this, they are significantly different with respect to some important provisions of their onto-epistemology and methodology.

Evolutionary cladistics partly conserves some features of classical phylogenetics [Hill and Crane 1982; Hill and Camus 1986; Pavlinov 1990, 1998, 2005, 2018; Wägele 2005]. It presumes prior elaboration of rather complex evolutionary scenarios for all characters, from which synapomorphic schemes are derived as a basis for reconstruction of the phylogeneric hypothesis. Acknowledgment of the importance of internal parallelisms between closely related groups is a characteristic of this school: they allow *underlying synapomorphies* to be recognized and, according to one of Darwin's principles, they serve as an additional support of monophyly [Tuomikoski 1967; Brundin 1972; Saether 1979, 1982, 1986]. In terms of cladistics, this actually means a "legalization" of at least some paraphyletic groups; this position is defended by some botanists actively using cladistic terminology [Cronquist 1987; Brummitt 1996; Hörandl 2006; Stuessy and Hörandl 2014; Willner et al. 2014; Lachance 2016]. It is proposed that classifications thus developed are called *para-* or *patrocladistic* [Stuessy and König 2008; Wiley 2009; Carter et al. 2015]. *Stratocladistics*, also in a traditional style, takes into account geochronological dating of fossil forms for clarification of their cladistic relations [Fisher 2008]. The zoologist Quentin Wheeler believes that the development of this approach means the formation of another "new taxonomy" [Wheeler 2008], but in fact this is a partial return to classical phylogenetic origins.

Parsimony cladistics develops the ideas of the "founding fathers" in a much more formalized way. In it, as in the previous version, the idea of evolution is initially accepted, with the main task being to reconstruct phylogenies as the basis for developing the respective classifications [Gaffney 1979; Farris 1983, 1986; Kluge 1984]. But this approach presumes minimizing the traditional speculative contents of the evolutionary scenarios: with reference to the methodological *principle of parsimony*, it is argued that such scenarios specified for particular cladistic characters are equivalent to adopting undesirable *ad hoc* prior judgments. Accordingly, all the latter are excluded from the prior analyses of characters accepted without predetermined polarities; such a background model corresponds to a unified hypothesis of undirected (random) evolution. The latter drastically distinguishes this version of cladistics from the Hennigian one, where irreversibility of character evolution is considered one of the basic assumptions [Hennig 1966]. So, the only prior assumption adopted is that of monophyly of the studied group relative to a certain outgroup; reference to the latter serves as a decisive means in determining the base and hence the entire hierarchy of the cladogram, from which all conclusions about synapomorphies are derived on a posterior basis [Watrous and Wheeler 1981; Farris 1982, 1983; Pavlinov 1990, 2005; Nixon and Carpenter 1993; Barriel and Tassy 1997; Härlin 1999; Sereno 2005]. This approach is based on analysis of as many characters as possible without their prior substantive weighting, indicating a noticeable element of the phenetic idea in its methodology. In order to emphasize its "phenetic bias," this school is sometimes called *neocladism* [Saether 1986] or *phenetic cladistics* [Wägele 2005].

Pattern cladistics develops the original ideas of W. Hennig in a very paradoxical way, and therefore it was called *transformed cladistics* by its leaders [Nelson 1978; Platnick 1979; Nelson and Platnick 1981; Patterson 1983]. It excludes any prior considerations of phylogeny from its background ontology, thus making the latter non-evolutionary [Beatty 1982; Brady 1985; Scott-Ram 1990; Ebach et al. 2008; Pavlinov 2018]; this gave a reason to call this approach *methodological cladistics* [Hill and Crane 1982]. Its ontic basis consists of an idea of the hierarchical structure (pattern) of the diversity of organisms as a consequence of the orderliness of their ontogenies [Patterson 1983, 1988]. The general argumentation scheme is almost directly derived from K. von Baer's epigenetic typology (see Section 2.4.3), according to which the more generalized characters appear earlier than the more specialized ones in animal ontogeny. Based on this, the principle of synapomorphy is reformulated into the *principle of the levels of generality of defining characters*: each of the latter (or their combination) defines a certain synapomorphic group, and their combined internested hierarchy determines the entire hierarchy of the cladogram and thus the cladistic classification inferred from it. As a result, this version of cladistics loses its phylogenetic roots and turns into one of the versions of typology of an epigenetic kind [Tatarinov 1977; Charig 1982; Riedley 1986; Scott-Ram 1990; Pavlinov 2018; Brower 2019].

5.8 THE EVOLUTIONARY ONTOGENETIC PROGRAM?

Generally speaking, the *evolutionary ontogenetic* research program is just beginning to take shape and is therefore is barely noticeable against the currently dominating cladistics based on the analysis of molecular data [Pavlinov 2019]. Its specificity is that it focuses on the evolutionarily interpreted disparity of ontogenetic patterns of multi-cellular organisms [Minelli 2015]. The basis for this is provided by a synthesis of the phylogenetics considered in its rather broadened sense, epigenetic typology, and the "evo-devo" concept (a now well-known abbreviation for evolutionary developmental biology). The concepts of dynamic archetype (phylocreod) and phylotype (phylotypic stage) mentioned above fit well into the context of this taxonomic theory, thus making a typological approach compatible with modern macroevolutionary theory [Lewens 2009b, Riegner 2013].

From a historical perspective, this theory goes back to the classical principle of trifold parallelism of the mid 19th century [Gould 1977]. Its fundamental novelty is incorporation of the evo-devo concept that focuses on the historical changes of the genetic mechanisms of regulation of ontogenesis [Minelli 2003, Laubichler and Maienschein 2008, Minelli and Pradeu 2014].

As can be seen, the evo-devo (or phylo-evo-devo) taxonomic theory and the respective research program are based on a rather rich biologically meaningful ontology, which distinguishes it positively from reductionist cladistics and molecular phylogenetics. This means another, newer version (along with biomorphics and evolutionary taxonomy) of the most recent biologization of systematics. At the same time, as far as phylogeny is considered one of the cornerstones of this program, it is possible to consider the latter as another branch of the phylogenetic program in its widest sense.

By focusing on the evolution of ontogenetic patterns and epigenetic mechanisms ensuring their historical stability and dynamics, this research program brings its own version of representing historical patterns of biodiversity and their respective classifications. The former can be represented by a phylo-ontogenetic tree, which is actually a phylogenetically interpreted "Baerian tree" (see Section 3.4 on the latter). This tree is transformed into a corresponding evo-devo classification in the same manner as the phylogenetic one, with its ranking scale being derived from a sequence of appearances of respective ontogenetic patterns in the evolution of multicellular organisms. The main characteristics of an evo-devo taxon become its specific ontogenetic pattern as a whole dynamic system, not reducible to any particular developmental stages [Orton 1955; Martynov 2011, 2012; Pavlinov 2013c; Minelli 2015]. All this provides biological systematics with a rich ontological basis and allows it to get rid of the overload reductionism brought in by the above-mentioned "new phylogenetics."

From an epistemic viewpoint, the research program under consideration faces a serious problem caused by its rich natural ontology. The latter presumes that the elaboration of evo-devo classifications should be based on a joint exploration of two complexly interacting multifaceted dynamic systems of phylogeny and ontogeny [Rieppel 1989, 1990; Pavlinov 2013c]. In such a knotty cognitive situation, the above-mentioned *NP-completeness problem* [Garey and Johnson 1979] (see Section 5.7.1) becomes quite relevant.

It is evident that the evo-devo research program is not universal: its application is limited to groups of multicellular organisms with sufficiently developed ontogenetic cycles. Accordingly, many protists and apparently all prokaryotes appear to be outside the scope of its competence. However, this circumstance should hardly be considered a serious disadvantage: as emphasized repeatedly above, any research program in systematics and each particular taxonomic theory underlying it are inevitably local with regard to its applications.

At the moment, those classifications that realize the evo-devo taxonomic theory most consistently and, thus, belong to the program in question are very few. The reason is that detailed studies on diversity and evolution of the mechanisms of regulation of ontogenesis in animals and plants on a modern epigenetic basis are just beginning. Therefore, as always happens with new disciplines, they involve analyses of only a few model organisms. So, it seems premature to consider how actively this research program will be developing, how productive it may turn out to be for systematics, how serious the alterations of taxonomic classifications may be, and which particular alterations will occur. Among the main tasks to be solved by evo-devo taxonomic theory, to make the program in question more promising, seem to be the following: (a) elaboration of a calculus for assessment of the relative significance (weight) of the differences in molecular sequences and ontogenetic rearrangements; (b) elaboration of the unified ranking scale for the evo-devo classifications of different groups of organisms; and (c) development of an optimal way to combine vertical and horizontal interrelations between groups with different ontogenetic patterns to reflect best both their primitive (ancestral) and derived features.

6 Taxonomic Puzzles

The world's a puzzle; no need to make sense out of it.

Socrates[1]

According to the philosophy of science by Thomas Kuhn, two more or less clearly defined phases are distinguished in the development of a scientific discipline [Kuhn 1962]. One of them is "revolutionary": it involves a *paradigm shift*, which is a fundamental change of one exploratory model to another with respective change of the entire cognitive situation. Another phase is "normal": it is associated with refining and solution of specific research tasks within the framework of the same paradigm with only a minimal change in the cognitive situation. These tasks are known as *scientific puzzles* [Girill 1973].

The taxonomic theories (TTs) functioning in biological systematics can be considered its paradigms; from a Kuhnian perspective, their development involves posing specific *taxonomic puzzles* and elaborating the means to solve them [Pavlinov 2019]. For instance, as far as general TT is concerned, the main conundrum is about understanding the Natural System: the various proposed ways of solving it led to scientific revolutions in systematics (see Chapter 2). For the aspect-based partial TTs embodying these different understandings, the respective puzzles are shaped by elaborating methodologies that are most consistent with certain understandings of the Natural System (typological, phylogenetic, etc.). For object-based and relational partial TTs, their puzzles are shaped by searching for answers to the questions of how to interpret species or homology, how to configure hierarchies, how to correlate kinship and similarity, etc.

In considering taxonomic puzzles, one of their important features should be pointed out. Unlike the "Kuhnian" ones, their main concern is the solution of specific tasks of a type similar to "find I don't know what," so they are *problem puzzles*. For example, it is clear that the classification sought should be natural, but it is not clear how to define it precisely and unambiguously: after all, we can only guess what the System of Nature is. Suppose we believe that such a classification should be phylogenetic, but it is not clear what precisely phylogeny is: we cannot observe it, so it remains only to assume some of its essential properties based on some other theoretical considerations and/or empirical observations. Thus, when considering the cognitive situation of systematics from this perspective, we have a kind of "taskbook" with a set of "tasks," each with many unknowns, that make up the respective taxonomic puzzles. Moreover, it is completed with a kind of "solutions textbook" in the form of certain TTs—but there is nothing like definite "answers" at its end, with which one could check the definitive solutions to the puzzles. Instead, a spectrum of many possible "answers" appears during the attempt to solve the taxonomic puzzles.

---

[1] Cited after Dan Milliman's "Way of the Peaceful Warrior."

This chapter explores some of the taxonomic puzzles associated with the basic concepts of systematics. Among them are problematic understandings of what the natural classification is and how it is structured; how taxon and character are interconnected by an iterative classification procedure; how similarity and kinship relate to each other; and how archetype, homology, and species can be treated.

6.1   BETWEEN NATURAL AND ARTIFICIAL CLASSIFICATIONS

According to the basics of the general TT set out in Chapter 4, the main studied object of systematics is taxonomic diversity (TD) as a specific taxonomic *Umwelt*, and the main task is to describe this object in the form of the taxonomic system. The latter is thought of as an adequate cognitive model of TD, whereas in traditional (Linnaean) terminology it is usually called *natural classification*. In classical systematics, the latter is thought of as "*globally natural*" by reflecting the TD either as a whole or in its presumably most essential manifestation. With this, the conceptual history of systematics provided various particular understandings of this general notion, so its non-classical version presumes that different aspectual manifestations of TD are best represented by the corresponding "*locally natural*" particular classifications.

In any cognitive situation centered around the general conception of natural classification, regardless of its understanding, it is acknowledged that an exhaustive description of TD is theoretically unattainable. In fact, systematics develops particular classifications as more or less close approximations of the natural one; the least close, in contrast to the latter, are traditionally called *artificial classifications*. In general, all particular classifications are to some extent natural and to some extent artificial; the difference between them is quantitative, so the TD structure, however it is understood, may be represented by a certain set of classifications with different degrees of naturalness.

The main problem related to this global task of systematics is caused by the impossibility of gaining an unequivocal and commonly adopted understanding of what the natural classification is and how it can and should be attended. Because of this, the most fundamental and most problematic *naturalness puzzle* appears in systematics: it is caused, at a theoretical level, by a multiplicity of ways to understand the Natural System, and at a practical level, by a multiplicity of working classifications with various degrees of their "local naturalness." It can be argued, without great exaggeration, that the conceptual history of systematics was driven by a search for ways to pose and solve this general puzzle. Obviously, these ways can be different depending on how specific cognitive situations are initially shaped onto-epistemically. Consideration of all questions related to this puzzle is within the competence of the TT in both general and particular understanding.

Some partial TTs refuse to consider this puzzle as loaded with a rich ontology. They develop an idea of *general-purpose classifications* interpreted quite pragmatically. Such classifications are the better, the wider the set of scientific or applied tasks they solve effectively [Gilmour 1940; Sokal and Sneath 1963]. However, they are sometimes called "*Gilmour-natural.*"

In classical natural science, the natural classification is likened to the law of nature [Duhem 1954; Rozova 1986; Zarenkov 1988; Zabrodin 1989; Vityaev and Kostin 1992; Rozov 1995; Mahner and Bunge 1997; Wilkins and Ebach 2014]. Such understanding is evidently inherited from the natural philosophy of the 17th and 18th centuries with its dominating idea of the law-like System of Nature; however, there is no universal understanding of what this "taxonomic law" might be. In one of its versions ascending to the scholastic tradition, classifications based on the essences are considered natural [Rozova 1986; Shatalkin 2012]. Close to this is a version according to which classification is natural if its taxa can be interpreted as *natural kinds* in the sense of Quine [Spencer 2016]. In phylogenetics, classification is natural if it represents phylogenetic pattern most adequately. In natural systematics of botanists, classification is natural if it is embodies many (essential) characters.

A generalization over different ways of solving this puzzle is offered by a rationalist approach, according to which it is necessary to elaborate and apply some explicit *criteria of naturalness*, or rather a system of coherent criteria of naturalness. Accordingly, it is presumed that a classification is more natural if it meets the conditions of a given criterion (system of criteria) of naturalness [Michener 1957; Rozova 1986; Pavlinov 2006, 2010b, 2018]. Each of them serves as a certain "calculus" for assessing the "local naturalness" of particular classifications. It presumes a certain "naturalness scale," relative to which working classifications can be ordered from the most to the least natural.

Each particular TT (PTT) elaborates its own criterion (system of criteria) as part of its quasi-axiomatics based on the respective ontic and/or epistemic premises. In the case of ontology-based PTTs, particular *Umwelts* (particular TD manifestations) are first defined and then the particular systems of naturalness criteria of classifications are developed with respect to just these *Umwelts*. In contrast, PTTs with epistemic bases infer their criteria from certain methodological requirements, such as logical rigor. Thus, all systems of criteria of naturalness of classifications are theory-dependent and therefore local as attributes of certain PTTs. Therefore, any conclusions about the naturalness of classifications based on such criteria are also theory-dependent and local: a classification considered natural in the context of one PTT may appear artificial in other PTTs.

This conclusion seems to be relevant to an understanding of the significance of the *criterion of predictivity* of classifications as the fundamental criterion of their naturalness. It is usually considered of paramount importance by the taxonomists adherent of the physicalist philosophy of science [Gilmour 1940, Crowson 1970, Lyubishchev 1972; Rozova 1986; Starobogatov 1989; Marradi 1990; Rozov 1995], but the preceding consideration indicates it is no less "local" than any others. This is because the possibilities of reliable predictions (extrapolations) inferred from the respective classifications are obviously limited to the specific *Umwelts* they reflect [Pavlinov 2018]. This conclusion is compatible with the treatment of taxon as natural kind (in the sense of [Quine 1996]).

One of the general approaches to the solution to this puzzle elaborated by the rational cognitive program is based on the idea of a *natural method*: according to it, for a classification to be natural, it must be elaborated by means of such a method. Thus, the emphasis is shifted from the criteria of the naturalness of the classification itself to the criteria of naturalness of the classification method. With this, as far as ontology-based PTTs are concerned, the *principle of methodological correspondence* comes into effect; accordingly, the naturalness of a method is determined not on its own but with reference to the respective *Umwelt*. Therefore, in the search for a solution to this general puzzle, one must proceed from the comprehension of Nature (what it "really is") to the comprehension of the respective natural method. It should be remembered that the most significant paradigmatic shifts in the conceptual history of systematics in the 18th and 19th centuries involved different treatments of the natural method following different treatments of Nature (see Sections 2.4 and 2.5).

Any system of criteria of naturalness of systematic methods includes an appropriate selection ("weighting") of the classifying characters and the classification algorithms; in both cases, it depends by and large on the basic onto-epistemology. With respect to the characters, the most significant are those that most consistently indicate either certain features or relations of organisms considered essential within a certain PTT. Thus, recalling classical terminology, it can be said that one of the main tasks of the natural method is to elaborate the criteria for the selection of proper characters. As to classification algorithms, their "naturalness" is determined by their effectiveness as a means of elaborating classifications considered the most natural according to certain criteria.

These considerations can be demonstrably illustrated by two examples. If an *Umwelt* is based on the natural-philosophical idea of the overall interconnectedness of the elements of Nature, the respective natural method should be based on the consideration of "all and any" properties of organisms. Accordingly, the natural classification is elaborated either based on all available characters to produce a kind of "omnispective" one [Blackwelder 1967] or on the criterion of the compatibility of the characters (this goes back to the natural method of Adanson; see Section 2.4.2), according to which natural classification should be substantiated by the largest possible number of mutually compatible characters [Estabrook 1972; Le Quesne 1982]. In contrast, if an *Umwelt* is defined as the phylogenetic pattern, most attention should be paid to the correct selection of characters as most reliable indicators of monophyly (morphological or molecular), and to the elaboration of classification algorithms (say, parsimony or likelihood) to infer most effectively the "best" (for the given dataset) phylogenetic classification.

Facing a seemingly irreducible variety of "locally natural" classifications, the following important question arises: how might it be possible to move from a set of local solutions to a global one? One possible answer to this question, and that claims to be a fairly general solution to the naturalness puzzle, may be development of the so-called *faceted classification system*, which is organized similarly to relational databases [Broughton 2006; Chan and Salaba 2016]. It allows different locally natural classifications to be combined into a single pool with the help of a certain meta-language that makes it possible to implement the *principle of mutual interpretability* of classifications. As a result, instead of searching for either a single, quite fuzzy "omnispective" classification that claims to reflect "all in one" or a certain "locally natural" classification proclaimed to be most significant according to a certain criterion (phenetic, phylogenetic, typological, etc.), we focus on principles of developing a faceted classification system that embraces all manifestations of TD and thus is capable of claiming to be considered "globally natural" [Pavlinov 2018, 2020].

6.2 BETWEEN TAXON AND CHARACTER

As shown in Chapter 4, TR is mentally divided into two "orthogonal" complementary aspects, taxonomic and partonomic (meronomic) [Meyen 1977, 1978; Lyubarsky 1996; Pavlinov 2011a, 2018; Pavlinov and Lyubarsky 2011]. Turning to the first aspect, systematics deals with the diversity of organisms and identifies taxa as units of taxonomic systems representing certain groups of organisms (vertebrate animals, herbaceous plants, etc.). Turning to the second aspect, systematics deals with the diversity of the properties of organisms and recognizes partons (merons) as particular classes of these properties (limbs, body coloration, sexual behavior, etc.). Based on the partons recognized in partonomic systems, classifying characters are fixed, though without a strict (one-to-one) correspondence between them. Organisms are then described by these characters and, based on the results of their comparison by the respective characters, taxa are distinguished in the taxonomic system. The elaboration of the latter constitutes an ultimate goal of systematics, but this aim can be rich only if there is a certain partonomic system with respective characters that have already been developed. On the other hand, the structure of partonomic diversity can only be uncovered after at least some preliminary information has been obtained about the structure of TD in order to differentiate partons not "in general" (which is meaningless enough), but with reference to particular groups of organisms. As a result, we have a kind of dual unity: one aspect of diversity does not exist without another, and the study of one of them is both a prerequisite and a certain result of the study of the other—therefore, the better (more detailed) we know one aspect of TR, the better we know another.

On this basis, the fundamental *taxon–character puzzle* arises: how to analyze correctly taxa and their characters, if the analyses of the respective aspects of the TR are mutually conditioned. So one part of this puzzle is a hazard of taxonomic research possibly devolving into circular reasoning by the bi-unity of these two aspects: they are involved in the same classification procedure, in which they are mutually interdependent.

In solving this puzzle, it should be clearly understood that both its components—a taxon and a character—are conceptual constructs that cannot be represented outside the conceptual space developed by the general TT. This means that the general solution to this puzzle belongs to the competence of the latter: it defines possible ways to solve the puzzle depending on how the classification units of both taxonomic and partonomic systems and the interrelations between them are defined.

It is to be noted that the frequently declared identity of characters with some observable properties of organisms [Gilmour 1940; Michener and Sokal 1957; Sneath and Sokal 1973; Wagner 2001] is fundamentally incorrect from a conceptualist standpoint. The latter presumes a character to be an information (cognitive) model of the corresponding parton or a certain set thereof, and a sample of characters is an information (cognitive) model of the corresponding partonomic "subspace" [Pavlinov 2018]. Characters are distinguished based on prior recognition of partons, which appear as the result of a certain conceptually motivated cognitive activity involving, among other things, certain concepts of homology (see Section 6.6).

For a general understanding of how this puzzle can be solved, the following formalisms are of use.

To begin with, let us define a character as a *variable*, with character states (modalities) as *variable values* arranged by a certain *order relation* [Sokal and Sneath 1963; Colless 1967, 1985; Estabrook 1971]. This variable serves as a means to compare organisms; in this comparison, the order relation established for the variable is reflected in a set of organisms and establishes a similar order relation between them according to the variable values (character modalities) attributed to them. Based on this order relation between organisms, they can be grouped into taxa.

Let us assume that taxa and characters recognized in respective classifications are related to each other by *taxon–character correspondence* [Pavlinov 2018]. The latter's meaning can be thought of as a reflection of a set of taxa on a set of characters, and vice versa. Accordingly, this correspondence establishes a kind of logical sequence of judgments about taxa and characters, which participates in the shaping of the classification algorithm. Taxon–character correspondence is always fulfilled: thanks to this, we can in practice define any taxon by certain characters comprising its specific diagnosis; however, this correspondence can be strict or non-strict.

Strict taxon–character correspondence yields the *monothetic* taxon definition expressed by the formula "one taxon—one character." It means that each taxon in classification can be exhaustively delineated by a single character or a single set of completely congruent characters. In such an ideal situation, all subsets of characters would provide completely overlapping classifications of taxa, formalized by the *principle of character interchangeability* [Meyen 1978]. In systematics dealing with natural aggregates and their features, this formula is never fulfilled because of the incongruence of the characters. This means *non-strict taxon–character corres-pondence* yielding the *polythetic* taxon definition. Accordingly, (a) certain elements (modalities) of any one character can be attributed to different taxa; (b) in any one taxon there is always at least one member that does not possess elements of at least one character in the taxon's diagnosis; and therefore (c) any taxon in a classification can be comprehensively specified by a combination of several characters only [Sokal and Sneath 1963].

The principle of taxon–character correspondence presumes the possibility of relating taxa and characters by a kind of *precedence*, according to which their logical and procedural relations can be represented as either "taxon→character" or "character→taxon." Based on this, the following stepwise classification algorithm of taxonomic research can be construed. The latter begins with a selection of particular sets of organisms and their characters based on certain previous results. At the next step, an analysis of the data may be conducted in two ways depending on a particular taxon–character precedence. In one case, it may start with a more detailed partonomic ordering, which results in the recognition of characters used to identify the respective taxa intensionally. In another case, it may start with an extensional delineation of taxa, to which particular characters are then attributed. Thus, in the first case, character recognition precedes taxa recognition, while in the second case the precedence is the opposite. The first option is most consistently implemented by scholastic genus–species scheme, by typology, and by biomorphics: most significant characters are first identified following certain criteria, and then taxa are recognized on their basis. The second option in its "pure form" may occur, when certain new (e.g., molecular) characters are selected to see what might be the outcome of employing them for the classification of organisms. Some of its elements are implemented by numerical taxonomy: taxa are first distinguished based on an analysis of the entire set of characters employed, and then particular diagnostic characters are attributed to each particular taxon.

A rigid dichotomy of these two schemes with opposite "precedences" represents the general classification procedure in a simplified and scarcely realistic form. In fact, practical systematic research is organized more complexly, in such a way that the precedences "taxon→character" and "character→taxon" alternate within a single sequential iterative procedure [Meyen 1984; Lyubarsky 1996; Pavlinov and Lyubarsky 2011; Pavlinov 2018]. In this algorithm, taxon–character correspondence is manifested dynamically as an interleaving of opposite "precedences" at different iteration steps. Suppose a certain group of organisms is first selected for which both taxonomic and partonomic order relations are preliminary, given by preceding research. As a result of its taxonomic analysis, particular characters are specified, on the basis of which taxonomic classification is carried out, with taxa being more precisely outlined and diagnosed. In the context of this taxonomic classification, the characters are studied in more detail, with their homology, weight, etc. being specified and changed by necessity. Then, based on specified characters, taxonomic analysis is again carried out, with parameters of taxa (composition, ranks, etc.) being specified in their turn, etc. As a result, in particular taxonomic research with a sufficient number of characters and taxa included, taxon–character correspondence, being less strict at its beginning, can be expected to become stricter at its end due to the action of the *principle of convergence*.

An idea of the successive iterative procedure of taxonomic research solves the puzzle in question in a rather general form. It "breaks" a closed logical circle of reasoning by replacing it with a sequence of internested hermeneutic circles. The latter means that, at each iteration step, a particular classification task does not "mirror" a previous one but rather is specified and solved in a less fuzzy context than the previous one.

Because of the non-strict nature of the taxon–character correspondence, employing different sets of weakly correlated characters yields unavoidably different classifications with various degrees of naturalness. Since the characters constitute the only basis for any classifications elaborated by a comparative method, the question of "naturalness" of both classifications and methods yielding them is thus largely a question of the choice of "natural" characters. This fact was recognized at the very beginning of rational classification activity, and it is formalized by the fundamental *principle of character inequality*. The choice of due characters for elaborating due classifications is routinely termed *character weighting*, so the principle just mentioned is specified as the *principle of differential character weighting*. It presumes the elaboration and application of a certain *weighting function*, by which specific *weights* are ascribed to characters; they are directly proportional to the contribution of the characters to the resulting classification. This function is operationalized by a set of *weighing criteria* which allows characters to be arranged according to a certain *weighting scale* by ascribing them particular weights.

Therefore, character weighting is a very important part of any version of the natural method: without much exaggeration, one may say that the latter is designed largely to interpret properly the general weighting function to make its particular interpretation most adequate to a particular understanding of what the natural classification should be. Its "adequacy" is provided in that it is developed by the PTT as part of its methodology based on its onto-epistemic premises, so the character weighting is theory-dependent. The latter means that there cannot be a single universal weighting function with the respective universal weighting scale: what is considered important for the character weighting in one PTT will not necessarily be so in the others. Thus, in typological systematics, characters should characterize archetypes, body plans, or ontogenetic patterns; in phylogenetic theory, they should indicate kinship relations; and in biomorphics they are selected so as to characterize biomorphs (life forms); and so on.

It follows from the above that, in order to elaborate natural classification with naturally defined taxa according to a certain system of criteria of naturalness, characters should be weighted "naturally" according to the respective weighting criteria. Therefore, proper application of the weighting function constitutes an important part of the above iterative classification algorithm, and its "partonomic step" includes proper character weighting. For instance, an iterative procedure is implemented in numerical phyletics by the method of successive character weighting [Farris 1969] and by Bayesian inference [Chen et al. 2014]. Thus, the development of the weighting function with respective weighting scales and criteria and its implementation in classification algorithms is a prerequisite for solving the taxon–character puzzle at an operational level.

6.3 HIERARCHIES: TO RANK OR NOT TO RANK?

In the structure of any sufficiently complex system, including the classification one, two basic components can be generally distinguished, hierarchical and non-hierarchical. They relate to each other in a complex manner, manifest themselves in different ways, and their consistent combination in a single classification system is fraught with serious methodological difficulties. All this creates a serious multifaceted *hierarchy puzzle* in systematics.

A *hierarchical component* of the structure of the classification system implies a certain subordination of its classification units (taxa, partons) set by a *hierarchization scale*, which establishes both the sequence and character of this subordination; this scale can be of two kinds. The *external* scale is borrowed from a certain ordering system independently of the classification system in question; it is *universal* with respect to its structure and determines an *absolute hierarchy* in it: classification units can be arranged in this hierarchy independently of each other with reference to this scale. The *internal* scale is part of the classification system in question; it is local with respect to its structure and determines a *relative hierarchy* in it: the position of the classification units in such hierarchy is mutually interdependent. These two scales are exemplified by the hierarchy of monophyletic groups in phylogenetic classifications: it can be determined either absolutely with reference to a universal geochronological scale or relatively by a branching sequence within each lineage (see Section 5.7).

Hierarchical ordering is twofold [Salthe 1985, 2012; Knox 1998; Pavlinov 2018]. A *linear hierarchy* arranges the classification units according to varying degrees of importance; sometimes it is called an *exclusive hierarchy*. A typical example is the social hierarchy of the "host–clave" kind; in systematics, it occurs in the character ranking (weighting) according to their contribution to the classification. An *inclusive* (encaptic, nested) *hierarchy* means subordination of the classification units according to their levels of generality: units of a higher level of generality include units of lower levels of generality. Accordingly, the inclusive hierarchy establishes the relations of the "set–subset" or "whole–part" kind between these units.

A *non-hierarchical component* of the structure of a classification system means that the classification units are linearly arranged by the gradient of a certain external variable. The latter sets the order of these units along its gradient, but, in contrast to the case of linear hierarchy, nothing like their subordination is presumed. An illustrative example is the Ladder of Nature, in which taxa are arranged according to the complexity of the organisms allocated to them.

In taxonomic terminology, classifications with inclusive hierarchy are customarily designated as "*vertical*," while those with a non-hierarchical arrangement are designated as "*horizontal*"; the latter are sometimes called *parametric* [Lyubishchev 1923, 1972], its not in the sense of [Subbotin 2001]. These components reflect substantially different aspects of the ordering of organisms and are weakly correlated with each other. As an illustrative example, it is enough to mention two classifications of tetrapods: classifying them by their kinship relation provides an inclusive hierarchy of monophyletic groups, while they are arranged linearly along a scale from cold-blooded to warm-blooded according to the properties of their thermoregulation.

Thus, these components can be reflected by different classifications of the same group of organisms, ordering their overall diversity in significantly different ways. Therefore, it is not possible to combine them consistently in a single taxonomic system; for each of them a particular system is to be construed. This collision gives rise to specific questions about the hierarchy puzzle: if both of these components are objectively inherent in the diversity of living beings and thus equally "natural," and at the same time they are mutually incompatible, how should a classification be built to make it "globally" natural? If it is impossible to combine them harmoniously, which one should be preferred as the most natural "locally"? This problem evidently relates to the one of different kinds of "locally" natural classifications considered above (see Section 6.1).

Different PTTs answer these questions differently depending on which component of the TD is given the greatest importance. Scholastic systematics and cladistics, aimed at developing strictly "vertical" hierarchical classifications, are at one extreme. In them, the only scale of hierarchization is the branching sequence of the initial tree, either a generic–species scheme or a cladogram, according to which the respective classification units are arranged. The already-mentioned version of onto-rational systematics developing periodic classification systems (see Section 5.2.1) is another extreme. Somewhere between these two extremes are PTTs that pay equal attention to both components of the ordered diversity. Examples include Oken's organismic natural philosophy, classical (Haeckelian) phylogenetics, and evolutionary taxonomy; they are usually called syncretic because they attempt to combine the uncombinable.

When working with inclusive hierarchies, another specific question about the hierarchy puzzle arises: an inclusive hierarchy can be either ranked or rankless. In the first case, clearly fixed levels of generality are distinguished and denoted by specific terms, and they are called *taxonomic ranks*: species, genus, family, etc. In the second case, the ranks are neither fixed nor denoted in any way.

Ranked hierarchy is established by a certain *ranking scale*, according to which both the sequence and number of fixed taxonomic ranks are determined. A set of taxa belonging to one rank constitutes the same *taxonomic category*, and it is termed the same as its corresponding rank (species, genus, etc.). According to the *principle of rank equivalence*, taxa of the same level of generality belong to the same rank as members of the same equivalence class, while those of different levels of generality belong to different ranks and are not equivalent in this sense. At an operational level, this principle is complemented by the *principle of rank coordination*, or the *rule of uniform level* [Starobogatov 1989; Vasilieva 1992, 1998; Shatalkin 1995; Lyubarsky 1996, 2018].

If the ranking scale is rigidly defined, it generates a *strict hierarchy*; this condition is usually met in the case of absolute ranking. In this case, the same sequence of taxonomic ranks is observed in different parts of a classification. In contrast, in the case of relative ranking, a *non-strict* (degenerate) *hierarchy* appears with its characteristic *rank uncertainty* [Zarenkov 1988; Pavlinov 2015, 2018]. The latter means that: (a) taxa cannot be unambiguously assigned to a specific rank; and (b) an equivalence of taxa of the same rank is not strict. With this, in the case of relative ranking, the more distantly taxa are located in a classification, the less it is possible to assess their equivalence [Pavlinov 2018]. The strict hierarchy is characteristic of most traditional classifications; typical examples of non-strict hierarchy are as follows. In traditional classifications, this happens in the case of extensional coincidence of the subordinate taxa of different ranks: for example, a monotypic family coincides in its species composition with its only genus [Gregg 1950, 1954; Sklar 1964; Gordon 1999]. In cladistics, *non-fixed ranks* are permissible, which makes it possible (a) to allocate the taxa with formally different ranks to the same level of generality, and (b) to use "out-of-rank" *plesions* [Wiley 1979, 1981; Gauthier et al. 1988; Pavlinov 1990, 2005; McKenna and Bell 1997].

In the *rankless hierarchy*, fixed categories are not distinguished and not designated terminologically. In such a hierarchy, a certain sequence of the levels of generality is set depending on a particular research task being solved. One of the most notable examples is the *fractal* with a scaling hierarchy set by different levels of detail in the description of a complexly structured object; this concept is relevant to systematics [Burlando 1990; Minelli et al. 1991]. A non-strict hierarchy occupies an intermediate position between strict ranked and rankless hierarchies.

The hierarchy of the genus–species scheme of scholastics, which was a starting point for systematics, was rankless (see Section 2.2.2). Fixed ranks with specific names were "invented" in the 18th century; at first there were 4–5 of them, and then during the 19th and 20th centuries their number increased significantly [Stevens 1994; Ereshefsky 1997, 2001a; Pavlinov 2015, 2018; Lyubarsky 2018]. With the fragmentation of the ranked hierarchy, the number of categories and their designations increased more and more, especially influenced by cladistics [Hennig 1966; Wiley 1981; de Queiroz and Gauthier 1992; Ereshefsky 2001a, 2001b, 2002], so a kind of "rank inflation" progressed [Stuessy 2008]. As a result, rank uncertainty increased, so the very meaningfulness of the over-complicated ranked hierarchy became doubtful. Based on this, it was proposed to abandon it and build rankless classifications in cladistics; in fact, this suggestion means formally a return to the genus–species scheme .

The coexistence of both traditional and cladistic approaches in contemporary systematics led to the fact that two hierarchical systems, ranked (classical) and rankless (cladistic), function in it almost on an equal footing. A contradiction between them, constituting another point of the hierarchy puzzle, is due to the lack of clear understanding of what fixed ranks might mean, whether they are needed or not, and if they are used, then what for?

To the extent that systematics is engaged in the development of natural classifications in their ontic understanding, the main question in this puzzle can be presented as follows: are there certain aggregations of organisms of different levels of generality really existing in the structure of Nature to which certain taxonomic ranks might correspond? If there are clearly delineated levels in this structure, the respective ranks are "real" in a classical sense; if not, they are nominal. With this, it should to be taken into account that different manifestations of the TD structure can be associated with different internested hierarchies (phylogenetic, biomorphic, typological, etc.), each with its own ranking scale; when considered together as components of the same faceted classification, they yield a kind of *polyarchic* system [Knox 1998].

Objectification of the ranked hierarchy implicitly presupposes a principal possibility of (or even need for) an absolute ranking scale; the latter is given different substantiations by different PTTs. In typology, it is inferred from an "objective" subdivision of Nature-as-superorganism into partons of different levels of generality [Beklemishev 1994; Lyubarsky 1996, 2018], which is rooted in the organismic natural philosophy of Oken. Similar to this in some respect is substantiation of an objective ranked hierarchy with a few fundamental ranks by reference to a "parallelism" between stages of evolutionary and ontogenetic differentiation of organisms [Ho 1988, 1998; Ho and Saunders 1993; Goodwin 1994; Shatalkin 1995, 2012]; this viewpoint ascends to the epigenetic typology of von Baer. In evolutionary taxonomy, the objectivity of ranks is substantiated by reference to adaptive zones and a kind of "quantum" shifting of evolving groups between them [Simpson 1961; Legendre 1972]. An objective nature of ranked hierarchy in biomorphics [Aleyev 1986] and in onto-rational systematics aimed at distinguishing various natural kinds [Webster and Goodwin 1996; Zakharov 2005] is substantiated in their own ways. The objective species rank is substantiated by reference to specific mechanisms of generating and integrity of species, from their divine creation to breeding mechanisms (see Section

6.7). However, according to the "Ladder" natural philosophy, and phenetic and cladistic PTTs, there is nothing in Nature that would justify a unified ranked taxonomic hierarchy [Brown 1810; Bentham 1875; Kozo-Polyansky 1922; Sokal and Sneath 1963; Løvtrup 1977; de Queiroz and Gauthier 1992; Zachos 2011].

Leaving aside a fundamental philosophical consideration of reality *vs.* nominality of taxonomic ranks as unresolvable under current circumstances, a critical question of the puzzle in question becomes more practical. It can be put as follows: does the ranked hierarchy bring anything important to the solution of some biologically meaningful tasks? A positive answer to it may go as follows. The system of fixed ranks gives classification a certain stability by serving as a kind of rigid frame that provides a specific basis for comparing taxa of different levels of generality [Simpson 1961; Vences et al. 2013]. A possible analogy here can be the Cartesian coordinate system: no one believes that such a system exists in reality—and nevertheless, it is very actively used in a variety of disciplines [Lyubarsky 2018]. Similarly, the ranked hierarchy makes it possible to bring different taxa from very distant fragments of classification into a certain relation to each other by establishing at least some "approximate" equivalence between them. Indeed, when experts speak of, say, orders of insects, mammals, monocotyledonous plants, etc., they usually mean an approximately similar (though hardly "the same") level of generality different from that of the genera.

From this point of view, the problem of the impossibility of strict rank equivalence loses its acuteness. Ranked hierarchy is used as a special technical tool: though not too precise, it nevertheless allows us to solve some scientifically significant tasks with some approximation. For example, in many ecological studies of natural communities, the species category is of particular significance: it is important for specialists to compare just the species, even if they are theoretically defined differently [Schwarz 1980]. When considering the global evolutionary dynamics of TD, the family rank is most often considered as the main reference level [Smith 1994; Sepkoski 1996]. Excluding ranked hierarchy, all these and similar studies remain without an important basis for comparison.

6.4   BETWEEN SIMILARITY AND KINSHIP

The content and structure of a taxonomic system are determined by the network of interrelations between its taxa, which are generally referred to as *taxonomic relations*. This system was defined above as an informational (cognitive) model of some manifestation of taxonomic reality, while taxa are similar models of the respective units of that reality (see Section 3.4); accordingly, taxonomic relations are also informational. To the extent that this system is meaningfully interpreted, the main purpose of these relations is to reflect somehow the real relations that exist in Nature between the units of the TD structure. Thus, the meaningful interpretation of the taxonomic system is largely determined by the meaningful interpretation of taxonomic relations shaping it; the *principle of taxonomic unity* serves as the basis for this interpretation.

In most research programs in systematics, taxonomic relations have two main substantive interpretations, *similarity* and *kinship*; in some approaches they are united by the general concept of *affinity* [Cain and Harrison 1958]. The latter ascends to its natural-philosophical interpretation applied also to, say, chemical elements, minerals, etc.; in biomorphics and partly in biosystematics, ecological relations are added to them. At an operational level, only *similarity relations* (similarity + difference) are actually studied as a kind of "primary" ones. Indeed, all that a researcher actually deals with is a sample of specimens with characters describing them. A researcher compares these specimens by these characters and infers similarity relations between them. On this basis, all other relations (kinship, etc.) are deduced as particular interpretations "superimposed" on similarities.

Proceeding from this argumentation scheme, ideologists of the positivist concept of systematics (and empiricism in general) insist that similarity relations should not be interpreted in any speculative way, so the classifications should be limited to this operational basic level. The rationale for this position is the conviction that it is only the similarity that is an observable manifestation of the affinity and that can therefore be considered "objective" [Gilmour 1940; Sokal and Sneath 1963; Colless 1967]; typologists and some evolutionists agree with them in this respect [Simpson 1961; Meyen 1977, 1978; Lyubarsky 1996; Epshtein 2003]. In contrast, other relations that can be taken into account in taxonomic research, including kinship, are not directly observable, so they are "conjectural" and therefore "subjective" to more or less degree. Therefore, classifications based on similarity relations are "objective," while all others are "subjective." However, for phylogeneticists, it is the kinship (genealogical) relation that is an immanent attribute of living nature and thus "objective"; the basis for this is an assumption that phylogeny is a real process that produces kinship network between species and monophyletic groups [Naef 1919; Zimmermann 1934; Hennig 1966; Wiley 1981; Pavlinov 2005].

The question of objectivity *vs.* subjectivity of similarity and kinship relations between organisms obviously belongs to the sphere of philosophy. It constitutes the *similarity vs. kinship puzzle* determined by the ambiguity of interpretations of both these relations themselves and interconnections between them. This puzzle is one of the most fundamental in systematics; it ascends to those times when the above-mentioned natural-philosophical affinity was considered as a fundamentum for elaborating essentialist classifications.

The *similarity relation* is quite paradoxical in a sense. It actually underlies all cognitive activity, including classifying [Tversky 1977; Quine 1996]; as noted above, it is usually considered by empiricists as "observable" and (literally) "obvious." However, if considered not from a common-sense standpoint, but as one of the key elements of the conceptual space systematics deals with, it loses any sign of being "obvious." To paraphrase Aurelius Augustine's reflection about time, it can be said: as long as I don't think about what similarity is, everything is clear about it; as soon as I begin thinking about it, my understanding ceases to be clear.

The paradoxicality of similarity begins with the absence of a clearly expressed onto-epistemic status that can be attributed to it. As was just noted, in approaches based on the empirical philosophy of science, it is considered objective and observable. However, from the point of view of conceptualism, this is not the case: "as philosophers have long realized, similarity without theory is empty" [Sober 1984: 336];

there are indeed quite a lot of such philosophers and philosophically minded biologists [Goodman 1972; Tversky 1977; Dupré 1993; Murphy 2002; Rieppel and Kearney 2002; Bartlett 2015; Pavlinov 2018; etc.]. In our case, one should speak not so much about a certain PTT, but rather about the conceptual space that configures the cognitive situation in systematics. Within its framework, a system of preferences is formed, which determines the selective pattern of an "observation": an observer "sees" only what is necessary to develop a sought classification. This means that a researcher does not simply "observe" the similarity itself in all its suchness, but rather compares objects selectively by certain characters within the context of a certain research task. Therefore, this "observing" is actually a pretty sophisticated specific cognitive act, and "similarity" appears as a result of a researcher's value judgment about commonality of certain characters in the objects being compared [Tversky 1977].

Thus, one may conclude (paradoxically enough) that similarity relation as such does not exist outside and apart from a specific cognitive situation that includes an obligatory observer [Goodman 1972; Tversky 1977]. This means that the similarity is not a "physical" attribute of aggregates of organisms connecting them by their direct or indirect interactions, therefore it is hardly possible to think of it as an "objective" in a traditional sense. Instead, it is an informational (essentially logical) relation imposed by a researcher according to the following cognitive model. To infer this relation, a researcher first observes and describes the compared objects using certain properties (characters) and thus turns these objects into their descriptive (information) models best suited to the researcher's task. It is these models, and not the objects themselves, that are actually compared based on the chosen characters, and an intellectual result of this mental (or machinery, or other) comparison is what is routinely called "similarity," or more correctly, similarity relation. And finally, the compared objects are classified based on this relation into groups to fulfill the general principle of taxonomic unity, i.e., to maximize similarities within each group and to minimize them between groups.

This conclusion is all the truer in cases where quantitative methods are used to assess similarity in numerical taxonomic research. As noted in Section 5.3, there are different quantitative measures of similarity relations providing their different estimates and, accordingly, different classifications. It was emphasized in the same section that there is nothing "objective" in these measures: they are invented by people to solve particular research tasks. So there is nothing especially "objective" in the classifications based on such estimates of similarity: they are just "intersubjective."

The *kinship relations* between organisms are generally defined as *relations by origin*: groups of organisms are kins if they are interconnected by the unity of origin. These groups and the kinship relations between them arise as a result of objective evolutionary processes: ancestors give birth to descendants, and they diverge over time, so that relations with both the ancestral form and among themselves gradually decrease. One can conclude from this that the kinship relations connecting the ancestors and their descendants exist "objectively" in a rather strict sense—that is, outside and apart from an observer. The opposite position consists in denying the reality (objectivity) of kinship: the "monophyletic groups [...] are not material systems, because no processes take place between the parts of a monophylum [...]

Obviously, monophyles are units of our thinking, but they are not coherent objects or systems of nature" [Wägele 2005: 71]. This thesis is certainly true if addressed to the judgments inferred by researchers as phylogenetic hypotheses; it is certainly wrong if addressed to the structural units of the phylogenetic pattern produced by the phylogenetic process. It denies the fact that the information flow between living organisms, including those between ancestors and their descendants, is no less "material" and thus "objective" than energy and matter flow between them [Brooks and Wiley 1986].

Does this conclusion imply that classifications based on kinship relations are more "objective" than those based on similarity relations as such? The answer is not clear cut. The point is that the phylogenetic chains of ancestors and descendants are not directly observable themselves—they are conjectured; moreover, in contrast to similarity, the researcher does not have any observable characteristics of objects that would directly indicate these chains. As a matter of fact, a certain value judgment about kinship is inferred in a certain cognitive situation based on certain empirical data as its indirect evidence. An extra element of subjectivity is added in that the definitions of kinship are different in different versions of phylogenetics. Therefore, kinship relation, though supposedly objective, cannot be "observed" and "measured" as such in order to place organisms along the gradient "less related–more related."

In this regard, a question arises: what is the basis for an indirect judgment about kinship and for any of its estimates? The answer is quite obvious this time: it is the similarity. This conclusion is deduced from the quasi-axiom of inherited similarity, which is fundamental for phylogenetic systematics: it states that, on a large scale and without going into detail, the greater the kinship of organisms, the greater their similarity. The *principle of similarity–kinship correspondence* is derived from it; this asserts (by a "reverse" reading of the quasi-axiom): the more similar organisms are, the closer their kinship is. And since the similarity is measurable, the kinship also becomes indirectly measurable.

But, as is usually said, "the devil is in the detail": the latter have to be dealt with in all those cases where the correspondence principle is applied in practice. These details are as follows.

Firstly, a fundamental difference between kinship and similarity must be taken into account. The similarity logically groups organisms into subsets, while the kinship really links the parts of the whole—the genealogical lineages of the evolving biota [Woodger 1952; Griffiths 1974a; Panova and Schreyder 1975; Mahner and Bunge 1997; Knox 1998; Pavlinov 2005, 2018; Wägele 2005; Lyubarsky 2018]. Thus the similarity is a kind of taxonomic relations, while the kinship is a kind of partonomic relations [Tversky 1989; Pavlinov 2005]. So, from the epistemology point of view, a transition from similarity to kinship is not a trivial task—there is a certain "logical gap" in it.

At an operational level, several specific uncertainties contribute to this non-triviality. One of them is that there are a lot of characters by which organisms can be compared; they are not strongly correlated with each other, therefore comparisons based on different characters provide different similarity relations. Another is that there are many ways to "measure" similarity, with different numerical approaches providing different estimates of similarity for the same set of characters. In turn, there are several interpretations of kinship relations, for each of which certain most fitting similarity estimates are to be adjusted. As a result, the above similarity–kinship correspondence is *not strict*: roughly speaking, no particular similarities are indicative of kinship. Or, if we take into account that a judgment about similarity is based on a researcher's judgment about characters in common to the groups compared, it turns out to be as follows: not every commonality in characters indicates a close kinship. This leads to the acknowledgment of the critical importance of character selection in order to obtain the most reliable indirect evidence of kinship relations.

The general argumentation scheme, which establishes a certain correspondence between similarity and kinship relations, can be imagined as follows [Pavlinov 2018].

In a cognitive situation, similarity and kinship actually appear "on an equal footing" in a form of value judgment: they are about the commonality of characters in the case of similarity and about the commonality of origin in the case of kinship. A formal basis for judging kinship by similarity is that they both are relations that link objects in the form of a ternary judgment: this means that, in both cases, two objects are more closely related to each other than each of them to a third object [Nelson and Platnick 1991]. An ontic basis for this judgment is shaped by reference to a certain model of the evolutionary process: it generates the diversity of organisms in a way that endows these two kinds of relations with some analogous properties and provides a certain objective correspondence between them. This correspondence is not strict; therefore, at an epistemic level, there is always more or less uncertainty in the similarity–kinship relationship. The main goal of phylogenetic research is to minimize this uncertainty as much as possible. This aim is achieved by choosing such characters in which similarity makes it possible to judge kinship most reliably. As a result, the whole scheme for the elaboration of the kinship-based classifications looks like this: (a) development of an evolutionary model as the basis of particular definitions of both similarity and kinship and interrelations between them; (b) selection of the characters as the most reliable indirect evidence of kinship; (c) assessment of similarity in these characters; (d) inference of a kinship relations network based on the similarity assessed; and (e) inference of classification from this network.

Thus, the problematic content of the similarity *vs.* kinship puzzle can be reduced at an operational level to two questions: (a) how the characters should be selected (weighted) so that (b) similarity between organisms in these characters would most reliably reflect the latter's kinship. Both these questions and their possible answers are formulated by certain PTTs: each of them determines in its specific way how to interpret similarity and kinship, which assessment of similarity is most consistent with a particular kinship definition and, accordingly, which characters are to be selected in order to obtain the required similarity assessment. The most striking example is the different weighting of homologs and analogs in phylogenetics: similarity in the former is considered evidence of kinship (inherited from a common ancestor), whereas similarity in the latter is not (a result of convergent evolution). In cladistics, in addition to this, monophyly in its "narrow" sense is evidenced by synapomorphies but not by symplesiomorphies (see Section 5.7.3). Another weighting approach involves not particular characters but particular sets thereof: these may be the largest set of mutually most compatible characters or a set with the largest possible number of characters that provides for overall similarity as the best measure of kinship (the above-mentioned total evidence).

6.5   WHAT IS THE (ARCHE)TYPE?

The notion of *type* in its general meaning is very broad: it appears in various scientific disciplines in quite different hypostases. This makes this notion very ambiguous, including many different meanings, and this is partly reflected in a rather rich terminology. This diversity provides a kind of *(arche)type puzzle*, which is studied by *metatypology*.

The main content of the *type concept* is determined by recognition of a common in particulars, a unity in the plural, a constant in what is transient [Voigt 1973; Vasilieva 1989b, 2003; Lyubarsky 1996; Mahner and Bunge 1997; Filatov et al. 2007; Plotnikov 2010; Pavlinov 2018]. This concept appears in Ancient times in two main versions that treat a type as either a specific *standard* of comparison or an ideal *prototype* of the real forms; the second interpretation is reinforced by the notion of *archetype* as a "primary," "original" for these forms [Koort 1936; Hammen 1981; Card 1996]. In Medieval theology, *divine archetypes* are thought of as prototypes of all things; this understanding permeates all natural theology and passes from it into natural philosophy [Amundson 2005]. In this natural-philosophical sense, the type is likened to a law of Nature: as the zoologist A. Naef says, "organisms relate to the type in the same way as events relate to the law they manifest" [Naef 1919: 7]. From this point of view, both the type and natural law (in its physicalist understanding) are quite comparable fundamental attributes of Nature: they are not directly observable but conjectural, and yet they are completely "material" as natural phenomena [Schindewolf 1969; Mahner and Bunge 1997]. Thus understood, the general concept of type is necessarily embedded in any scientific discipline and, in particular, serves as one of the organizing principles of "the biological way of thought" [Beckner 1959; Grene 1974; Bunge 1979; Richards 1992; Lyubarsky 1996; Amundson 2005; Shatalkin 2012; Pozdnyakov 2015].

From a more concrete and still quite natural-philosophical understanding, the type represents a certain generalized structural characteristic of an organism considered in an equally generalized (idealized) form. Such a type expresses a general body plan (Cuvier) or a developmental plan (Baer) or metamorphosis of parts of an archetype (Goethe); it belongs to partonomy and can be designated as the *organismic type*. With this understanding, the (arche)type concept is applied to any organismal attributes (structures, functions, processes) the diversity of which can be represented as an ordered unity of certain constituents. It played a key role in the initial development of the homology concept, without which no taxonomic (and indeed no comparative) study is possible (see Section 6.6).

Thus "quasi-ontologically" understood, the (arche)type is characterized by a rather complex structure, which is adequate for the structure of the organismal disparity that it is designed to generalize. In fairly advanced typologies, it appears as a hierarchically ("vertically") organized whole of parts (partons) of different levels of generality [Kälin 1945; Kaspar 1977; Vasilieva 1992, 1998, 2003; Beklemishev 1994;

Shatalkin 1994, 2012; Lyubarsky 1996]. Another manifestation of its complexity is its differentiation by "horizontal": less typical and more typical manifestations can be distinguished in it. Accordingly, the general archetype (*alpha archetype*) possesses its core (*beta archetype*) and periphery shaped by a set of *styles* [Lyubarsky 1996]. Thus, it makes sense to ascribe fuzzy character to the (arche)type understood in this way [Sattler 1996; Hermon and Niccolucci 2002; Pavlinov 2018].

From an epistemic standpoint, the type can be considered a specific informational (cognitive) model of a group of organisms capturing its certain unity and/or specificity according to a certain chosen basis of comparison [Jurgens and Vogel 1965; Voigt 1973; Kaspar 1977]. For this, the *principle of typicality* is introduced, which presumes that the type as a representation must have a certain special quality, a *typehood*: it is the latter that allows peculiar properties of a group of organisms to be represented by its (arche) type [Hermon and Niccolucci 2002; Murphy 2002]. Such (arche)type may most appropriately be called a *representational type* [Love 2009; Pavlinov and Lyubarsky 2011; Moskvitin 2013; Pavlinov 2018]; this conceptualist interpretation is widely adopted in studies based on the typological method as it is generally understood [Weber 1904; Hempel 1965; Bertalanffy 1968; Voigt 1973; Oderberg 2009]. This is a mental construct, so it is an *ideal type* [Weber 1904] that appears in a cognitive situation as a derivative of all three of its basic components—ontic, epistemic, and subjective [Pavlinov 2018]. The latter's selective cognitive activity yields a *constructed type* depending on the choice of certain characters to resolve a particular research task set by a particular researcher [Becker 1940; Bailey 1992]. The above typicality can be understood in different more particular ways. On the one hand, the type can express the most characteristic (literally typical) properties of a group of organisms, so it is a *central type* [Remane 1956]; it summarizes the disparity of this group, so it is a *synthetic type* [Smirnov 1925]. On the other hand, the type may reflect a certain extreme property of a group; this is its *extreme type* [Hempel 1965; Sattler 1996; Filatov et al. 2007].

Most pertinent to the particular tasks of systematics seems to be two basic versions of the (arche)type concept elaborated by the respective versions of typology; these are dynamic and stationary ones (see Section 2.4.3) [Pavlinov 2018]. When considered at an organismal level, they are generalized by the above-mentioned organismic (arche)type as a totality of both structural and transformation relations between partons (parts, organs, etc.) of a generalized (idealized) organism [Meyen 1973, 1978, Kaspar 1977; Lyubarsky 1996]. For the purposes of systematics, these versions are summarized by the concept of *classification type* as a totality of the typical features of a taxon which determines relations between its members through their relations to its type. The organismic (arche)type serves as the basis for recognizing general and serial homologs, while the classification type allows special homologs to be recognized.

If the historical time dimension is included in the characterization of the (arche) type, it receives an evolutionary interpretation [Hammen 1981; Stevens 1984b]. According to this, a dynamic supra-organismic (arche)type can be considered as a canalized trajectory of the historical development of a complex biological structure, which is called *phylocreod* [Waddington 1962; Meyen 1984; Wagner and Stadler 2003; Pavlinov 2005, 2018]. An evolutionary shift from one to another phylocreod is caused by a critical reorganization of the "design" (*phylotype*) of this structure [Sander 1983; Slack et al. 1993; Hall 1996; Richardson et al. 1998]; such reorganizations provide evolutionary-typological definitions of the respective taxa.

A variant of the classification type, understood in quite a simplified form, is an *empirical type*—a set of diagnostic characters used to identify a certain group of organisms [Read 1974; Klein 1991]. This group can be recognized as a taxon in its typological understanding, as a monophyletic group, as a life form, etc., so its empirical type depends completely on its prior context-dependent definition. For instance, in cladistics, a set of synapomorphies distinguishing a clade can be considered its empirical type [Nelson 1979; Pavlinov 1990; Shatalkin 1994]. Empirical typology, based on the use of quantitative methods, recognizes *computed type* defined as a sample centroid [Smirnov 1924, 1925]. In its broadest interpretation, empirical (arche) type is suggested to encompass all characters of a group, including ecological ones [Meyen and Shreyder 1976]. It seems that such an "all-encompassing" interpretation of (arche)type deprives it of its original meaning and brings typology developing this version closer to classification phenetics.

Finally, mention should be made of a purely applied concept of the type as a concrete physical object taken as an *example* of a certain set of objects. In systematics, this variant corresponds to the *collection type*; the Codes of Nomenclature assign reference function to it and thus make it the *nomenclature type* [Farber 1976; Ogilvie 2006; Pavlinov 2015].

The epistemic significance of the (arche)type concept is undeniable: it is a necessary element of the synthetic phase of the cognitive activity, allowing the studied phenomena to be presented in a generalized form. An ontical aspect of this concept is seriously criticized as being groundless: it is indicated that there is nothing in Nature that would correspond to an (arche)type from any understanding. This criticism is based on an extremely simplified understanding of type, which is particularly characteristic of empiricism. The general concept of contemporary essentialism [Ellis 2001; Walsh 2006; Oderberg 2009; Rieppel 2010b] makes many of the negative assessments addressed to this concept untenable: it implies that certain natural phenomena are indeed endowed with some specific, fairly stable emergent features that may be denoted as their (arche)types. Such a standpoint presumes that the (arche) type puzzle can be reasonably resolved in a pluralistic manner only.

6.6 HOMOLOGY, AN UNRESOLVED PROBLEM

Let us recall that the taxonomic reality studied by systematics is divided at its basic level into two general aspects, viz. taxonomic and partonomic. In the former, diversity of organisms is investigated by the totality of their characters; this diversity is ordered by grouping organisms into taxa and building taxonomic systems. In the latter, disparity of the properties of organisms is investigated, and this is ordered by grouping the organismal properties into partons (= merons) and building partonomic systems. The "puzzle" issues concerning the taxonomic aspect were considered in previous sections of this chapter; an issue of this kind is considered here concerning the partonomic aspect. This consideration is facilitated by the fact that the problems associated with the structure of taxonomic and partonomic aspects of the diversity of organisms is similar in many ways.

Each parton represents a class of equivalence of certain organismal properties considered indistinguishable with reference to the respective class-forming parameter. As units of partonomic classifications, the partons can be of different kinds. In systematics and related disciplines, of paramount importance is recognition of their status as either natural or artificial; in one of the most commonly used traditional terms they are designated as *homologies* and *analogies*. The procedure of separating partons is *partonomization*; accordingly, two components are distinguished in it, viz. *homologization* and *analogization*.

Without going into detail, the basis for a general classic understanding of the homologies and analogies can be presented as follows. The homologies are formed by certain "internal" causes of an "essential" kind: they can be interpreted as manifestations of the same essences of organisms, as elements of their general body plans or as derivatives of a certain archetypal (ancestral) structure. In contrast to this, the analogies are formed by the "external" causes which are accidental with respect to the organisms: the influence of environmental conditions is most usually meant, and consideration aspects fixed by a researcher also belong to this class of causes. Thus, homologies express a certain "deep" affinity of partons, while analogies reflect their "superficial" similarity. This explains the great attention that is traditionally paid to the task of revealing homologies and distinguishing them from analogies in solving particularly both taxonomic and partonomic tasks.

For systematics, the significance of partonomization is determined by the fact that partons serve as the common basis for identifying characters by which organisms are described, compared, and classified. In the simplest version, there is a one-to-one correspondence between them: one parton corresponds to one character, and elements of this parton correspond to the modalities of the respective character. In a more complex variant, characters may correspond to several partons—for example, describing the ratio of different parts of an organism. From a formal point of view, all partons are equivalent as the bases for distinguishing characters; with this, homologies yield *homologous characters*, while analogies yield *analogous characters*.

Such a differentiated evaluation of the homologies and analogies and characters associated with them is rooted in the natural-philosophical understanding of the Natural System as a network of the natural affinity of organisms manifested in their "essential similarity" by "essential characters." This affinity was understood either natural-philosophically (scholastics, taxonomic "esotericism") or genealogically (Darwin, Haeckel), while the "essential characters" became understood as homologies, in contrast to the "accidental characters" equated with analogies. In contemporary taxonomic schools of thought, such a distinction is inherited by typology, phylogenetics, and partly phenetics. However, in biomorphics and partly biosystematics, their assessment is different: analogies may be of greater importance than homologies for identifying life forms.

Contemporary understandings of the ways of defining the *homology concept* in its general meaning are so diverse that there seems to be no sign of a common agreement: this gave the zoologist Gavin de Beer reason to declare that "homology is an unresolved problem in biology" [de Beer 1971]. This *homology problem*, which is comparable in its importance and unresolvedness with the species problem (discussed in Section 6.7), shapes the *homology puzzle* in systematics. Its main content, as in the case of the species problem, lies in the impossibility of defining homology in a trivial unified way [Bock 1989; Brigandt 2002; Hossfeld and Olsson 2005; Jamniczky 2005; Kleisner 2007; Pavlinov 2012, 2018; Pavlinov and Lyubarsky 2011; Minelli 2016]. Several special books are devoted to this problem [de Beer 1971; Voigt 1973; Hall 1994; Sanderson and Hufford 1996; Bock and Cardew 1999; Timonin 2001; Wagner 2014].

The general metaphysical context of an understanding of the partonomic structure of organisms, and thus the homology problem, is provided by a hierarchical whole–part relationship [Woodger 1952; Jardine 1967; Bertalanffy 1968; Ghiselin 1981, 2005; Rieppel 1988b, 1992; Lyubarsky 1996; Pavlinov 2012, 2018]. It assumes (a) the subdivision of a whole (organism, archetype) into parts and the existence of certain relations between both (b) different parts of the same whole and (c) the respective parts belonging to different wholes, with the latter (d) being, in their turn, parts of a higher-order whole.

The beginning of the contemporary concepts of homology and analogy was laid by the zoologist Richard Owen [Hall 1994] (see Section 2.4.4). His concept is based on an idea of the *archetype* (partly in the sense of Goethe) as a general principle of the structural organization of an ideal super-organism, which is partitioned into basic structural elements, *homotypes* [Owen 1848]. Owen defined homology as the correspondence of elements of the same homotype, which made them "the same," and analogy as the absence of such correspondence. This general scheme, as applied to particular organisms, is detailed in three ways: (a) *general homology* is a correspondence of the main structural elements (parts, organs, etc.) in a given organism to archetypal homotypes; (b) *serial homology* is a correspondence of repeating elements along a segmented organism that realize sequentially the same homotype; and (c) *special homology* is a correspondence of structural elements in different organisms that realize the same archetypal homotype. Taking the vertebrate body plan as an instance, we have: its different parts (fins, limbs, glands, blood vessels, etc.) as different general homologs, paired limbs as serial homologs, and variants of the forelimb in fishes, birds, and mammals as special homologs.

According to Owen, whose natural philosophy was close to "Biblical Platonism" (see Section 2.4.4), an ideal archetype is single and all-encompassing for a certain group of organisms, therefore distinguishing its structural elements (partons) as homologs and analogs is also the only possible one. The Natural System of organisms is a result of different modes of realizing this ideal archetype and the embodiment of its homotypes into the corresponding special homologs of particular organisms. This implies the above-mentioned priority status of the homology: Owen himself paid most attention to homology, while he lost interest in analogy. Such an attitude was inherited by most of the explorers of the homology problem: it is believed that all comparative morphology is the "science of homology" [Remane 1956].

Owen's typological concept subsequently underwent significant changes caused by its different interpretation. The phylogenetic one gave an idea of *phylogenetic homology* and led to dividing it into inherited from ancestors (*homophyly* of Haeckel, *homogeny* of Lancaster, *complete homology* of Gegenbaur, *genetic homology* of E. Wilson) and acquired as a result of parallel evolution (*homoplasy* of Lancaster, *incomplete homology* of Gegenbaur, *latent homology* of Osborn, *homoiology* of Plate). A special phylogenetic interpretation of anatomical correspondences was proposed by the zoologist Edward Cope, who borrowed his concepts of *homologous* and *heterologous series* from chemistry and "inscribed" them into his theory of polyphyletic evolution [Cope 1887]. Simultaneously with the phylogenetic view, an *ontogenetic* understanding of homology was developed based on the ideas of K. Baer's epigenetic typology. On this basis, a detailed interpretation of serial homologies developed during ontogenesis was elaborated [Bronn 1858; Haeckel 1866]; this general understanding was reinforced by the term *ontogenetic homology* [Mivart 1870]. As a result, by the end of the 19th century, three main concepts of homology existed, viz. typological, phylogenetic (in several versions), and ontogenetic (embryological). This fragmentation continued further: in the 1920s–1930s, up to five main "homologisms" appeared, two decades later there were about ten of them, and in two more decades their number increased to several dozen [Blacher 1976]. In phylogenetics, special attention was and is paid to the delineation of homogenies and homoplasies,[2] which turns out to be more than problematic [Lankester 1870; Osborn 1902; Spemann 1915; Söderström 1925; Hubbs 1944; Sanderson and Hufford 1996; Pavlinov 2012; Minelli and Fusco 2013].

In contemporary ideas of homologies, two main generalized interpretations can be distinguished: *structural* (stationary) and *transformational* (dynamic, generative). They differ mainly by the ontic interpretation of the nature of interrelations between homologous elements: they are united either through their structural "sameness" or through their sequential transformation, respectively. Some experts contrast these two interpretations; in an extreme version, it is considered impossible to include the structural consideration of homology in studies of transformations of biological forms [Naef 1931; Kälin 1945; Borhvardt 1988; Shatalkin 1990b]. Others attempt to consider them not mutually exclusive but complementary and to combine them in one way or another [Rieppel 1985, 1988b; Brigandt and Griffiths 2007; Pavlinov 2012, 2018]. This intention is fixed by *biological* or *synthetic* concepts of homology: the latter is interpreted as a canalized development of structural correspondences [Wagner 1989; Szucsich and Wirkner 2007].

An emphasis on the transformational aspect suggests an important transition from the traditional homology of definitive structures to the homology of the processes [Bertalanffy 1968; Laubichler 2000; Gilbert and Bolker 2001; Scholtz 2005, 2010; Minelli and Fusco 2013]. In this case, the homology of structures can be correctly defined in terms of certain processes that generate them: this approach leads to the recognition of partons as *process homologs* [Hall 1992, 1995, 1996; Gilbert and Bolker 2001; Minelli 2003; Kleisner 2007]. Extending homology to include both developmental processes and structures as their results allows the introduction of the general

---

[2]   In contemporary cladistics, homogeny is incorrectly identified with the whole homology, thus ignoring other particular interpretations of the latter.

concept of *organizational homology* as a correspondence between morphoprocesses and morphostructures [Müller 2003; Kleisner 2007]. Two types of process that generate such homologies have been considered since the end of the 19th century, phylogeny and ontogeny. Homologies defined by them are sometimes contrasted and considered incompatible [Wagner 1989, 1994; Shatalkin 1990b; Rieppel 1992]. More promising seems to be their joint consideration within the framework of the general concept of "evo-devo," which gives an understanding of a homolog as a structural unit capable of separate evolutionary development due to its ontogenetic (quasi)autonomy [Wagner 1994, 2014; Laubichler 2000; Amundson 2005; Brigandt 2007; Suzuki and Tanaka 2017]. The correspondence of definitive structures resulting from either the same or different generative (both ontogenetic and evolutionary) pathways was proposed to designate *syngeny* and *allogeny*, respectively [Butler and Saidel 2000].

In the most recent studies on process homology, special emphasis is given to the regulator genes (*Hox*, *MADS*, etc.) that affect the formation of basic morphostructures in the early stages of ontogenesis [Holland et al. 1996; Schierwater and Kuhn 1998; Galis 1999; Shatalkin 2003, 2012; Scholtz 2005; Davis 2013]. Accordingly, it was proposed that *genetic homology* should be recognized as a special category [Hossfeld and Olsson 2005]. It is assumed that homology of these genes in animals makes it possible to homologize structures traditionally considered paradigmatic analogs, such as wings in insects and birds [Shatalkin 2003]. Taking a less optimistic view, especially taking into account that *MADS* genes are also found in plants [Niklas 1997; Ng and Yanofsky 2001), things are not as simple and unambiguous as they seem. It has been shown that homologous genes are responsible for non-homologous (in a traditional sense) morphostructures, while homologous (in the same sense) morphostructures can be regulated by non-homologous genes [Striedter and Northcutt 1991; Wray and Abouheif 1998; Wray 1999].

At a molecular level of organization, there are specific subtleties in the understanding of homology: in this case, a distinction between *orthology* and *paralogy* is of importance [Williams 1993; Hillis 1994; Wheeler 2001, Sonnhammer and Koonin 2002; Freudenstein 2005; Wagner 2007, 2014]. Orthology is a kind of special homology that corresponds to the homogeny of morphological structures: orthologous genes are regions of the same macromolecule in two organisms (groups of organisms) inherited from their closest ancestor. Paralogous genes in molecular biology (as opposed to morphology; see above) are duplicated regions; if they occur within the same organism, they correspond to the serial homology; those in different phyletic lineages superficially correspond to homoplasy. In addition, *xenology* is distinguished to designate alien DNA or RNA fragments included in the genome as a result of horizontal transfer [Gogarten 1994].

Modern research is characterized by an expansion of the application of the general concept of homology to include consideration of functions of organisms [Roth 1982, 1988, 1991; Wake 1992; Gilbert and Bolker 2001; Love 2007; Matthen 2007; Tetenyi 2013]. The following examples are worth mentioning here: biochemical reactions constituting the Krebs cycle; viviparity that repeatedly appeared in the evolution of animals; and behavioral stereotypes. Such interpretations bring specific problems in the homology puzzle: for example, in the case of behavioral stereotypes, it is not clear whether they should be considered conjointly with the morphological structures that perform them or independently of these structures.

An important part of the modern content of the homology puzzle, in contrast to its classical versions, is an understanding that distinctions between homologies and analogies are context-dependent and thus relative. These contexts may be set conceptually by different definitions of homology (which is self-evident); by higher-order hypotheses, within which the respective structures and functions are homologized; by specific aspects considering the structural and functional organization of biological objects; etc. [Roth 1988, 1991; de Pinna 1991; Brigandt 2002; Pavlinov 2012, 2018]. For instance, in phylogenetics, a distinction between homogenies and homoplasies depends on the contexts of particular phylogenetic hypotheses [Hennig 1966; Wiley 1981; Pavlinov 2005]. In the "new typology," homologies and analogies are distinguished cognitively depending on a particular research task [Meyen and Shreyder 1976; Meyen 1978; Lyubarsky 1996].

Another important aspect of the relative character of the boundaries between homologies and analogies is determined by the recognition of their quantitative and fuzzy character. This means that, instead of their strict delineation, it may be more correct to consider varying "degrees of homology/analogy," according to which there may be *complete* and *partial* homologies [Gegenbaur 1865]. This is most characteristic, for example, of ontogenetic interpretations: the more similar developmental trajectories are, the more homologous are both corresponding morphoprocesses themselves and morphostructures generated by them [Sattler 1994; Minelli 2003]. Specific methods for determining homologies in numerical taxonomy make them strictly quantitative: the mutual similarity of the structures is considered to be a measure of the degree of their homology [Smirnov 1959; Sneath and Sokal 1973]. The same is true for the homologization of macromolecules through alignment and assessment of the similarity of their sequences [Hillis 1994; Doyle and Davis 1998; Wheeler 2001, 2016; Morrison et al. 2015]. A quantitative character distinguishing between homologies and analogies also occurs when considering judgments of them as hypotheses that may be more or less plausible.

An important part of the homology puzzle is an elaboration of *homology criteria* for solving the particular tasks of homologization. For anatomical structures, a sufficiently consistent system of criteria is developed on a typological basis [Remane 1956], which goes back to the theory of analogs by E. Geoffroy de Saint-Hilaire [Geoffroy Saint-Hilaire 1830]. This includes three main criteria: of special quality, of position, and of intermediate forms. The first two are primary and allow distinction between two descriptive forms of structural homology, *compositional* and *positional*, respectively [Jardine 1967; Sluys 1996; Minelli and Fusco 2013]; the third is secondary to them. According to the *criterion of special quality*, structures are homologous if they coincide in some essential "internal" properties (for instance, tissue composition). The *criterion of position* (connectivity) implies that structures in different organisms are homologous if they occupy the same position among the same structures that have already been proved to be homologous. The *criterion of transitional forms* (continuity) is addressed if two previous criteria do not provide quite clear recognition of homology; it is especially relevant for the transformational homology concept. An auxiliary *criterion of congruence* gives preference to the hypothesis of homology of a structure that is more consistent with other hypotheses of homologies of other structures within the context of a particular higher-order (phylogenetic, etc.) hypothesis [Patterson 1982; de Pinna 1991]; this is a kind of logical judgment by analogy (see Section 3.6 on the latter).

Currently, several ideas of how to solve the homology problem and puzzle are considered, and these ideas can be divided between two opposites. One of them presumes rejection of the very general concept of homology, inventing different particular concepts and corresponding terms [Borhvardt 1988; Shatalkin 1990b]; this position goes back to early attempts to interpret homology in a phylogenetic manner (see above). It is sometimes justified by reference to scientific pluralism [Heather and Jamniczky 2005; Kleisner 2007], by which the homology problem is quite comparable with the species problem [Ereshefsky 2001b; Pavlinov 2012] (see next section). On the other hand, it is proposed that we should think of a totality of structural, functional, and developmental correspondences of partons, which might be represented in the form of a complex hierarchical system integrated by the concept of *hierarchical* (deep), or *combinatorial* (factorial) homology [Minelli 1998, 2016; Scotland 2010; Minelli and Fusco 2013]. According to this, at different levels of hierarchically stratified supra- and intra-organismal diversity, different but interconnected particular concepts of homology can operate, shaping a single conceptual system of *metahomology* [Kleisner 2007]; this viewpoint agrees in general with the idea of a conceptual pyramid (see Section 3.2.3).

6.7   AN UNDISCOVERABLE ESSENCE OF SPECIES?

At the turn of the 5th to 6th centuries, the Neoplatonist Boëthius expressed his concern about the foundations of the logical generic–species scheme by declaring that "if we do not know what species is, nothing would secure us from delusion" (cited after [Boëthius 1906]). This became a kind of testament for many subsequent generations of thinkers: first were scholastics, then natural philosophers, and finally biologists. The main question that determined the content of the *species puzzle* then was this: if we designate something as species, is there any real essence hidden behind it in Nature, or is it only a nominal category? Over time, this puzzle preserved its fundamental status, though its content changed in one way or another.

In scholasticism, a nominal treatment of species seemed to dominate. Among the naturalists of the 17th and 18th centuries, a belief in the reality of species began to take hold: for example, J. Ray and after him C. Linnaeus believed that species appeared in the days of the Divine creation (see Section 2.3.2). Buffon, agreeing with them, wrote in one of his treatises that "Species are the only creatures of Nature, eternal and unchanging like itself" (cited after [Buffon 1843: 52]). At that time, the monistic concept of the overall System of Nature dominated, with one of its basic elements being just the species. And since the System of Nature was imagined to be unified, both the species unit and category were also considered real and unified for all Nature; this implied that a true understanding of species must also be unified. This general position is well reflected in the unified taxonomic species of systematics.

Starting in the second half of the 19th century, Darwinian evolutionary theory has significantly shaken (for some time, even subverted) belief in an exclusive status of the species category: the latter began to be understood as just a final stage in the gradual differentiation of local geographical races. And it was Charles Darwin who expressed an idea inserted into the title of this section: he believed that biological science had to become free "from the vain search for the undiscovered and undiscoverable essence of the term species" [Darwin 1859: 485]. Darwin's "species nihilism" gave rise to a "species crisis" [Mayr 1965; Skvortsov 1967; Zavadsky 1968] and was taken up by biosystematics of the first half of the 20th century [Turesson 1922; Lotsy 1931; Mayr 1942]. With this, the *species question* emerged, which was basically about how to distinguish between species and intraspecific categories [Britton 1908; Turrill 1925]. In the synthetic theory of evolution that emerged, speciation was recognized as a nodal point in the evolutionary process; this led to the restoration of the priority status of the species category in those branches of biology that were in one way or another oriented to that theory, including several research programs of systematics [Mayr 1965]. A kind of apotheosis of this return became several proposals to establish a special "science of the species," called *hexonomy* [Skvortsov 1967], *eidology* [Zavadsky 1968], or *eidonomy* [Dubois 2011].

Most recently, interest in species increased with the emergence of the concept of biodiversity as a natural resource that requires special study and conservation [Wilson 1988]. The reason is that, in the currently dominating noticeably simplified version of biodiversity, the species is usually given the same key place as in "Linnaean" systematics: it is considered the main "counting" unit of the diversity of organisms [Claridge et al. 1997; Sigwart 2018].

A gradual return of interest in species led to the fact that, throughout most of the 20th century, there was an active search for its "undiscoverable essence." By the end of the 1930s, it became clear that, in different groups of organisms, a structural unit commonly called species manifests itself in quite different manners. These differences were primarily associated with various mechanisms (ecological, behavioral, epigenetic, etc.) that ensured species integrity. Interpretation of species as a *singameon* [Lotsy 1931] led to the fundamentally important conclusion that different breeding systems might presume different *kinds of species* [Faegri 1937; Turrill 1938; Cain 1954]. Depending on particular mechanisms, different *species concepts* were formulated based on the general notion of species. These concepts, each with its own rationale, appeared to be scarcely generalized by any single "superconcept": that was how the *species problem* in its original version emerged [Robson 1928; Faegri 1937; Clark 1956].[3] Its further development led to the fact that a general intuitive understanding of the "natural species" (in the sense of [Kunz 2012]) was supplanted by its conceptualization and related formalizations [McOuat 2001; Pavlinov 2013a, 2018; Bartlett 2015]. Thus, *species pluralism* appeared to come into conflict with the classical *species monism*. As this problem was developing, the situation did not become clear but worsened because the species concepts got multiplied: theorists

---

[3]   Actually, this word combination appeared earlier in the sense of the "species question" (e.g., [Turesson 1922]).

"established a minor industry devoted to the production of new definitions for the term species" [de Queiroz 1988: 57]. Hundreds of articles and more than a dozen books devoted to this problem and its possible solutions appeared; among the books the most significant seem to be: [Ghiselin 1997; Wilson 1999; Wheeler and Meier 2000; Hey 2001; Stamos 2003; Wilkins 2009; Richards 2010; Pavlinov 2013d; Slater 2013; Zachos 2016].

Thus, the main reason for the ascendancy and persistence of the species problem in its traditional sense is due to the existence of different species concepts. However, it seems to be only part of this problem. The entire cognitive situation construed around the general species concept is problematic, not just because of the diversity of particular concepts as such, but rather due to a contradiction between aspiration and inability to reduce this diversity to a single species concept, be it the most general ("omnispective") or a particular "most principal" one [Pavlinov 2009b, 2013a, 2018]. So, in more general terms, this problem is caused by a *contradiction between species monism and species pluralism*: as can be assumed, it is this contradiction that accounts for the current content of the species puzzle.

Part of the latter is an "erosion" of the species category, which is reflected in the specific terminology denoting different degrees of species differentiation: hypospecies, semispecies, "almost species," allospecies, ex-conspecies, superspecies [Mayr 1965, 1969; Graybeal 1995; Dubois 2006; Mallet 2013]. As a result, the species category turned out to be vague, and it was suggested that the pool of taxonomic studies addressed to it should be called "around-species systematics" [Mikhailov 2003].

Currently, there are about two dozen particular species concepts; they differ in what is considered the key parameter in the definition of the species unit [Mayden 1997; Hey 2001; Mallet 2001; Wilkins 2009; Zachos 2016; Pavlinov 2018]. These concepts can be classified into various categories; of the "ontically motivated," two seem to be the most general—*structural* (static, synchronous) and *processual* (dynamic, diachronic) [Dobzhansky 1935; Lee and Wolsan 2002; Stamos 2003]. Besides, it is important to distinguish between theoretical and operational concepts. The following are lower-level categories to which most of the species concepts can be referred:

- species as a *unit defined by similarity*, including the following concepts: phenetic (similarity of phenotypes), genetic (similarity of genotypes), typological, as well as all concepts based on operational definitions of species units
- species as a *least reproductively isolated unit*, according to the biological concept, the self-recognition concept
- species as a *historical unit*, according to the classical generative concept ("like gives birth to like"), phylogenetic, and genealogical (including lineage) concepts
- species as an *ecological unit*, including the ecospecies concept
- species as an *ontological unit*, including understanding of species as a natural kind.

There is a number of complex concepts trying to combine several criteria, of which the *evolutionary species concept* is worth mentioning here [Simpson 1961; Wiley and Mayden 1981; Mayden 1997].

Looking at the species problem from a "non-classical" point of view does not allow us to consider the diversity of species concepts as "seditious." Each concept fixes a certain particular manifestation of a particular structural unit of taxonomic reality, which is usually called species (or "around-species"). This position legitimizes species pluralism and pledges not so much to fight against it but rather to look for its natural causes. A general prerequisite for such a search is provided by the structure of the cognitive situation in which the species problem is considered [Pavlinov 2013a, 2018].

Taking into account the hierarchical ("pyramidal") structure of the conceptual space shaping this situation (see Section 3.2.3), let us first assume that any particular species concept of a certain level of generality can be rationally defined within the context of a higher, more inclusive level of the respective conceptual pyramid. In this regard, a question inevitably arises as to how to develop some "most general" species concept that would serve as a kind of "umbrella" to integrate ("shelter") particular concepts [Reydon 2005; Yuichi 2017]; in a way it is similar to the question posed by David Hull about an "ideal" species concept [Hull 1997]. It is evident that, for such a general concept to be properly developed, it is necessary to raise the respective pyramid's highest level of generality so that it permits consideration of the general species concept within the context of a certain more general biological conception/ theory. In more formal terms, it means the need to denote such a "logical genus," by which division it would become possible to get a "logical species" containing a general definition of the natural species phenomenon biology deals with. In order to avoid the logical "genus–species" tautology, it is necessary to define, in the same, more general context ("logical genus"), some other biological units ("logical species") of the same level of generality as the species phenomenon proper, but which are not species. Such an approach is aimed at understanding what the natural species phenomenon is and how it differs from any "non-species" phenomena (such as life forms, syntaxa, guilds, etc.); otherwise, it seems to be impossible to decide positively why we think of a phenomenon as of species and not anything else.

Taking into consideration that natural "species" and "non-species" units are different elements of biota, it seems reasonable to fix the latter's certain conception at the "top" of the entire conceptual pyramid. In this respect, treating biota as a developing non-equilibrium system seems quite attractive; it allows the main emphasis to be placed on those natural causes (factors) that operate at the level of biota and structure it, as it develops and functions, thus generating and individualizing its various structural units. The joint co-action of these causes yields dynamic stability of both biota and any of its units, including species, as one of their most fundamental properties to be comprehended [Brooks and Wiley 1986].

So, to get a metaphysically consistent understanding (and eventually definition) of the general concept of species as a natural phenomenon, the entire conceptual pyramid should be construed as a descending cascade of those causes that structure biota and provide dynamic stability of species as a natural phenomenon. There are several of them (external, internal, historical, etc.) which interact in a complex manner with each other, so no one of them, taken separately, can explain the species phenomenon exhaustively. If this is so, then the desired definition (or at least understanding) of the natural "species in general" should not be reductionist but as exhaustive (omnispective) as possible to incorporate all causes ensuring its existence [Sluys 1991; Pavlinov 2013a]. On the one hand, this would provide a certain comprehension of what makes the natural "species in general" just species differing from other units of the structure of developing and functioning biota. On the other hand, this would allow clearer recognition of the particular causes responsible for the particular manifestations of the natural species reflected in the respective particular concepts [Ereshefsky 1992; Pavlinov 2018].

According to the considerations given above, attempts to find a metaphysically sound general understanding (and eventually definition) of the natural species phenomenon seems to plunge the species problem into the context of "new essentialism" [Ellis 2001; Rieppel 2010b; LaPorte 2017; Maxwell et al. 2020]. This means acknowledging that, if a species is not reducible to just a sum of its constituent organisms, it must be endowed with some kind of emergent essential property [Sober 1984, 2000]. The latter may be informally designated as a *specieshood* of whatever content, distinguishing it from any "non-species" with their own essences or "-hoods" [Pavlinov 1992, 2009b, 2013a, 2018; Griffiths 1999; Wilkins 2007; LaPorte 2017]. A part of this metaphysics is the assumption that specieshood, however conceptualized, is an integrated part of the overall natural history of organisms. It incorporates, in a certain unobvious way, particular mechanisms responsible for the dynamic stability of the particular species—their self-reproduction and mutual isolation, their place in the niche structure of ecosystems, their persistence as genealogical successions, etc. Such an understanding seemingly presumes the dependence, to a greater or lesser degree, of specieshood on other aspects of an integrated natural history of organisms. This inevitably makes it group-specific: even if we suppose that some natural species considered in general may be endowed with certain emergent properties common to all (or to the vast majority of) living beings, it may have different manifestations depending on the specific biological properties of particular groups of organisms. One of the outcomes of such group-specific specieshood manifestations is the above-mentioned existence of different "kinds of species" associated with different breeding systems.

From this, the assumption follows that specieshood manifestations change together with other biological properties of organisms in the course of the evolutionary development of the functional and structural organization of biota in general and its various elements in particular. So we have here something like an "evolving essence" which might be pertinent to so-called "historical essentialism" [Griffiths 1999; Pedroso 2012; Maxwell et al. 2020]. More particularly, this means the supposition that biological mechanisms responsible for the dynamic stability of the species and constituting their respective specieshoods were quite loose (fuzzy) at the beginning of evolution, which resulted in *quasispecies* or *pseudospecies* in viruses, prokaryotes, and lower eukaryots [Dobzhansky 1970; Eigen 1983; Nowak 1992; Cohan 2001, 2002; Domingo 2002; Stamos 2003; Hanage et al. 2005; Wilkins 2006; Pavlinov 2013a; Andino and Domingo 2015]. Finally, those mechanisms shaping specieshood became more perfect in more advanced organisms, thus making their species more cohesive and discrete, so they became *euspecies* ("real species") described by the reproductive (or self-recognition) species concept [Dobzhansky 1970; Eldredge 1985; Pavlinov 2013a]. Thus, it might be an important task of comparative eidology/eidonomy to study the distribution of various kinds of species, with their specific specieshoods, over groups of organisms with different natural histories to reveal the causal intercorrelations of these "-hoods"and "histories."

Were it to appear possible, contrary to Darwin, to discover a certain essence of species ("specieshood"), the probable theoretical verdict regarding the species problem might be as follows: the natural (biological in its most general sense) species as a natural phenomenon manifested itself in a universal unit of the structure of biota of a certain level of generality most likely exists, but it is implemented in different ways in different groups of organisms with different natural histories. This will serve as justification for the use of a unified *taxonomic species concept* in systematics and, at the same time, a certain motivation for a more thorough elaboration of different species concepts. Otherwise, it will be necessary to acknowledge that there is no such universal natural unit in its general meaning; instead, there are actually different natural phenomena (units), each with its own essence, inherent in different groups of organisms. And systematics will most possibly be obliged to respond appropriately to such a conclusion by inventing local classification units to replace the currently used universal one (as a legacy of scholastics) in order to correspond better to what occurs "in reality" [Dubois 2011; Nathan 2017].

The latter position corresponds to the proposal to exclude the general species concept from the biological thesaurus, since it is both outdated and cannot be rigorously and unambiguously defined [Burma 1954; Michener 1962; Sokal and Sneath 1963; Riddle and Hafner 1999; Hendry et al. 2000; Kober 2008]. However, when considering such a far-reaching suggestion, one should take into consideration that both the general species concept and notion are sturdily embedded in the thesaurus of many fundamental and applied biological disciplines. So, realization of this suggestion may lead to a substantial reorganization of a large part of the conceptual space of all of biology. The reason is quite obvious; such a rejection would necessarily entail rejection or serious correction of other concepts and notions that are associated, in one way or another, with the notion of species.

For instance, in ecology, this notion is not essential for describing the structure of the local ecosystems, as far as such descriptions are based on biomorphs [Krivolutsky 1971; Chernov 1991]. However, in comparative analyses of different ecosystems, there is the evident need for a certain unified basis of comparison (the above "denominator") that would allow the biomorphs recognized within each of the local ecosystems to be linked to each other. It happens that currently it is the species that fulfills such a function; as a matter of fact, for the evolutionary ecologists, biomorphs exist in local ecosystems not by themselves but as manifestations of local populations of widespread species [Schwarz 1980].

This particular aspect of the species puzzle is not only a consequence of conservatism of the conceptual apparatus of biology but also reflects one of the universally valid epistemological principles. According to the latter, in order to explore any differences between objects, one must have some more or less solid unified basis of comparison making these objects components of a certain unity (elements of the same class, tokens of the same natural kind, etc.) by possessing some fundamental feature(s) in common. For many research tasks in biology, such a currently acknowledged basis refers to *conspecificity*, i.e., to organisms, differing from each other in some details, belonging to the same species as a real existing natural unit. Such reference function of species faces the serious problem of the substantive incommensurability of the species units recognized in biologically different groups of organisms, with their "fuzzy" treatment providing a rather weak solution to this problem. And yet, it is clear from this perspective that, in order to get rid of the notion of species in biology, it would be necessary to introduce some other no less general notion as a solid unified basis of comparison, substantiating such a replacement by reference to some biologically meaningful and sufficiently comprehensive theory. With this, it should be kept in mind that any new concept replacing that of species will certainly face the same ontic and epistemic limitations and pitfalls outlined in the philosophical chapter of the book: ontic reduction, incompleteness, fuzziness, etc. So the species problem will not actually be "closed," but will simply turn into another, no less challenging, problem.

References

Abachiev, S. K. 2004. [*Evolutionary theory of knowledge. An experience of systemic construction.*] Moscow: Editorial URSS. ISBN 978-5354007240 (in Russian).

Abbot, L. A., F. A. Bisby, and D. J. Rogers. 1985. *Taxonomic analysis in biology. Computers, models, and databases.* New York: Columbia University Press. ISBN 978-0231049269.

Adami, C. 2002. What is complexity? *BioEssays* 24:1085–94. DOI 10.1002/bies.10192.

Adanson, M. 1763. *Familles des plantes*, Pt. 1. Paris: Vincent. ISBN 978-0913196250 (1964 reprint).

Agassiz, L. 1859. *An essay on classification.* London: Longman. ISBN 978-0486435800 (2004 reprint).

Agnarsson, I., and M. Kuntner. 2007. Taxonomy in a changing world: Seeking solutions for a science in crisis. *Systematic Biology*, 56:531–39. DOI 10.1080/10635150701424546.

Akhutin, A. V. 1988. [*The concept of Nature in Antiquity and New time.*] Moscow: Nauka. ISBN 978-5020080164 (in Russian).

Albert, V. A. (ed.). 2005. *Parsimony, phylogeny, and genomics.* New York: Oxford University Press. ISBN 978-0199297306.

Alello, T. 2003. Pierre Magnol: His life and works. *Magnolia* (the Journal of the Magnolia Society) 38:1–10. ISSN 0738-3053.

Aleyev, Yu. G. 1986. [*Ecomorphology.*] Kiev: Naukova Dumka. www.read.in.ua/book111767/ (in Russian).

Aliseda, A. 2006. *Abductive reasoning. Logical investigations into discovery and explanation.* Berlin: Springer-Verlag. ISBN 978-1402039065.

Amo, A., J. Montero, and V. Cutello. 1999. On the principles of fuzzy classification. In *Proceedings of the North American Fuzzy Information Processing Society Conference*, 675–9. New York: NAFIPS. ISBN 978-0780344532.

Amundson, R. 1998. Typology reconsidered: Two doctrines on the history of evolutionary biology. *Biology and Philosophy* 13:153–77. DOI 10.1023/A:1006599002775.

Amundson, R. 2005. *The changing role of the embryo in evolutionary thought roots of evo-devo.* Cambridge (UK): Cambridge University Press. ISBN 978-0521806992.

Anderson, F. J. 1977. *The illustrated history of the herbal.* New York: Columbia University Press. ISBN 978-0231040020.

Andino, R., and E. Domingo. 2015. Viral quasispecies. *Virology* 479–480:46–51. DOI 10.1016/j.virol.2015.03.022.

Antipenko, L. G. 1986. [*Problem of the theory incompleteness and its gnoseological meaning.*] Moscow: Nauka (in Russian).

Antonov, A. S. 2006. [*Genosystems of plants.*] Moscow: Akademkniga. http://msu-botany.ru/gallery/antonov_2006_genosistematika_rastenii.pdf (in Russian).

Arber, A. 1938. *Herbals: Their origin and evolution. A chapter in the history of botany, 1470–1670.* Cambridge (UK) Cambridge University Press. ISBN 978-1108016711.

Arber, A. 1950. *The natural philosophy of plant form.* Cambridge (UK): Cambridge University Press. ISBN 978-1108045056.

Archibald, J. D. 1994. Metataxon concepts and assessing possible ancestry using phylogenetic systematics. *Systematic Biology* 43(1):27–40. DOI 10.1093/sysbio/43.1.27.

Archibald, J. D. 2014. *Aristotle's ladder, Darwin's tree. The evolution of visual metaphors for biological order.* New York: Columbia University Press. ISBN 978-0231164122.

Ariño, A. 2010. Approaches to estimating the universe of natural history collections data. *Biodiversity Informatics* 7:81–92. DOI 10.17161/bi.v7i2.3991.

Aristotle. 1910. *The history of animals*, I–IX. Oxford: Clarendon Press. ISBN 978-0674994812.

Aristotle. 2001. *On the parts of animals*, I–IV. Oxford: Clarendon Press. ISBN 978-1419138805.

Armand, A. D. 2008. [*Two in one: A law of complementarity.*] Moscow: LKI Press. ISBN 978-5382005263 (in Russian).

Arnoldi, K. V. 1939. [On the continuous geographical variability in its general and taxonomic meaning.] *Zoological Journal* 13:685–710. ISSN 0044-5134 (in Russian).

Ashlock, P. D. 1971. Monophyly and associated terms. *Systematic Zoology* 20:63–9. DOI 10.1093/sysbio/20.1.63.

Ashlock, P. D. 1984. Monophyly: Its meaning and importance. In *Cladistics: Perspectives on the reconstruction of evolutionary history*, ed. T. Duncan, and T. F. Stuessy, 39–46. New York: Columbia University Press. ISBN 0-231054300.

Atkin, A. 2013. Peirce's theory of signs. In *The Stanford Encyclopedia of Philosophy*, ed. E. N. Zalta. https://plato.stanford.edu/archives/sum2013/entries/peirce-semiotics/.

Atran, S. 1981. Natural classification. *Social Science Information* 20:37–91. DOI 10.1177/053901848102000102.

Atran, S. 1990. *The cognitive foundations of natural history: Towards an anthropology of science*. New York: Cambridge University Press. ISBN 978-0521438711.

Atran, S. 1998. Folk biology and the anthropology of science: Cognitive universals and cultural particulars. *Behavioral and Brain Sciences* 21:547–609. ISSN 10.1017/S0140525X98001277.

Atran, S. 1999. The universal primacy of generic species in folk biological taxonomy: Implications for human biological, cultural, and scientific evolution. In *Species: New interdisciplinary essays*, ed. R. A. Wilson, 231–61. Cambridge (MA): The MIT Press. ISBN 978-0262731232.

Atran, S., and D. Medin. 2008. *The native mind and the cultural construction of nature*. Cambridge (MA): The MIT Press. ISBN 978-0262514088.

Aubert, D. 2015. A formal analysis of phylogenetic terminology: Towards a reconsideration of the current paradigm in systematics. *Phytoneuron* 66:1–54. www.phytoneuron.net/2015Phytoneuron/66PhytoN-PhylogeneticTerminology.pdf.

Avise, J. C., and G. C. Johns. 1999. Proposal for a standardized temporal scheme of biological classification for extant species. *Proceedings of the National Academy of Sciences* 96: 7358–63. DOI 10.1073/pnas.96.13.7358.

Avise, J. C., and J-X. Liu. 2011. On the temporal inconsistencies of Linnean taxonomic ranks. *Biological Journal of the Linnean Society* 102:707–14. DOI 10.1111/j.1095-8312.2011.01624.x.

Avise, J. C., and D. Mitchell. 2007. Time to standardize taxonomies. *Systematic Biology* 56:130–3. DOI 10.1080/10635150601145365.

Ax, P. 1987. *The phylogenetic system. The systematization of organisms on the basis of their phylogenesis*. Chichester: John Wiley. ISBN 978-0471907541.

Ayala, F. J. 1999. Molecular clock mirages. *BioEssays* 21:71–5. DOI 10.1002/(SICI)1521-1878(199901)21:1<71::AID-BIES9>3.0.CO;2-B.

Baer, K. E., von. 1828. *Über die Entwicklungsgeschichte der Thiere. Beobachtungen und Reflexion*, Vol. 1. Konigsberg: Gebrudern Borntrager. ISBN 3487108712 (1999 reprint).

Baer, K. 1959. [On artificial and natural classifications of animals and plants.] *Annaly Biologii* 1:367–405. ISSN 0970-0153 (in Russian).

Bailey, K. D. 1992. Typologies. In *Encyclopedia of sociology*, ed. E. F. Borgatta, and M. L. Borgatta, 2188–94. New York: Macmillan. ISBN 9780028648538.

Bailey, L. H. 1896. The philosophy of species-making. *Botanical Gazette* 22:454–62. ISBN 978-0332049809.

Baldwin, J. T. 1987. Classification theory. In *Classification theory*, ed. J. T. Baldwin, 1–23. Heidelberg: Springer-Verlag. ISBN 978-3540186748.

Balme, D. M. 1962. Γενοσ and ειδοσ in Aristotle's biology. *Classical Quarterly* 12:81–98. DOI 10.1017/S0009838800011642.

Balme, D. M. 1987. The place of biology in Aristotle's philosophy. In *Philosophical issues in Aristotle's biology*, ed. A. Gotthelf and J. G. Lennox, 9–20. Cambridge (UK): Cambridge University Press. ISBN 978-0511552564.

Barantsev, R. G. 1983. [System triads and classification.] In [*Theory and methodology of biological classifications.*] ed. Yu. A.Shreyder, and B. S. Sornikov, 81–9. Moscow: Nauka (in Russian).

Barantsev, R. G. 2003. [*Synergetics in modern natural science.*] Moscow: Editoria URSS. ISBN 535400201X (in Russian).

Barriel, V., and P. Tassy. 1997. Rooting with multiple outgroups: Consensus versus parsimony. *Cladistics* 14:193–200. DOI 10.1111/j.1096-0031.1998.tb00332.x.

Barsanti, G. 1992. *La Scala, la mappa, l'albero: immaginie classificazioni della natura frasei e ottocento*. Florence: Sansoni. ISBN 978-8838313844.

Bartha, P. 2013. Analogy and analogical reasoning. In *The Stanford encyclopedia of philosophy*, ed. E. N. Zalta. https://plato.stanford. edu/entries/reasoning-analogy/.

Bartlett, H. 1940. The concept of the genus. 1. History of the generic concept in botany. *Bulletin of the Torrey Botanical Club* 67:319–62. DOI 10.2307/2481068.

Bartlett, S. J. 2015. *The species problem and its logic*. http://philsci-archive.pitt.edu/11655/1/Bartlett_The_Species_Problem_and_Its_Logic.pdf

Bauhin, G. 1623. *ΠΙΝΑΞ Theatri botanici Caspari Bauhini [...] sive Index in Theophrasti, Dioscoridis, Plinii et botanicorum qui à seculo scripserunt opera: plantarum circiter sex millium ab ipsis exhibitarum nomina cum earundem synonymiis & differentiis methodicè secundùm earum & genera & species proponens*. Basileae Helvet.: Ludovici Regii. ISBN 978-5881618476 (2012 reprint).

Baum, D. A., and S. D. Smith. 2013. Tree thinking: An introduction to phylogenetic biology. Greenwood Village (CO): Roberts. ISBN: 978-1936221165

Bazhanov, V. A. 2009. [*Vasil'ev and his imaginary logic. Resurrection of one forgotten idea.*] Moscow: Canon+. ISBN 978-5883731968 (in Russian).

Beatty, J. 1982. Classes and cladists. *Systematic Zoology* 31:25–34. DOI 10.2307/2413234.

Becker, G., R. Heim, H. Bianchi, et al. 1957. *Tournefort*. Paris: Museum National d'Histoire Naturelle. ISBN 978-2011784216.

Becker, H. 1940. Constructive typology in the social sciences. *American Sociological Review* 5:40–55. DOI 10.2307/2083940.

Beckner, M. 1959. *The biological way of thought*. New York: Columbia University Press. ISBN 978-0231931403.

Beklemishev, V. N. 1994. [*Methodology of systematics.*] Moscow: KMK Sci. Press. ISBN 5873170053 (in Russian).

Bentham, G. 1875. On the recent progress and present state of systematic botany. *Report of the British Association for the Advancement of Science (1874)*:27–54.

Berendsohn, W. G. (ed.). 2007. *Access to biological collection data. ABCD schema 2.06 ratified TDWG standard*. Berlin: TDWG Task Group on Access to Biological Collection Data, BGBM. www.bgbm.org/TDWG/CODATA/Schema/default.htm.

Berkov, V. F., and Ya. S. Yaskevich. 2001. [*History of logic.*] Minsk: Novoe Znanie (in Russian).

Berlin, B. 1973. Folk systematics in relation to biological classification and nomenclature. *Annual Review of Ecology and Systematics* 4:259–71. DOI 10.1146/annurev.es.04.110173.001355.

Berlin, B. 1992. *Ethnobiological classification: Principles of categorization of plants and animals in traditional societies*. Princeton: Princeton University Press. ISBN 978-0691631004.

Berlin, B. 2004. How a folk botanical system can be both natural and comprehensive: One Maya Indian's view of the plant world. In *Nature knowledge: Ethnoscience, cognition, and utility*, ed. S. Glauco, and O. Gherardo, 38–46. New York: Berghan Books. ISBN 978-1571818232.

Berlin, B., D. E. Breedlove, and P. H. Raven. 1973. General principles of classification and nomenclature in folk biology. *American Anthropologist* 75:214–42. DOI 10.1525/aa.1973.75.1.02a00140.

Bertalanffy, L., von. 1968. *General system theory. Foundations, development, applications*. New York: G. Braziller. ISBN 978-0807604533.

Berti, E. 2016. Aristotle's concept of Nature: Traditional interpretation and results of recent studies. In *Evolving concepts of nature*. Vatican City: Libreria Editrice Vaticana. www.pas.va/content/ dam/accademia/pdf/acta23/acta23-berti.pdf.

Bessey, C. E. 1915. Phylogenetic taxonomy of flowering plants. *Annals of the Missouri Botanical Garden* 2:5–155. DOI 10.2307/2990030.

Bezdek, J. C. 1974. Numerical taxonomy with fuzzy sets. *Journal of Mathematical Biology* 1:57–71. DOI 0.1007/BF02339490.

Bigelow, R. S. 1958. Classification and phylogeny. *Systematic Zoology* 7:49–59. DOI 10.2307/2411792.

Blacher, L. Ya. 1976. [*Problems of animal morphology. Historical essays.*] Moscow: Nauka (in Russian).

Blackwelder, R. E. 1967. *Taxonomy. A text and reference book*. New York: John Wiley. ISBN 978-0471078005.

Blackwelder, R. E., and A. Boyden. 1952. The nature of systematics. *Systematic Zoology* 1:26–33. DOI 10.2307/2411851.

Blumenbach, J. F. 1782. *Handbuch der Naturgeschichte*. Göttingen: Johann Christian Dieterich. ISBN 978-0274455188 (2018 reprint).

Bobrov, E. G. 1970. [*Carl Linné.*] Leningrad: Nauka (in Russian).

Bock, G. R., and G. Cardew (eds.). 1999. *Homology*. Chichester: John Wiley. ISBN 978-0123195838.

Bock, H. 1546. *New Kreuterbuch von Underscheidt, Würckung und Namen der Kreuter, so in teutschen Landen wachsen*. Strasbourg.

Bock, W. 1974. Philosophical foundations of classical evolutionary classification. *Systematic Zoology* 11:375–92. DOI 10.2307/2412945.

Bock, W. J. 1977. Foundations and methods of evolutionary classification. In *Major patterns of vertebrate evolution*, ed. M. K. Hecht, P. C. Goody, and B. M. Hecht, 851–95. New York: Plenum Press. ISBN 978-0306356148.

Bock, W. J. 1989. The homology concept: Its philosophical foundation and practical methodology. *Zoologische Beiträge* 32:327–53. ISSN 0044-5150.

Bock, W. J. 1994. Concepts and methods in ecomorphology. *Journal of Biosciences* 19:403–13. DOI 10.1007/BF02703177.

Boëthius. 1906. *Anicii Manlii Severini Boethii in Isagogen Porphyrii commentorum editionis secundae*. In *Corpus Scriptorum Ecclesiasticorum Latinorum*, ed. S. Brandt. Vienna/ Leipzig: Tempsky/Freitag. www.logicmuseum.com/wiki/Authors/Boethius/isagoge/CPorII.

Boger, H. 1989. The stem-group problem. In *Phylogeny and the classification of fossil and resent organisms*, ed. N. Schmidt-Kittler, and R. Willmann, 45–52. Hamburg: Verlag Paul Prey. ISBN 978-3490144966.

Bonde, N. 1976. Cladistic classification as applied to vertebrates. In *Major patterns of vertebrate evolution*, ed. M. K. Hecht, P. C. Goody, and B. M. Hecht, 741–804. New York: Plenum Press. ISBN 978-1468488531.

Bonnet, C. 1769. *Contemplation de la Nature*, 2nd ed., Vol. 1. Amsterdam: Marc-Michel Rey. ISBN 978-2013032414 (2017 reprint).

Borhvardt, V. G. 1988. Homology, live concept or dogma? *Vestnik Liningradskogo Gosudarstvennogo Universiteta, Series Biological.* 3, 4(24):3–7. ISSN 1025-8604 (in Russian).

Botha, M. E. 1989. Theory development in perspective: The role of conceptual frameworks and models in theory development. *Journal of Advanced Nursing* 14:49–55. DOI 10.1111/j.1365-2648.1989.tb03404.x.

Bowler, P. J. 1973. Bonnet and Buffon: Theories of generation and the problem of species. *Journal of the History of Biology* 6:259–81. DOI 10.1007/BF00127610.

Bowler, P. J. 1975. The changing meaning of "evolution". *Journal of the History of Ideas* 36:95–114. DOI 10.2307/2709013.

Brady, R. H. 1985. On the independence of systematics. *Cladistics* 1:113–26. DOI 10.1111/j.1096-0031.1985.tb00416.x.

Breidbach, O., and M. Ghiselin. 2006. Baroque classification: A missing chapter in the history of systematics. *Annals of the History and Philosophy of Biology* 11:1–30. ISBN ISBN: 978-3941875974.

Bremekamp, C. E. B. 1931. The principles of taxonomy and the theory of evolution. *South African Biological Society* 4:1–8.

Bremekamp, C. E. B. 1953. A re-examination of Cesalpino's classification. Acta *Botanica Neerlandica* 1: 580–93. DOI 10.1111/j.1438–8677.1953.tb00033.x.

Brigandt, I. 2002. Homology and the origin of correspondence. *Biology and Philosophy* 17:389–407. DOI 10.1023/A:1020196124917.

Brigandt, I. 2007. Typology now: Homology and developmental constraints explain evolvability. *Biology and Philosophy* 22:709–25. DOI 10.1007/s10539-007-9089-3.

Brigandt, I., and P. E. Griffiths. 2007. The importance of homology for biology and philosophy. *Biology and Philosophy* 22:633–41. DOI 10.1007/s10539-007-9094-6.

Brisseau-Mirbel, C. F. 1802 *Histoire naturelle, générale et particulière, des plantes*, Vol. 1. Paris: F. Dufart. ISBN 9781013192203 (2019 reprint).

Britton, N. L. 1908. The taxonomic aspect of the species question. *American Naturalist* 42:225–42. DOI 10.1086/278927.

Bronn, H. 1858. *Morphologische Studien über die Gestaltungs-gesetze der Naturkorper überhaupt und der organische Insbesondere.* Leipzig: C. F. Winter. ISBN 978-0260926937 (2018 reprint).

Brooks, D. R., and E. O. Wiley. 1986. *Evolution as entropy.* Chicago: University of Chicago Press. ISBN 978-0226075747.

Broughton, V. 2006. The need for a faceted classification as the basis of all methods of information retrieval. *Aslib Proceedings* 58:49–72. DOI 10.1108/00012530610648671.

Brower, A. V. Z. 2019. Background knowledge: The assumptions of pattern cladistics. *Cladistics* 35:717–31. DOI 10.1111/cla.12379.

Brown, C. H. 1986. The growth of ethnobiological nomenclature. *Current Anthropology* 27:1–19. DOI 10.1086/203375.

Brown, C. H., J. Kolar, B. J. Torrey, et al. 1976. Some general principles of biological and non-biological folk classification. *American Ethnologist* 3:73–85. DOI 10.1525/ae.1976.3.1.02a00050.

Brown, D. E. 2004. Human universals, human nature, human culture. *Daedalus* 133:47–54. DOI 10.1162/0011526042365645.

Brown, R. 1810. *Prodromus florae Novae Hollandiae et Insulae Van Diemen...* London: R. Taylor.

Brummitt, R. K. 1996. In defence of paraphyletic taxa. In *The biodiversity of African plants.*, ed. L. J. G. van der Maesen, X. M. van der Burg, and J. M. van Medenbach de Rooy, 371–84. Dordrecht: Kluwer Academic Publ. DOI 10.5962/bhl.title.3678.

Brundin, L. 1972. Evolution, causal biology, and classification. *Zoologica Scripta* 1:107–20. DOI 10.1111/j.1463–6409.1972.tb00670.x.

Brunfels, O. 1530. *Herbarum vivae eicones ad naturae imitationem summa cum diligentia et artificio effigiatse, una cum effectibus earundem, in gratiam veteris*, T. I. Argentorati. Strasbourg: Ioannem Scottü. ISBN 978-9333327398 (2015 reprint).

Bryan, M. 2005. *John Ray (1627–1705), pioneer in the natural sciences. A celebration and appreciation of his life and work.* Braintree (UK): The John Ray Trust. ISBN 978-0955015007.

Buck, R. C., and D. L. Hull. 1966. The logical structure of the Linnaean hierarchy. *Systematic Zoology* 15:97–111. DOI 10.2307/2411628.

Buffon. 1835. Premier discours. De la manière d'étudier et de traiter l'histoire naturelle. In *Oeuvres complètes de Buffon. Théorie de la Terre*, 37–79. Paris: Pourrat Frères.

Buffon. 1843. De la Nature. Seconde vue. In *Oeuvres choisies de Buffon, contenant les discours académiques, des extraits de la théorie de la terre, les époques de la nature, la génésie des minéraux, l'histoire naturelle de l'homme etdes animaux*, Vol. 1:52–64. Paris: Pourrat Frères.

Bulmer, R. 1974. Folk biology in the New Guinea highlands. *Social Science Information* 13:9–28. DOI 10.1177/053901847401300402.

Bunge, M. 1979. Some topical problems in biophilosophy. *Journal of Social and Biological Structures* 2:155–72. DOI 10.1016/0140-1750(79)90006-X.

Burlando, B. 1990. The fractal dimension of taxonomic systems. *Journal of Theoretical Biology* 146:99–114. DOI 10.1016/S0022-5193(05)80046-3.

Burma, B. H. 1954. Reality, existence, and classification: A discussion of the species problem. *Madroño* 12:193–209. ISSN 0024-9637.

Burtt, B. L. 1966. Adanson and modern taxonomy. *Notes from the Royal Botanical Garden, Edinburgh* 26:427–31. ISSN: 0080-4274.

Butler, A. B., and W. M. Saidel. 2000. Defining sameness: Historical, biological, and generative homology. *BioEssays* 22:846–53. DOI 10.1002/1521–1878(200009)22:9<846::AID-BIES10>3.0.CO;2-R.

Cain, A. J. 1954. *Animal species and their evolution.* London: Hutchinson. ISBN 978-0691020983.

Cain, A. J. 1958. Logic and memory in Linnaeus' system of taxonomy. *Proceedings of the Linnean Society of London* 169:144–63. DOI 10.1111/j.1095–8312.1958.tb00819.x.

Cain, A. J. 1959a. Taxonomic concepts. *Ibis* 101:302–18. DOI 10.1111/j.1474-919X.1959.tb02387.x.

Cain, A. J. 1959b. Deductive and inductive methods in post-Linnaean taxonomy. *Proceedings of the Linnean Society of London* 170:185–217. DOI 10.1111/j.1095–8312.1959.tb00853.x.

Cain, A. J. 1994. Rank and sequence in Caspar Bauhin's Pinax. *Botanical Journal of the Linnean Society* 114:311–56. DOI 10.1111/j.1095–8339.1994.tb01839.x.

Cain, A. J. 1999. John Ray on the species. *Archives of Natural History* 26:223–38. DOI 10.3366/anh.1999.26.2.223.

Cain, A. J., and G. A. Harrison. 1958. An analysis of the taxonomist's judgment of affinity. *Proceedings of the Zoological Society of London* 131:85–98. DOI 10.1111/j.1096–3642.1958.tb00634.x.

Calosi, C., and P. Graziani (eds.). 2014. *Mereology and the sciences: Parts and wholes in the contemporary scientific context.* Dordrecht: Springer. ISBN 978-3319053561.

Camardi, G. 2001. Richard Owen, morphology and evolution. *Journal of the History of Biology* 34:481–515. DOI 10.1023/A:1012946930695.

Camin, J. H., and R. R. Sokal. 1965. A method for deducing branching sequences in phylogeny. *Evolution* 19:311–26. DOI 10.2307/2406441.

Camp, W. H. 1951. Biosystematy. *Brittonia* 7:113–27. DOI 10.2307/2804701.

Camp, W. H., and C. L. Gilly. 1943. The structure and origin of species, with a discussion of intraspecific variability and related nomenclatural problems. *Brittonia* 4:323–85. DOI 10.2307/2804896.

Card, C. R. 1996. The emergence of archetypes in present-day science and its significance for a contemporary philosophy of Nature. *Dynamical Psychology.* https://goertzel.org/dynapsyc/1996/natphil.html.

Carnap, R. 1969. *The logical structure of the world: And pseudoproblems in philosophy.* Chicago: Open Court Publ. ISBN 978-0812695236.

Carr, D. C. 1923. *Linnaeus and Jussieu, or, The rise and progress of systematic botany.* Charleston (SC): Nabu Press. ISBN 978-1287413240 (2013 reprint).

Carrier, M. 1994. *The completeness of scientific theories.* Dordrecht: Kluwer Academic Publ. ISBN 978-0792324751.

Carter, J. G., C. R. Altaba, L. C. Anderson, et al. 2015. The paracladistic approach to phylogenetic taxonomy. *Paleontological Contributions* 12:1–9. DOI 10.17161/PC.1808.17551.

Caruel, T. 1883. Pensées sur la taxinomie botanique. *Botanische Jahrbucher fur Systematik, Pflanzengeschichte und Pflanzengeographie* 4:549–616. ISSN 0006-8152.

Cavalli-Sforza, L. L., and A. W. F. Edwards. 1964. Reconstruction of evolutionary trees. In *Phenetic and phylogenetic classification*, ed. V. H. Heywood, and J. McNeilm, 67–76. London: Syst. Assoc. ISBN 978-0902551015.

Cesalpino, A. 1583. *De plantis libri XVI Andreae Cesalpini Aretini.* Florentiae: Georgium Marescottum. ISBN 9789333481038 (2015 reprint).

Chambers, R. 1860. *Vestiges of the natural history of creation.* New York: Wiley & Putnam. ISBN 978-0226100739 (1994 reprint).

Chan, L. M., and A. Salaba. 2016. *Cataloging and classification: An introduction*, 4th ed. Lanham (MD): Rowman & Littlefield Publ. ISBN 978-0810860001.

Charig, A. J. 1982. Systematics in biology: A fundamental comparison of some major schools of thought. In *Problems of phylogenetic reconstruction*, ed. K. A. Joysey, and A. E. Friday, 363–440. London: Academic Press. ISBN 978-0123912503.

Chebanov, S. V. 2007. [To what extent did Linnaeus deal with classification?] In *Archives of the Zoological Museum of Moscow State University* 48:437–54. ISSN 0134-8647 (in Russian).

Chebanov, S. V. 2016. [The importance of biological diversity for the typological turn of the 20th century.] In *Aspects of biodiversity*, ed. I. Ya. Pavlinov, M. V. Kalyakin, and A. V. Sysoev, 629–54. Moscow: KMK Sci. Press. ISBN 978-5990841666 (in Russian).

Chebanov, S. V., and G. Ya. Martynenko. 2008. [From the history of typological notions.] In [*Structural and applied linguistics*] 6:328–90. ISSN 0202-2400 (in Russian).

Chen, M.-H., L. Kuo, and P. O. Lewis (eds.). 2014. *Bayesian phylogenetics. Methods, algorithms, and applications.* Boca Raton (FL): CRC Press. ISBN 978-1466500792.

Chernov, Yu. I. 1991. [Biological diversity, its essence, and related problems.] *Uspehi Sovremennoj Biologii* 111:499–507. ISSN 0042-1324 (in Russian).

Claridge, M. F., H. A. Dawah, and M. R. Wilson (eds.). 1997. *Species: The units of biodiversity.* London: Chapman & Hall. ISBN 978-0412631207.

Clark, R. B. 1956. Species and systematics. *Systematic Zoology* 5:1–10. DOI 10.2307/2411648.

Coggon, J. 2002. Quinarianism after Darwin's origin: The circular system of William Hincks. *Journal of the History of Biology* 35:5–42. DOI 10.1023/A:1014582710287.

Cohan, F. M. 2001. Bacterial species and speciation. *Systematic Biology* 50:513–24. DOI 10.1080/10635150118398.

Cohan, F. M. 2002. What are bacterial species? *Annual Review of Microbiology* 56:457–87. DOI 10.1146/annurev.micro.56.012302.160634.

Coley, J. D., D. L. Medin, and S. Atran. 1997. Does rank have its privilege? Inductive inferences within folkbiological taxonomies. *Cognition* 64:73–112. DOI 10.1016/S0010-0277(97)00017-6.

Colless, D. H. 1967. An examination of certain concepts in phenetic taxonomy. *Systematic Zoology* 16:6–27. DOI 10.2307/2411512.

Colless, D. H. 1970. The phenogram as an estimation of phylogeny. *Systematic Zoology* 19:352–62. DOI 10.2307/2412276.

Colless, D. H. 1985. On "character" and related terms. *Systematic Zoology* 34:229–33. DOI 10.2307/sysbio/34.2.229.

Collingwood, R. G. 1994. *The idea of history*. Oxford: Oxford University Press. ISBN 978-0192853066.

Cope, E. D. 1887. *The origin of the fittest. Essays of evolution*. New York: Appleton. ISBN 978-0387964959.

Corazzon, R. 2019. Parmenides of Elea. Annotated bibliography of the studies in English (02/16/2019). www.academia.edu/38365125/Parmenides_of_Elea_Annotated_Bibliography_of_the_studies_in_English_02_16_2019_.

Costello, M. J. 2020. Taxonomy as the key to life. *Megataxa* 1:105–113. DOI 10.11646/megataxa.1.2.1.

Cotterill, F. 2016. The Tentelic Thesis, interdisciplinarity, and Earth System science: How natural history collections underpin geobiology. In *Aspects of biodiversity*, ed. I. Ya. Pavlinov, M. V. Kalyakin, and A. V. Sysoev, 692–732. Moscow: KMK Sci. Press. ISBN 978-5990841666.

Cotterill, F. P. D. 2002. The future of natural science collections into the 21st century. In *Conferencia de clausura. actas del i simposio sobre el patrimonio natural en las colecciones públicas en espaca*, 237–82. Vitoria: Museo de Ciencias Naturales de Álava. ISBN 978-8479932251.

Cracraft, J., and M. J. Donoghue (eds.). 2004. *Assembling tree of life*. Oxford: Oxford University Press. ISBN 978-0195172348.

Cronquist, A. 1987. A botanical critique of cladism. *Botanical Review* 53:1–52. DOI 10.1007/BF02858181.

Crowson, R. A. 1970. *Classification and biology*. London: Heinemann Education. ISBN 978-0202309057.

Cuvier, G. 1801. *Leçons d'anatomie comparée*, Vol. 1. Paris: Crochard. ISBN 978-2329388854 (2020 reprint).

Cuvier, G. 1840. *Cuvier's animal kingdom*. London: Orr & Smith. ISBN 978-1169367708 (2010 reprint).

Darlington, P. J. 1971. Modern taxonomy, reality, and usefulness. *Systematic Zoology* 20:341–65. DOI 10.2307/2412346.

Darwin, C. 1859. *On the origin of species by means of natural selection, or the preservation of favoured races in the struggle for life*. London: John Murray. ISBN 978-1615340378 (2009 reprint).

Daston, L. and P. Galison. 2007. *Objectivity*. New York: Zone Books. ISBN 978-1890951795.

Datta, S. K. 2020. Horto-taxonomy in context of classical taxonomy. *International Journal of Life Sciences* 9:85–114. DOI 10.5958/2319-1198.2020.00006.8.

Daudin, H. 1927. De Linne à Jussieu. Méthode de la classification et l'idée de la série en botanique et en zoologie (1740–1790). *Revue Philosophique de la France et de l'Etranger* 103:469–73. DOI 0.1038/121085a0.

Davis, M. C. 2013. The deep homology of the autopod: Insights from hox gene regulation. *Integrative Comparative Biology* 53:224–32. DOI 10.1093/icb/ict029.

Davis, P. H., and V. H. Heywood. 1963. *Principles of angiosperm taxonomy*. London: Oliver & Boyd. ISBN 978-8170193838.

Dayrat, B. 2005. Towards integrative taxonomy. *Biological Journal of the Linnean Society* 85:407–15. DOI 10.1111/j.1095-8312.2005.00503.x.

de Beer, G. 1971. *Homology, an unsolved problem*. Oxford: Oxford University Press. ISBN 978-0199141111.

de Blainville, H. 1816. Prodrome d'une nouvelle distribution du règne animal. *Bulletin de la Société Philomathique* 8:113–24. ISSN 1153-6470.

de Candolle, A.-P. 1819. *Théorie élémentaire de la botanique, ou, Exposition des principes de la classification naturelle et de l'art de décrire et d'étudier les végétaux*, 2nd ed. Paris: Deterville. ISBN 978-1142242220 (2010 reprint).

de Carvalho, M. R., F. A. Bockmann, D. S. Amorim, et al. 2007. Taxonomic impediment or impediment to taxonomy? A commentary on systematics and the cybertaxonomic-automation paradigm. *Evolutionary Biology* 34:140–43. DOI 10.1007/s11692-007-9011-6.

de Pinna, M. C. C. 1991. Concepts and tests of homology in the cladistic paradigm. *Cladistics* 7:367–94. DOI 10.1111/j.1096-0031.1991.tb00045.x.

de Queiroz, K. 1988. Systematics and the Darwinian revolution. *Philosophy of Science* 55:238–59. DOI 10.1086/289430.

de Queiroz, K. 1992. Phylogenetic definitions and taxonomic philosophy. *Biology and Philosophy* 7:295–313. DOI 10.1007/BF00129972.

de Queiroz, K. 2005. Linnaean, rank-based, and phylogenetic nomenclature: Restoring primacy to the link between names and taxa. *Symbolae Botanicae Upsalienses* 33:127–40. https://pdfs.semanticscholar.org/cb4d/912a9e082d9b2e405b8a26429462ff565c88.pdf.

de Queiroz, K., and P. D. Cantino. 2001. Phylogenetic nomenclature and the PhyloCode. *Bulletin of Zoological Nomenclature* 58:254–71. http://citeseerx.ist.psu.edu/viewdoc/download?doi=10.1.1.483.4425&rep=rep1&type=pdf.

de Queiroz, K., and J. Gauthier. 1990. Phylogeny as a central principle in taxonomy: Phylogenetic definitions of taxon names. *Systematic Zoology* 39:307–22. DOI 10.2307/2992353.

de Queiroz, K., and J. Gauthier. 1992. Phylogenetic taxonomy. *Annual Review of Ecology, Evolution and Systematics* 23:449–80. DOI 10.1146/annurev.es.23.110192.002313.

Dean, J. 1979. Controversy over classification: A case study from the history of botany. In *Natural order. Historical studies of scientific culture*, ed. B. Barnes, and S. Shapin, 211–30. London: Sage ISBN 978-0803909595.

DeCandolle, A.-P., and K. Sprengel. 1821. *Elements of the philosophy of plants, containing the principles of scientific botany*. Edinburgh: Blackwood; London: T. Cadell. ISBN 9781108037464 (2011 reprint).

Devitt, M. 2005. Scientific realism. In *The Oxford handbook of contemporary philosophy*, ed. F. Jackson, and M. Smith. Oxford: Oxford University Press. ISBN 978-0199234769.

Di Gregorio, M. A. 2008. Zoology. In *The Oxford handbook of contemporary philosophy*, ed. P. J. Bowler, and J. Pickstone, 205–24. Oxford: Oxford University Press. ISBN 978-0199234769.

Dobzhansky, T. 1935. A critique of the species concept in biology. *Philosophy of Science* 2:344–55. DOI 10.1086/286379.

Dobzhansky, T. 1970. *Genetics of evolutionary process*. New York: Columbia University Press. ISBN 978-0231083065.

Dobzhansky, T. 1973. Nothing in biology makes sense except in the light of evolution. *American Biological Teacher* 35:125–9. DOI 10.2307/4444260.

Domingo, E. 2002. Quasispecies theory in virology. *Journal of Virology* 76:463–5. DOI 10.1128/JVI.76.1.463–465.2002.

Dompere, K. K. 2009. *Fuzzy rationality. A critique and methodological unity of classical, bounded and other rationalities.* Berlin: Springer-Verlag. ISBN 978-3540880837.

Doolittle, W. F. 1999. Phylogenetic classification and the universal tree. *Science* 284:2124–8. DOI 10.1126/science.284.5423.2124.

Doyle, J. J., and J. I. Davis. 1998. Homology in molecular phylogenetics: A parsimony perspective. In *Molecular systematics of plants, Pt. II. DNA sequencing*, ed. D. E. S. Soltis, P. Soltis, and J. J. Doyle, 101–31. Boston: Kluwer Academic Publ. ISBN 978-1461554196.

Drew, J. 2011. The role of natural history institutions and bioinformatics in conservation biology. *Conservation Biology* 25:1250–2. DOI 10.1111/j.1523-1739.2011.01725.x.

Driesch, H. 1899. Von der Methode der Morphologie. *Biologisches Zentralblatt* 19:33–58. ISSN 0006-3304.

Driesch, H. 1908. *The science and philosophy of the organism.* Aberdeen: Print. University. ISBN 978-1428640924 (2006 reprint).

Drouin, J.-M. 2001. Principles and uses of taxonomy in the works of Augustin-Pyramus de Candolle. *Studies in History and Philosophy of Biological & Biomedical Sciences* 32:255–75. DOI 10.1016/S1369-8486(01)00002-4.

Du Rietz, G. E. 1930. The fundamental units of biological taxonomy. *Svensk Botanisk Tidskrift* 24:333–428. https://ru.scribd.com/doc/248585975/Du-Rietz-1930-the-Fundamental-Units-of-Biological-Taxonomy.

Du Rietz, G. E. 1931. Life forms of terrestrial flowering plants. *Acta Phytogeographica Suecica* 3:1–95. www.diva-portal.org/smash/get/diva2:565457/FULLTEXT01.pdf.

Dubois, A. 2006. New proposals for naming lower-ranked taxa within the frame of the International Code of Zoological Nomenclature. *Comptes Rendus Biologies* 329:823–40. DOI 10.1016/j.crvi.2006.07.003.

Dubois, A. 2011. Species and "strange species" in zoology: Do we need a "unified concept of species"? *Comptes Rendus de l'Académie des Sciences, Series IIA, Earth and Planetary Science* 10:77–94. DOI 10.1016/j.crpv.2011.01.002.

Duhem, P. M. M. 1954. The aim and structure of physical theory. Princeton: Princeton University Press. ISBN 978-0691025247.

Dunn, G., and B. S. Everitt. 1982. *An introduction to mathematical taxonomy.* New York: Cambridge University Press. ISBN 978-0486435879.

Dupré, J. 1993. *The disorder of things. Metaphysical foundations of the disunity of science.* Cambridge (MA): Harvard University Press. ISBN 978-0674212619.

Dwyer, P. D. 2005. Ethnoclassification, ethnoecology and the imagination. *Le Journal de la Société des Océanistes* 120–121:11–25. DOI 10.4000/jso.321.

Eades, D. C. 1970. Theoretical and procedural aspects of numerical phyletics. *Systematic Zoology* 19:142–71. DOI 10.2307/2412451.

Ebach, M. C., J. J. Morrone, and D. M. Williams. 2008. A new cladistics of cladists. *Biology and Philosophy* 23:153–6. DOI 10.1007/s10539-007-9069-7.

Ebach, M. C., D. M. Williams, and J. J. Morrone. 2006. Paraphyly is bad taxonomy. *Taxon* 55:831–2. DOI 10.2307/25065678.

Edwards, J. G. 1954. A new approach to infraspecific categories. *Systematic Zoology* 3:1–20. DOI 10.2307/2411490.

Eernisse, D. J., and A. G. Kluge. 1993 Taxonomic congruence versus total evidence, and amniote phylogeny inferred from fossils, molecules, and morphology. *Molecular Biology and Evolution* 10:1170–95. DOI 10.1093/oxfordjournals.molbev.a040071.

Efremov, V. A. 2009. [A theory of concept and the conceptual space.] *Izvestia, Herzen University Journal of Humanities & Sciences*, 104:96–106. ISSN 1992–6464 (in Russian).

Eigen, M. 1983. Viral quasispecies. *Scientific American* 269:42–9. DOI 10.1128/MMBR. 05023-11.

Eldredge, N. 1985. *Unfinished synthesis: Biological hierarchies and modern evolutionary thought.* New York: Oxford University Press. ISBN 978-0195036336.

Eldredge, N., and J. Cracraft. 1980. *Phylogenetic patterns and the evolutionary process.* New York: Columbia University Press. ISBN 0 231 03802 X.

Ellen, R. F. 1993. *The cultural relations of classification. An analysis of Nuaulu animal categories from Central Seram.* Cambridge (UK): Cambridge University Press. ISBN 978-0521025737.

Ellen, R. F. 2008. *The categorical impulse: Essays on the anthropology of classifying behavior.* Oxford: Berghahn Books. ISBN 978-1845450175.

Ellis, B. 2001. *Scientific essentialism.* Cambridge (UK): Cambridge University Press. ISBN 978-0521800945.

Endler, J. A. 1977. *Geographic variation, speciation and clines.* Princeton (NJ): Princeton University Press. ISBN 978-0691081922.

Engelmann, G. F., and E. O. Wiley. 1977. The place of ancestor-descendant relationships in phylogeny reconstruction. *Systematic Zoology* 26:1–11. DOI 10.1093/sysbio/26.1.1.

Engler, A. 1898. *Syllabus der Pflanzenfamilien: eine Übersicht über das gesamte Pflanzensystem mit Berücksichtigung der Medicinal- und Nutzpflanzen nebst einer Ubersicht uber die Florenreiche und Florengebiete der Erde zum Gebrauch.* Berlin: Gebruder Borntraeger. ISBN 978-0331613391 (2018 reprint).

Envall, M. 2008. On the difference between mono-, holo-, and paraphyletic groups: A consistent distinction of process and pattern. *Biological Journal of the Linnean Society* 94:217–20. DOI 10.1111/j.1095-8312.2008.00984.x.

Epshtein, V. M. 1999. [*Systematic philosophy.*] Karkiv: Ranok. ISBN 5873171408 (in Russian).

Epshtein, V. M. 2002. [*Systematic philosophy, book 3. Contemporary problems of the theory of systematics.*] Moscow: KMK Sci. Press. ISBN 5873171173 (in Russian).

Epshtein, V. M. 2003. [*Systematic philosophy, book 2. Principles of construing the theory of systematics and the problem of organism integrity in the history of biology.*] Moscow: KMK Sci. Press. ISBN 5873171408 (in Russian).

Epshtein, V. M. 2004. [*Systematic philosophy, book 3. Theoretical systematics. Ideographic systematics. Assertions and comments.*] Donetsk: Nordpress. ISBN 9668085418 (in Russian).

Epshtein, V. M. 2009. [*Systematic philosophy, book 5. Theoretical systematics. Part 2. Nomothetical systematics. Assertions and comments.*] Gelsenkirchen: Edita Gelsen. ISBN 978-3941464414 (in Russian).

Ereshefsky, M. 1992. Eliminative pluralism. *Philosophy of Science* 59:671–90. DOI 10.1086/289701.

Ereshefsky, M. 1997. The evolution of the Linnaean hierarchy. *Biology and Philosophy* 12:493–519. DOI 1 0.1023/A:1006556627052.

Ereshefsky, M. 2001a. *The poverty of the Linneaean hierarchy: A philosophical study of biological taxonomy.* New York: Cambridge University Press. ISBN 978-0521038836.

Ereshefsky, M. 2001b. Philosophy of biological classification. In *Encyclopedia of life sciences.* http://mrw.interscience.wiley.com/emrw/9780470015902/els/article/a0003447/current/abstract.

Ereshefsky, M. 2008. Systematics and taxonomy. In *A companion to the philosophy of biology,* ed. S. Sarkar, and A. Plutynski, 99–118. Oxford: Blackwell Publ. ISBN 978-1405125727.

Ereshefsky, M. 2011. Mystery of mysteries: Darwin and the species problem. *Cladistcis* 27:167–79. DOI 10.1111/j.1096-0031.2010.00311.x.

Estabrook, G. F. 1971. Some information theoretic optimality criteria for general classification. *Mathematical Geology* 3:203–7. DOI 10.1007/BF02045962.

Estabrook, G. F. 1972. Cladistic methodology: A discussion of the theoretical basis for the induction of evolutionary history. *Annual Review of Ecology, Evolution and Systematics* 3:427–56. DOI 10.1146/annurev.es.03.110172.002235.

Faegri, K. 1937. Some fundamental problems of taxonomy and phylogenetics. *The Botanical Review* 3:400–23. DOI 10.1007/BF02870489.

Fairthorne, R. A. 1969. Empirical hyperbolic distributions (Bradford–Zipf–Mandelbrot) for bibliometric description and prediction. *Journal of Documentation* 25:319–43. DOI 10.1108/eb026481.

Faith, D. P. 2003. Biodiversity. In *The Stanford encyclopedia of philosophy*, ed. E. N. Zalta. http://plato. stanford.edu/archives/sum2003/entries/ biodiversity/.

Faith, D. P. 2006. Science and philosophy for molecular systematics: Which is the cart and which is the horse? *Molecular Phylogenetics and Evolution* 38:553–7. DOI 10.1016/j.ympev.2005.08.018.

Falcon, A. 1997. Aristotle's theory of division. *Bulletin of the Institute of Classical Studies*, Suppl. 68:127–46. DOI 10.1111/j.2041-5370.1997.tb02267.x.

Farber, P. L. 1972. Buffon and the concept of species. *Journal of the History of Biology* 5:259–84. DOI 10.1007/BF00346660.

Farber, R. L. 1976. The type-concept in zoology in the first half of the nineteenth century. *Journal of the History of Biology* 9:93–119. DOI 10.1007/BF00129174.

Farris, J. S. 1969. A successive approximations approach to character weighting. *Systematic Zoology* 18:374–85. DOI 10.2307/2412182.

Farris, J. S. 1982. Outgroup and parsimony. *Systematic Zoology* 31:328–34. DOI 10.2307/2413239.

Farris, J. S. 1983. The logical basis of phylogenetic analysis. *Advances in Cladistics* 2:7–36. https://citeseerx.ist.psu.edu/viewdoc/download?doi=10.1.1.1078.6366&rep=rep1&type=pdf.

Farris, J. S. 1986. On the boundaries of phylogenetic systematics. *Cladistics* 2:14–27. DOI 10.1111/j.1096-0031.1986.tb00439.x.

Farris, J. S. 2008. Parsimony and explanatory power. *Cladistics* 24:825–47. DOI 10.1111/j.1096-0031.2008.00214.x.

Farris, J. S., A. G. Kluge, and M. J. Eckardt. 1970. A numerical approach to phylogenetic systematics. *Systematic Zoology* 19:72–91. DOI 10.2307/2412452.

Feliner, G. N., and I. A. Fernandez. 2000. Biosystematics in the 90s: Integrating data from different sources. *Portugaliae Acta Biologica* 19:9–19. DOI 10.1111/aen.12158.

Felsenstein, J. 1982. Numerical methods for inferring evolutionary trees. *The Quarterly Review of Biology* 57:379–404. DOI 10.1086/412935.

Felsenstein, J. 1983. Parsimony in systematics: Biological and statistical issues. *Annual Review of Ecology, Evolution and Systematics* 14:313–33. DOI 10.1146/annurev. es.14.110183.001525.

Felsenstein, J. 1988. The detection of phylogeny. In *Prospects in systematics*, ed. D. L. Hawksworth, 112–27. Oxford: Clarendon Press. ISBN 978-0198577072.

Felsenstein, J. 2004. *Inferring phylogenies*. Sunderland (MA): Sinauer Assoc. ISBN 978-0878931774.

Ferris, G. F. 1928. *The principles of systematic entomology*. Stanford (CA): Stanford University Press. ISBN 978-0598947185.

Filatov, V. P., A. P. Ogurtsov, V. G. Fedotov, et al. 2007. [Discussing the topic "typological method".] *Epistemology & Philosophy of Science*, 11:157–68. ISSN: 1811-833X (in Russian).

Fisher, D. C. 2008. Stratocladistics: integrating temporal data and character data in phylogenetic inference. *Annual Review of Ecology, Evolution, and Systematics* 39:365–85. DOI 10.1146/annurev.ecolsys.38.091206.095752.

Fisher, H. 1966. Conrad Gesner (1516–1565) as bibliographer and encyclopedist. *The Library* 5:269–381. DOI 10.1093/library/s5-XXI.4.269.

Fisher, R. A. 1925. *Statistical methods for research workers*. Edinburgh: Oliver & Boyd. ISBN 978-0028447308.

Fitch, W. M., and E. Margoliash. 1967. Construction of phylogenetic trees. *Science* 155:279–84. DOI 10.1126/science.155.3760.279.

Fitzhugh, K. 2006. The abduction of phylogenetic hypotheses. *Zootaxa* 1145:1–110. DOI 10.11646/zootaxa.1145.1.1.

Fitzhugh, K. 2016. Dispelling five myths about hypothesis testing in biological systematics. *Organisms, Diversity, & Evolution* 16: 443–65. DOI 10.1007/s13127-016-0274-6.

Forey, P. L., R. A. Fortey, P. Kenrick, et al. 2004. Taxonomy and fossils: A critical appraisal. *Philosophical Transactions of the Royal Society, Series B* 359:639–53. DOI 10.1098/rstb.2003.1453

Forey, P. L., C. J. Humphries, I. J. Kitching, et al. 1992. *Cladistics: A practical course in systematics*. Oxford: Clarendon Press. ISBN 978-0198577669.

Foucault, M. 1970. *The order of things. An archaeology of the human sciences*. New York: Pantheon Books. ISBN 978-0679753353.

Fraassen, B. C., van. 2008. *Scientific representation: Paradoxes of perspective*. New York: Oxford University Press. ISBN 978-09278220.

French, R. 1994. *Ancient natural history. Histories of nature*. London: Routledge. ISBN 978-0415115452.

Freudenstein, J. V. 2005. Characters, states and homology. *Systematic Biology* 54:965–73. DOI 10.1080/10635150500354654.

Frigg, R., and J. Nguyen. 2016. Scientific representation. In *The Stanford encyclopedia of philosophy*, ed. E. N. Zalta. https://plato.stanford. edu/entries/scientific-representation/.

Gadamer, H.-G. 1960. *Truth and method*, 2nd ed. London: Continuum. ISBN 978-1780936246.

Gaffney, E. S. 1979. An introduction to the logic of phylogeny reconstruction. In *Phylogenetic analysis and paleontology*, ed. J. Cracraft, and N. Eldredge, 79–111. New York: Columbia University Press. ISBN 978-0231046930.

Galis, F. 1999. On the homology of structures and *Hox* genes: The vertebral column. In *Homology*, ed. G. Bock, and G. Cardew, 81–94. Chichester: John Wiley. ISBN 0471984930.

Gärdenfors, P. 2000. *Conceptual spaces*. Cambridge (MA): MIT Press. ISBN 978-0262071994.

Garey, M. R., and D. S. Johnson. 1979. *Computer and intractability: A guide to the theory of NP-completeness*. San Francisco: W.H. Freeman. ISBN 978-0716710455.

Gauthier, J., R. Estes, and de K. Queiroz. 1988. A phylogenetic analysis of Lepidosauromorpha. In *The phylogenetic relationships of the lizard families*, ed. R. Estes, and G. Pregill, 15–98. Stanford (CA): Stanford University Press. ISBN 978-0804714358.

Gaydenko, P. P. 1980. [*Evolution of the notion of science (formation and development of the first scientific programs)*.] Moscow: Nauka (in Russian).

Gaydenko, P. P. 1987. [*Evolution of the notion of science (formation of scientific programs of the new time of the 17th–18th centuries)*.] Moscow: Nauka. 487 (in Russian).

Gaydenko, P. P. 2003. [*Scientic rationality and philosophical mind*.] Moscow: Progress-Traditsia. ISBN 5898261427 (in Russian).

Gaydenko, V. P., and G. A. Smirnov. 1989. [*Western European science in the Middle Ages.*] Moscow: Nauka (in Russian).

Gegenbaur, C. 1865. *Untersuchungen zur vergleichenden Anatomie der Wirbeltiere.* Leipzig: Wilgelm Eidelmann. ISBN 978-3743371484 (2016 reprint).

Geoffroy Saint-Hilaire, E. 1830. *Principes de philosophie zoologique, discutés en mars 1830 au sein de l'Académie royale des sciences.* Paris: Pigeon et Didier. ISBN 978-2013555630 (2018 reprint).

Gesner, C. 1560. *Nomenclator aquatilium animantium. Icones animalium aquatilium in mari & dulcibus aquis degentium, plus quam DCC cum nomenclaturis singulorum.* Zurich: Christoph. Froschoverum. ISBN 978-5519152907 (2015 reprint).

Ghiselin, M. T. 1981. Categories, life, and thinking. *The Behavioral and Brain Sciences* 4:269–313. DOI 0.1017/S0140525X00008852.

Ghiselin, M. T. 1997. *Metaphysics and the origin of species.* New York: State University of New York Press. ISBN 978-0791434680.

Ghiselin, M. T. 1998. Folk metaphysics and the anthropology of science. *Behavioral and Brain Sciences* 21:573–4. DOI 10.1017/S0140525X98261279.

Ghiselin, M. T. 2005. Homology as a relation of correspondence between parts of individuals. *Theory in Biosciences* 124:91–103. DOI 10.1007/BF02814478.

Giampietro, M. 2002. Complexity and scales: The challenge for integrated assessment. *Integrated Assessment* 3:247–65. DOI 10.1076/iaij.3.2.247.13568.

Gibbs, R. W. (ed.). 2008. *The Cambridge handbook of metaphor and thought.* New York: Cambridge University Press. ISBN 9780511816802.

Gilbert, S. F., and J. A. Bolker. 2001. Homologies of process and modular elements of embryonic construction. *Journal of Experimental Zoology* 291:1–12. DOI 10.1002/jez.1.

Gillis, M., P. Vandamme, P. De Vos, et al. 2001. Polyphasic taxonomy. In *Bergey's manual of systematic bacteriology*, Vol. 1, ed. D. R. Boone, and R. W. Castenholz, 43–8. Baltimore: Williams & Wilkins. ISBN 978-0683041088.

Gilmour, J. S. L. 1940. Taxonomy and philosophy. In *The new systematics*, ed. J. Huxley, 461–74. Oxford (UK): Oxford University Press. ISBN 978-0403017867.

Gilmour, J. S. L. 1961. Taxonomy. In *Contemporary botanical thought*, ed. A. M. MacLeod, and L. S. Cobley, 27–45. Chicago: Quadrangle Book. ASIN B0000CL6NB.

Gilmour, J. S. L., and W. B. Turrill. 1941. The aim and scope of taxonomy. *Chronica Botanica* 6:217–19. DOI 10.5852/ejt.2017.712.

Gilmour, J. S. L., and S. M. Walters. 1963. Philosophy and classification. In *Vistas in botany. Recent researches in plant taxonomy*, ed. W. B. Turrill, 1–22. London: Pergamon Press. ISBN 978-1483210223.

Girill, T. R. 1973. The logic of scientific puzzles. *Zeitschrift für Allgemeine Wissenschaftstheorie* 4:25–40. DOI 10.1007/BF01801063.

Giseke, P. D. 1792. *Caroli a Linne [...] Praelectiones in ordines naturales plantarum.* Hamburg: Benj. Gottl. Hoffmanni. ISBN 978-1343182158 (2015 reprint).

Gmelig-Nijboer, C. A. 1977. *Conrad Gessner's "Historia animalium": An inventory of Renaissance zoology.* Meppel (Netherlands): Krips Repro. ASIN B0024TMTF4.

Godfray, H. C. J. 2002. Challenges for taxonomy. *Nature* 417:17–9. DOI 10.1038/417017a.

Goethe, J. W., von. 1790. *Versuch die Metamorphosen der Pflanzen zu erklären.* Gotha: Carl Wilhelm Ettinger. ISBN 978-3337783570 (2019 reprint).

Goethe, J. W., von. 1957. [*Selected issues on natural history.*] Moscow: Acad. Sci. Publ. (in Russian).

Gogarten, J. P. 1994. Which is the most conserved group of proteins? Homology-orthology, paralogy, xenology, and the fusion of independent lineages. *Journal of Molecular Evolution* 39:541–3. DOI 10.1007/bf00173425.

Gontier, N. 2011. Depicting the tree of life: The philosophical and historical roots of evolutionary tree diagrams. *Evolution, Education, and Outreach* 4:515–38. DOI 10.1007/s12052-011-0355-0.

Goodin, S. 1999. Locke and Leibniz and the debate over species. In *New essays on the rationalists*, ed. R. J. Gennaro, C. Huenemann, 163–77. New York: Oxford University Press. ISBN 978-0195124880.

Goodman, N. 1972. *Problems and projects.* Indianapolis (IN): Bobs-Merrill. ISBN 978-0814416822.

Goodwin, B. 1994. Homology, development, and heredity. In *Homology: The hierarchical basis of comparative morphology*, ed. B. K. Hall, 229–47. New York: Academic Press. ISBN 978-0123195838.

Gordon, A. D. 1999. *Classification*, 2d ed. London: Chapman & Hall/CRC. ISBN 978-1584880134.

Gordon, M. S. 1999. The concept of monophyly: A speculative essay. *Biology and Philosophy* 14:331–48. DOI 10.1023/A:1006535524246.

Górska, E. 2002. On partonomy and taxonomy. *Studia Anglica Posnaniensia* 39:103–11. ISSN 2082-5102.

Gotthelf, A. 2012. *Teleology, first principles, and scientific method in Aristotle's biology.* Oxford: Oxford University Press. ISBN 978-0199287956.

Gould, Ch. 1886. *Mythical monsters.* London: W.H. Allen. ISBN 978-1605204062 (2000 reprint).

Gould, S. J. 1977. *Ontogeny and phylogeny.* Cambridge (MA): Belknap Press. ISBN 97-80674639416.

Goulding, T. C., and B. Dayrat. 2016. Integrative taxonomy: Ten years of practice and looking into the future. In *Aspects of biodiversity*, ed. I. Ya. Pavlinov, M. V. Kalyakin, and A. V. Sysoev, 116–33. Moscow: KMK Sci. Press. ISBN 978-5990841666.

Graham, C. H., S. Ferrier, and F. Huettman. 2004. New developments in museum-based informatics and applications in biodiversity analysis. *Trends in Ecology and Evolution* 19:497–503. DOI 10.1016/j.tree.2004.07.006.

Grant, T., and A. G. Kluge. 2004. Transformation series as an ideographic character concept. *Cladistics* 20:32–41. DOI 10.1111/j.1096-0031.2004.00003.x.

Gray, A. 1858. *Introduction to structural and systematic botany, and vegetable physiology*, 5th ed. New York: Ivision & Phinney; Chicago: S.C. Griggs. ISBN 978-3337367800 (2017 reprint).

Gray, A. 1876. *Darwiniana. Essays and reviews pertaining to Darwinism.* New York: D. Appleton. ISBN 978-1108001960 (2009 reprint).

Graybeal, A. 1995. Naming species. *Systematic Biology* 44:237–50. DOI 10.1093/sysbio/44.2.237.

Greene, E. L. 1909. *Landmarks of botanical history. A study of certain epochs in the development of the science of botany. Pt 1. Prior to 1562.* Washington (DC): Smithsonian Institute Press. ISBN 978-1343207356.

Greene, J. C. 1992. From Aristotle to Darwin: Reflections on Ernst Mayr's interpretation in *The Growth of Biological Thought. Journal of the History of Biology* 25:257–84. DOI 10.1007/BF00162842.

Gregg, J. R. 1950. Taxonomy, language and reality. *American Naturalist* 84:419–35. DOI 10.1086/281639.

Gregg, J. R. 1954. *The language of taxonomy.* New York: Columbia University Press. ISBN 978-0804728058.

Gregor, J. W. 1942. The units of experimental taxonomy. *Chronica Botanica* 7:193–6. DOI 10.1007/BF01556761.

Grene, M. 1974. *The understanding of nature: Essays in the philosophy of biology.* Dordrecht: D. Reidel. ISBN 978-9401022248.

Grene, M. 1987. Historical realism and contextual objectivity: A developing perspective in the philosophy of science. In *The process of science,* ed. N. J. Nersessian, 69–81. Dordrecht: Nijhoff. ISBN 978-9400935198.

Grene, M. 1990. Evolution, typology, and population thinking. *American Philosophical Quarterly* 27:237–44. DOI 10.1002/jez.b.22796.

Griffiths, G. C. D. 1974a. On the foundations of biological systematics. *Acta Biotheoretica* 23:85–131. DOI 10.1007/BF01556343.

Griffiths, G. C. D. 1974b. Some fundamental problems in biological classification. *Systematic Zoology* 22:338–43. DOI 10.2307/2412942.

Griffiths, P. E. 1999. Squaring the circle: Natural kinds with historical essences. In *Species: New interdisciplinary essays,* ed. R. A. Wilson, 209–28. Cambridge: The MIT Press. ISBN 978-0262731232.

Grizkevich, V. P. 2004. [*History of museum activity till the end of the 18th century.*] Saint Petersburg: Saint Petersburg State University of Culture and Arts Publ. ISBN 5947080354 (in Russian).

Grushin, B. A. 1961. [*Essays on the logic of historical research.*] Moscow: Vysshaia Shkola (in Russian).

Guedes, M. 1967. La méthode taxonomique d'Adanson. *Revue d'Histoire des Sciences et de leurs Applications* 20:361–86. DOI 10.3406/rhs.1967.2543.

Gumperz, J. J., and S. C. Levinson. 1996. *Rethinking linguistic relativity.* Cambridge (UK): Cambridge Univ. Press. ISBN 978-0521448901.

Hacking, I. 1983. *Representing and intervening, Introductory topics in the philosophy of natural science.* Cambridge (UK): Cambridge University Press. ISBN 978-0521282468.

Haeckel, E. 1866. *Generelle Morphologie der Organismen. Allgemeine Grundzuge der organischen Formen-Wissenschaft, mechanisch begrundet durch die von Charles Darwin reformirte Descendenz-Theorie,* Vols. 1 and 2. Berlin: Georg Reimer. ISBN 978-3-11-010185-0 (1988 reprint).

Haeckel, E. 1868. *Natürliche Schöpfungsgeschichte: Gemeinverständliche wissenschaftliche Vorträge über die Entwickelungslehre im Allgemeinen und diejenige von Darwin, Goethe, und Lamarck im Besonderen.* Berlin: Georg Reimer.

Haeckel, E. 1894–1896. *Systematische Phylogenie,* Vols. 1–3. Berlin: Georg Reimer. DOI 10.5962/bhl.title.3947.

Haeckel, E. 1917. *Kristallseelen. Studien uber das anorganishe Leben.* Leipzig: Alfred Kroner Verlag.

Hagen, J. 2003. The statistical frame of mind in systematic biology from quantitative zoology to biometry. *Journal of the History of Biology* 36:353–84. DOI 10.1023/A:1024479322226.

Hall, B. K. 1992. *Evolutionary developmental biology.* London: Chapman & Hall. ISBN 978-0412785900.

Hall, B. K. (ed.). 1994. *Homology, the hierarchical basis of comparative biology.* San Diego (CA): Academic Press. ISBN 978-0123189202.

Hall, B. K. 1995. Homology and embryonic development. In *Evolutionary biology,* Vol. 28, ed. M. K. Hecht, R. J. MacIntyre, and M. T. Clegg, 1–37. New York: Plenum Press. ISBN 978-1461357490.

Hall, B. K. 1996. *Baupläne,* phylotypic stages, and constraint: Why there are so few types of animals? *Evolutionary Biology* 29:251–61. DOI 10.1007/s12052-012-0424-z.

Hall, H. M., and F. E. Clements. 1923. *The phylogenetic method in taxonomy: The North American species of Artemisia, Chrysothamnus, and Atriplex.* Washington (DC): Carnegie Institute Publ. ISBN 978-1332305636.

Hamilton, A. 2014. Historical and conceptual perspectives on modern systematics: Groups, ranks, and the phylogenetic turn. In *The evolution of phylogenetic systematics*, ed. A. Hamilton, 89–115. Berkeley (CA): University of California Press. ISBN 978-0520276581.

Hammen, L., van der. 1981. Type-concept, higher classification and evolution. *Acta Biotheoretica* 30:3–48. DOI 10.1007/BF00116071.

Hanage, W. P., C. Fraser, and B. G. Spratt. 2005. Fuzzy species among recombinogenic bacteria. *BMC Biology* 3:6. www.biomedcentral.com/1741-7007/3/6.

Härlin, M. 1999. The logical priority of the tree over characters and some of its consequences. *Biological Journal of the Linnean Society* 68:497–503. DOI 10.1006/bijl.1999.0348.

Harris, E. S. J., and B. D. Mishler. 2009. The delimitation of phylogenetic characters. *Biological Theory* 4:230–4. DOI 10.1162/biot.2009.4.3.230.

Harrison, P. 2006. The Bible and the emergence of modern science. *Science & Christian Belief* 18:115–32. DOI 10.1017/CHO9781139048781.029.

Harrison, P. 2009. Linnaeus as a second Adam? Taxonomy and the religious vocation. *Zygon* 44:879–93. DOI 10.1111/j.1467-9744.2009.01039.x.

Hayes, R. L., and R. Oppenheim. 1997. Constructivism: Reality is what you make it. In *Constructivist thinking in counseling practice, research, and training*, ed. T. L. Sexton, and B. L. Griffin, 19–40. Danvers (MA): Columbia University Press. ISBN 978-0807736104.

Hays, T. E. 1982. Utilitarian/adaptationist explanations of folk biological classification: Some cautionary notes. *Journal of Ethnobiology* 2:89–94. https://anthrosource.onlinelibrary. wiley.com/doi/pdf/10.1525/aa.1982.84.4.02a00070.

Hays, T. E. 1983. Ndumba folk biology and general principles of ethnobotanical classification and nomenclature. *American Anthropologist* 85:592–611. DOI 10.1525/aa.1983.85.3.02a00050.

Hecht, M. K., and J. L. Edwards. 1977. The methodology of phylogenetic inference above the species level. In *Major patterns in vertebrate evolution*, ed. M. K. Hecht, P. C. Goody, and B. M. Hecht, 3–51. New York: Plenum Press. ISBN 978-1-4684-8853-1.

Hedberg, O. 1997. Progress in biosystematics: An overview. *Lagascalia* 19:307–16. http://institucional.us.es/revistas/lagascalia/19/Progress%20Hedberg.pdf.

Heisenberg, W. 1959. *Physics and philosophy. The revolution in modern science.* London: George Allen & Unwin. ISBN 978-0061209192.

Hempel, G. 1965. *Aspects of scientific explanation and other essays in the philosophy of science.* New York: Free Press. ISBN 978-0029143407.

Hempel, C. G. 1966. *Philosophy of natural science.* Englewood Cliffs: Prentice-Hall. ISBN 978-0136638230.

Hendry, A. P., S. M. Vamosi, S. J. Latham, et al. 2000. Questioning species realities. *Conservation Genetics* 1:67–76. DOI 10.1023/A:1010133721121.

Hennig, W. 1950. *Grundzüge einiger Theorie der phylogenetische Systematik.* Berlin: Deutscher Zentralverlag. ISBN 978-3874291880.

Hennig, W. 1965. Phylogenetic systematics. *Annual Review of Entomology* 10:97–116. DOI 10.13140/RG.2.1.5013.3203.

Hennig, W. 1966. *Phylogenetic systematics.* Urbana (IL): University Illinois Press. ISBN 978-0252068140.

Hermon, S., and F. Niccolucci. 2002. Estimating subjectivity of typologists and typological classification with fuzzy logic. *Archeologia e Calcolatori* 13:217–32. www.archcalc. cnr.it/indice/PDF13/16Niccolucci.pdf.

Heslop-Harrison, J. 1960. *New concepts in flowering-plant taxonomy.* Cambridge (MA): Harvard University Press. ISBN 978-0435613907.

Hey, J. 2001. *Genes, categories, and species. The evolutionary and cognitive cause of the species problem.* New York: Oxford University Press. ISBN 978-0195144772.

Hill, C. R., and J. M. Camus. 1986. Pattern cladistics or evolutionary cladistics? *Cladistics* 2:362–75. DOI 10.1111/j.1096-0031.1986.tb00459.x.

Hill, C. R., and P. R. Crane. 1982. Evolutionary cladistics and the origin of angiosperms. In *Problems of phylogenetic reconstruction*, ed. K. A. Joysey, and A. E. Friday, 269–361. London: Academic Press. ISBN 978-0123912503.

Hillis, D. M. 1994. Homology in molecular biology. In *Homology, the hierarchical basis of comparative biology*, ed. B. K. Hall, 339–68. London: Academic Press. ISBN 978-0123195838.

Hillis, D. M., C. Moritz, and B. K. Mable (eds.). 1996. *Molecular systematics*, 2nd ed. Sunderland: Sinauer Associates. ISBN 978-0878932825.

Ho, M. W. 1988. How rational can rational morphology be? A post-Darwinian rational taxonomy based on a structuralism of process. *Theoretical Biology Forum* 81:11–55. ISSN: 2283-7175.

Ho, M. W. 1990. An exercise in rational taxonomy. *Journal of Theoretical Biology* 147:43–57. DOI 10.1016/S0022-5193(05)80251-6.

Ho, M. W. 1992. Development, rational taxonomy and systematics. *Theoretical Biology Forum* 85:193–211. ISSN: 2283-7175.

Ho, M. W. 1998. Evolution. In *Comparative psychology, a handbook*, ed. G. Greenberg, and M. M. Haraway, 107–19. London: Taylor & Francis. ISBN 978-1138971332.

Ho, M. W., and P. T. Saunders. 1993. Rational taxonomy and the natural system, with particular reference to segmentation. *Acta Biotheoretica* 41:289–304. DOI 10.1007/BF00709367.

Ho, M. W., and P. T. Saunders. 1994. Rational taxonomy and the natural system—segmentation and phyllotaxis. In *Models in phylogeny reconstruction*, ed. R. W. Scotland, D. J. Siebert, and D.M. Williams, 113–24. Oxford: Systematics Association. ISBN 978-0198548249.

Hoch, P. C., and A. G. Stephenson (eds.). 1995. *Experimental and molecular approaches to plant biosystematics.* St. Louis (IL): Missouri Botanical Garden. ISBN 978-0915279302.

Hofsten, N., von. 1958. Linnaeus's conception of Nature. *Kungliga Vetenskaps-societetens Årsbok från 1957/1958*:65–105. ISBN 978-9198240528.

Holbach, P. H. (M. Mirabaud). 1770. *Système de la nature ou des loix du monde physique et du monde moral.* London. ISBN 978-2051023351 (2011 reprint).

Holland, L. Z., P. W. Holland, and N. D. Holland. 1996. Revealing homologies between body parts of distantly related animals by in situ hybridization to developmental genes: Amphioxus versus vertebrates. In *Molecular zoology: Advances, strategies and protocols*, ed. J. D. Ferraris, and S. R. Palumbi, 267–95. New York: Wiley-Liss. ISBN 978-0471144618.

Holopainen, T. 1996. *Dialectic and theology in the eleventh century.* Leiden: Brill. ISBN 978-9004105775.

Hołyński, R. B. 2005. Philosophy of science from a taxonomist's perspective. *Genus* 16:469–502. ISSN 0016-6987.

Hopwood, N., S. Schaffer, and J. Secord. 2010. Seriality and scientific objects in the nineteenth century. *History of Science* 48:251–85. DOI 10.1177/007327531004800301.

Hoque, T. 2008. Buffon: from natural history to the history of nature? *Biological Theory* 2:413–19. DOI 10.1086/353225.

Hörandl, E. 2006. Paraphyletic versus monophyletic taxa. *Taxon* 55:564–70. DOI 10.2307/25065631.

Hörandl, E. 2010. Beyond cladistics: Extending evolutionary classifications into deeper time levels. *Taxon* 59:345–50. DOI 10.1002/tax.592001.

Hossfeld, U., and L. Olsson. 2005. The history of the homology concept and the "Phylogenetisches Symposium". *Theory in Biosciences* 124:243–53. DOI 10.1007/BF02814486.

Hubbs, C. L. 1944. Concepts of homology and analogy. *American Naturalist* 78:289–307. DOI 10.1086/281202.

Huelsenbeck, J. P., B. Rannala, and J. P. Masly. 2000. Accommodating phylogenetic uncertainty in evolutionary studies. *Science* 288:2349–50. DOI 10.1126/science.288.5475.2349.

Hull, D. 1985. Linne as an Aristotelian. In *Contemporary perspectives on Linnaeus*, ed. J. Weinstock, 37–54. Lanham (MD): University Press of America. ISBN 978-0819146984.

Hull, D. L. 1965. The effect of essentialism on taxonomy: Two thousand years of stasis. *British Journal for the Philosophy of Science* 15:314–26. DOI 10.1093/bjps/XV.60.314.

Hull, D. L. 1970. Contemporary systematic philosophies. *Annual Review of Ecology, Evolution and Systematics* 1:19–54. DOI 10.1146/annurev.es.01.110170.000315.

Hull, D. L. 1978a. A matter of individuality. *Philosophy of Science* 45:335–60. http://mechanism.ucsd.edu/teaching/philbiology/readings/hull.materofindividuality.1978.pdf.

Hull, D. L. 1978b. The principles of biological classification: The use and abuse of philosophy. In *Proceedings of the Biennial Meeting of the Philosophy of Science Association* 2:130–53. Chicago (IL): University of Chicago Press. ISBN 9780917586057.

Hull, D. L. 1988. *Science as a process*. Chicago (IL): University of Chicago Press. ISBN 978-0226360508.

Hull, D. L. 1997. The ideal species concept—and why we can't get it. In *Species. The units of biodiversity*, ed. M. F. Claridge, A. H. Dawah, and M. R. Wilson, 357–80. London: Chapman & Hall. ISBN 978-0412631207.

Hull, D. L. 1999. The use and abuse of Sir Karl Popper. *Biology and Philosophy* 14:481–504. DOI 10.1023/A:1006554919188.

Hull, D. L. 2001. The role of theories in biological systematics. *Studies in History and Philosophy of Biological and Biomedical Sciences* 32:221–38. DOI 10.1016/S1369-8486(01)00006-1.

Humberstone, I. L. 1996. A study in philosophical taxonomy. *Philosophical Studies* 83:121–69. DOI 10.1007/BF00354286.

Humboldt, A. 1806. Ideen zur einer Physiognomik der Gewächse. Tubingen: Cotta. ISBN 9783337198732 (2017 reprint).

Hunn, E. 1977. *Tzeltal folk zoology: The classification of discontinuities in nature.* New York: Academic Press. ISBN 978-0123617507.

Hunn, E. 1982. The utilitarian factor in folk biological classifications. *American Anthropologist* 84:830–47. DOI 0.1525/aa.1982.84.4.02a00070.

Hunn, E. S., and D. H. French. 2000. Alternatives to taxonomic hierarchy: The Sahapin case. In *Ethnobotany: A reader*, ed. P. E. Minnis, 118–28. Norman (OK): University of Oklahoma Press. ISBN 978-0806131801.

Huxley, J. (ed.). 1940a. *The new systematics*. London: Oxford University Press. ISBN 978-0403017867.

Huxley, J. 1940b. Introductory: Towards the new systematics. In *The new systematics*, ed. J. Huxley, 1–46. London: Oxford University Press. ISBN 978-0403017867.

Huxley, J. S. 1939. Clines: An auxiliary method in taxonomy. *Bijdragen tot de Dierkunde* 27:491–520. DOI 10.1038/142219a0.

Huxley, J. S. 1958. Evolutionary process and taxonomy, with special reference to grades. *Uppsala Universitets Arsskrift* 6:21–39. ISSN 0372-4654.

Huxley, T. H. 1864. *Lectures on the elements of comparative anatomy*, Vol. 1. London: John Churchill. ISBN 978-1332057917 (2017 reprint).

Hübner, K. 1988. *Critique of scientific reason*. Chicago (IL): University of Chicago Press. ISBN 978-0226357096

Ilyin, V. V. 2003. [*Philosophy of science.*] Moscow: Moscow University Publ. (in Russian).

Impey, O., and A. MacGregor (eds.). 2001. *The origins of museums: The cabinet of curiosities in sixteenth and seventeenth-century Europe*, 2nd ed. London: House of Stratus. ISBN 978-1842321324.

Innis, R. E. 2020. Semiotic framing of thresholds of sense. In: *Between philosophy and cultural psychology*, ed. R. E. Innis, 31–48. Cham (Switzerland): Springer International Publ. DOI 10.1007/978-3-030-58190-9_3.

Ipatiev, A. N. 1971. [*Differential systematics and differential geography of plants.*] Minsk: Vysshaia Shkola (in Russian).

Ivanova-Kazas, O. M. 2004. [*Mythological zoology.*] Saint Petersburg: Saint Petersburg State University Publ. ISBN 5846502121 (in Russian).

Ivin, A. A. 2005. [*Contemporary philosophy of science.*] Moscow: Vysshaia Shkola. ISBN 5060053091 (in Russian).

Ivin, A. A. 2016. [*The logic of assessments and norms. Philosophical, methodological, and applied aspects.*] Moscow: Prospect. ISBN 978-5392195930 (in Russian).

Jamniczky, H. A. 2005. Biological pluralism and homology. *Philosophy of Science* 72:687–98. DOI 10.1086/508108.

Janick, J. 2003. Herbals: The connection between horticulture and medicine. *Hort Thechnology* 13:229–38. DOI 10.21273/HORTTECH.13.2.0229.

Jardine, N. 1967. The concept of homology in biology. *British Journal for the Philosophy of Science* 18:125–39. DOI 10.1093/bjps/18.2.125.

Jardine, N. 1969. A logical basis for biological classification. *Systematic Zoology* 18:37–52. DOI 10.1093/sysbio/18.1.37.

Jardine, N., and R. Sibson. 1971. *Mathematical taxonomy (probability and mathematical statistics)*. New York: J. Wiley. ISBN 978-0471440505.

Jeffrey, C. 1992. *Biological nomenclature*, 3rd ed. Cambridge (UK): Cambridge University Press. ISBN 978-0713129830.

Jensen, R. J. 2009. Phenetics: Revolution, reform or natural consequence? *Taxon* 58:1–11. DOI 10.1002/tax.581008.

Jones, J.-E. 2006. Leibniz and Locke and the debate over species. In *Leibniz selon les nouveaux essais sur l'entendement humain*, ed. F. Duchesneau, and J. Girard. Paris: Bellarmin & Vrin. ISBN 978-2711683499.

Jung, J. (1662) 1747. *Ioachimi Iungi lubecensis* […] *Opuscula botanico-physica*. Coburg: typus Georgii Ottonis. ISBN 978-1173363604 (2011 reprint).

Jurgens, H. W., and C. Vogel. 1965. *Beiträge zur menschlichen Typenkunde*. Stuttgart: Ferdinand Enke. ASIN B0000BJVGZ.

Jussieu, A. L. 1773. Mémoire sur de la famille des renonculacées. *Histoire de l'Academie Royale des Sciences. Année 1773*: 214–40. Paris: l'Imprimerie Royale. https://gallica.bnf.fr/ark:/12148/bpt6k3572b.

Jussieu, A. L. 1789. *Genera plantarum secundum ordines naturales disposita, juxta methodum in horto regio parisiensi exaratam*. Paris: Viduam Herissant et Theophilum Barrois. ISBN 978-0332255293 (2018 reprint).

Jussieu, A. L. 1824. *Principes de la method naturelle des végétaux*. Paris: F. G. Levrault. ISBN 978-2329434711 (2020 reprint).

Kälin J. A. 1945. Die Homologie als Ausdruck ganzheitlicher Baupläne von Typen. *Bulletin de la Société Fribourgeoise des Sciences Naturelles* 37:135–61. ISSN 0366-3256.

Kamelin, R. V. 2004. [*Lectures on plant systematics. Chapters of theoretical systematics of plants.*] Barnaul: Azbuka. ISBN 5939571166 (in Russian).

Kanaev, I. I. 1963. [*Essays from the history of comparative anatomy before Darwin.*] Moscow: Acad. Sci. Publ. (in Russian).

Kanaev, I. I. 1970. [*Goethe as a naturalist.*] Leningrad: Nauka (in Russian).

Kant, I. 2009. The metaphysical foundations of natural science. In *Philosophy of science: An historical anthology*, ed., T. McGrew, M. Alspector-Kelly, and F. Allho, 232–7. Oxford (UK): Blackwell Publ. ISBN 978-1405175425.

Kashkarov, D. N. 1938. [*Basics of the animal ecology.*] Moscow: Medgiz (in Russian).

Kaspar, R. 1977. Der Typus—Idee und Realitat. *Acta Biotheoretica* 26:81–95. DOI 10.1007/BF00048426.

Kaup, J. J. 1844. *Classification der Säugethiere und Vögel*. Darmstadt: C. W. Leske. DOI 10.5962/bhl.title.51492.

Keet, C. M., and A. Artale. 2008. Representing and reasoning over a taxonomy of part–whole relations. *Applied Ontology* 3:91–110. DOI 10.3233/AO-2008-0049.

Kellert, S. H., H. E. Longino, and C. K. Waters (eds.). 2006. Scientic pluralism. Minneapolis (MN): University of Minnesota Press. ISBN 978-0816647637.

Kemp, T. S. 2016. *The origin of higher taxa: Palaeobiological, developmental and ecological perspectives*. Chicago (IL): University of Chicago Press. ISBN 978-0226335957.

Khalidi, M. A. 2016. Mind-dependent kinds. *Journal of Social Ontology* 2:223–46. DOI 10.1515/jso-2015-0045.

Kiriakoff, S. G. 1947. Le cline, une nouvelle catégorie systématique intraspecifique. *Bulletin et Annales de la Société Royale d'Entomologie Belgique* 83:130–40. ISSN: 0374-603.

Kirpotin, S. N. 2005. [Life forms of organisms as organization patterns and spatial environmental factors.] *Zhurnal Obshchei Biologii* 66:239–50. ISSN 0044-4596 (in Russian).

Kitching, I. J., P. L. Forey, C. J. Humphries, et al. 1998. *Cladistics: The theory and practice of parsimony analysis*, 2nd ed. Oxford: Oxford University Press. ISBN 978-0198501381.

Kitts, D. B. 1977. Karl Popper, verifiability, and systematic zoology. *Systematic Zoology* 26:185–94. DOI 10.1093/sysbio/26.2.185.

Klein, L. S. 1991. [*Archaeological typology.*] Leningrad: LNIAO (in Russian).

Kleinman, K. 2009. Biosystematics and the origin of species. Edgar Anderson, W. H. Camp, and the evolutionary synthesis. In *Descended from Darwin: Insights into the history of evolutionary studies, 1900–1970*, ed. J. Cain, and M. Ruse, 73–91. Philadelphia: Amer. Philos. Soc. ISBN 978-1606189917.

Kleisner, K. 2007. The formation of the theory of homology in biological sciences. *Acta Biotheoretica* 55:317–40. DOI 10.1007/s10441-007-9023-8.

Kline, M. 1980. *Mathematics: The loss of certainty*. Oxford: Oxford University Press. ISBN 978-0195030853.

Klix, F. 1983. *Erwachendes denken. Eine Entwicklungsgeschichte der menschlichen Intelligenz*. Berlin: VEB Deutscher Verlag der Wissenschaften. ISBN: 401-4-6635-8210-6.

Kluge, A. G. 1984. The relevance of parsimony to phylogenetic inference. In *Cladistics: Perspectives on the reconstruction of evolutionary history*, ed. T. Duncan, and T. F. Stuessy, 24–38. New York: Columbia University Press. ISBN 0-231054300.

Kluge, A. G. 1989. Metacladistics. *Cladistics* 5:291–4. DOI 10.1111/j.1096-0031.1989.tb00491.x.

Kluge, A. G. 1997. Testability and the refutation and corroboration of cladistic hypotheses. *Cladistics* 13:81–96. DOI 10.1111/j.1096-0031.1997.tb00242.x.

Kluge, A. G. 1998. Total evidence or taxonomic congruence: Cladistics or consensus classification. *Cladistics* 14:151–8. DOI 10.1111/j.1096-0031.1998.tb00328.x.

Kluge, A. G. 2009. Explanation and falsification in phylogenetic inference: Exercises in Popperian philosophy. *Acta Biotheoretica* 57:171–86. DOI 10.1007/s10441-009-9070-4.

Knox, E. B. 1998. The use of hierarchies as organizational models in systematics. *Biological Journal of the Linnean Society* 63:1–49. DOI 10.1111/j.1095–8312.1998.tb01637.x.

Knyazeva, E. N. 2006. [*Epistemological constructivism.*] *Filosofia Nauki & Tekhniki* 12:133–53. ISSN 2413–9084 (in Russian).

Knyazeva, E. N. 2015. [The concept of "Umwelt" by Jacob von Uexküll and its significance for modern epistemology.] *Voprosy Filosofii* 5:30–44. ISSN: 0042-8744 (in Russian).

Kober, G. 2008. *Biology without species: A solution to the species problem.* PhD thesis, Boston University. https://philpapers.org/rec/KOBBWS.

Kolchinsky, E. I., A. K. Sytin, and G. I. Smagina. 2004. [*Natural history in Russia.*] Saint Petersburg: Nestor-Istoria (in Russian).

Kozhara, V. L. 2006. *Classification movement.* Borok: Papanin Institute of Biology RAS (in Russian).

Komarov, V. L. 1940. [*Conception of species in plants.*] Moscow: Acad. Sci. Publ. 1940 (in Russian).

Koort, A. 1936. Beitrage zur Logik des Typusbegriffs, Vol. I. *Acta et Commentationes Universitatis Tartuensis* 38:1–138. ISSN 1406–2283.

Kosko, B. 1993. *Fuzzy thinking: The new science of fuzzy logic.* New York: Hyperion. ISBN 978-0006547136.

Kozo-Polyansky, B. M. 1922. [*Introduction to phylogenetic systematics of higher plants.*] Voronezh: Priroda i Kultura (in Russian).

Krell, F.-T. 2004. Parataxonomy vs. taxonomy in biodiversity studies—pitfalls and applicability of 'morphospecies' sorting. *Biodiversity and Conservation* 13:795–812. DOI 10.1023/B:BIOC.0000011727.53780.63.

Kripke, S. A. 1972. *Naming and necessity.* Cambridge (MA): Harvard University Press. ISBN 9780631128014.

Krivolutskiy, D. A. 1999. Live forms and biological diversity of animals. *Bulleten Moskovskogo obshestva ispytatelej prirody, Biol.* 104:61–7 (in Russian).

Krivolutsky, D. A. 1971. [Modern ideas about animal life forms.] *Ekologia* 3:19–25. ISSN: 0367-0597 (in Russian).

Kuhn, T. S. 1962. *The structure of scientific revolutions.* Chicago: University of Chicago Press. ISBN 978-0226458083.

Kull, K. 2009. Umwelt and modelling. In *The Routledge companion to semiotics*, ed. P. Cobley, 43–56. London: Routledge. ISBN 978-0415440738.

Kunz, W. 2012. *Do species exist? Principles of taxonomic classification.* Weinheim (Germany): Wiley-VCH Verlag. ISBN. 978-3527332076.

Kupriyanov, A. V. 2005. [*Prehistory of biological systematics.*] Saint Petersburg: Saint Petersburg European University (in Russian).

Kuraev, V. I., and F. V. Lazarev. 1988. [*Precision, truth, and growth of knowledge.*] Moscow: Nauka (in Russian).

Kusakin, O. G. 1995. [The crisis of megataxonomy and ways to overcome it.] *Biologia Moria* 21:236–62. ISSN 0134-3475 (in Russian).

Kuznetsova, N. I. 2009. [Presentism and antiquarism—two pictures of the past.] *Arbor Mundi* 15:164–96 (in Russian).

Kwok, H., and Bing, R. 2011. Phylogeny, genealogy and the Linnaean hierarchy: A logical analysis. *Journal of Mathematical Biology* 63:73–108. DOI 10.1007/s00285-010-0364-6.

Lachance, M.-A. 2016. Paraphyly and (yeast) classification. *International Journal of Systematic and Evolutionary Microbiology* 66:4924–9. DOI 10.1099/ijsem.0.001474.

Lakatos, I. 1978. *The methodology of scientific research programmes.* Cambridge (UK): Cambridge University Press. ISBN 978-0521280310.

Lam, K. H. 2020. The realism of taxonomic pluralism. *Metaphysics* 3:1–16. DOI https://doi. org/10.5334/met.32.

Lamarck, J.-B. 1809. *Philosophie zoologique: ou Exposition des considerations relative a l'histoire naturelle des animaux*, Vols. 1 and 2. Paris: Dentu (Libr.), (Mus. Hist. Nat.). [Lamarck, J.-B. 1963. *Zoological philosophy. An exposition with regard to the natural history of animals*. New York: Hafner Publ. ISBN 9780226468105]

Lankester, E. R. 1870. On the use of the term homology in modern zoology, and the distinction between homogenetic and homoplastic agreements. *Annals and Magazine of Natural History Series 5* 6:34–43. DOI 10.1080/00222937008696201.

LaPorte, J. 2017. Modern essentialism for species and its animadversions. In *The Routledge handbook of evolution and philosophy*, ed. R. Joyce, 182–93. New York: Routledge. ISBN 9781138789555.

Larson, J. L. 1967. Linnaeus and the natural method. *Isis* 58:304–20. www.journals.uchicago. edu/doi/pdf/10.1086/350265.

Larson, J. L. 1971. *Reason and experience: The representation of natural order in the work of Carl von Linné*. Berkeley (CA): University of California Press. ISBN 978-0520018341.

Laubichler, M. D. 2000. Homology in development and the development of the homology concept. *American Zoologist* 40:777–88. DOI 10.1093/icb/40.5.777.

Laubichler, M. D., and J. Maienschein (eds.) 2008. *Evolving pathways: Key themes in evolutionary developmental biology*. New York: Cambridge University Press. ISBN 978-0521880244.

Laudan, L. 1981. A confutation of convergent realism. *Philosophy of Science* 48:19–49.

Le Quesne, W. J. 1982. Compatibility analysis and its applications. *Zoological Journal of the Linnean Society* 74:267–75. DOI 0.1111/j.1096–3642.1982.tb01151.x.

Lee, M. S. Y. 2004. The molecularization of taxonomy. *Invertebrate Systematics* 18:1–6. DOI 10.1071/IS03021.

Lee, M., and M. Wolsan. 2002. Integration, individuality and species concepts. *Biology and Philosophy* 17 :651–60. DOI 10.1023/A:1022596904397.

Lefevre, W. 2001. Natural or artificial systems? The eighteen-century controversy on classification of animals and plants and its philosophical contents. In *Between Leibniz, Newton, and Kant. Philosophy and science in the eighteenth century*, ed. W. Lefevre, 191–209. Dordrecht: Kluwer Academic. ISBN 978-9048157747.

Legendre, P. 1972. The definition of systematic categories in biology. *Taxon* 21:381–406. DOI 10.2307/1219102.

Leibniz, G. W. 1900. *Système nouveau de la Nature et de la communication des substances, aussi bien que de l'union qu'il y a entre l'âme et le corps*. Paris: Félix Alcan. ISBN 978-2080707741 (1994 reprint).

Lektorsky, V. A. 2001. [*Classical and non-classical epistemology.*] Moscow: Editorial URSS (in Russian).

Lennox, J. G. 1980. Aristotle on genera, species, and "the more and the less". *Journal of the History of Biology* 13:321–46. DOI 10.33134/rds.314.

Lenoir, T. 1988. Kant, von Baer, and causal-historical thinking in biology. *Poetics Today* 9:103–15. DOI 10.1002/bewi.19850080208.

Leontiev, D. V., and A. Yu. Akulov. 2004. [Ecomorphema of the organic world: An experience of reconstruction.] *Zhurnal Obshchei Biologii* 65:500–26. ISSN 0044-4596 (in Russian).

Lesch, J. E. 1990. Systematics and the geometrical spirit. In *The quantifying spirit in the 18th century*, ed. T. Frangsmyr, J. L. Heilbron, and R. E. Rider, 73–111. Berkeley (CA): University of California Press. ISBN 978-0585071985.

Lévi-Strauss, C. 1966. *The savage mind*. London: Weidenfeld & Nicolson. ISBN 978-0226474847.

Levit, G. S., and K. Meister. 2006. The history of essentialism vs. Ernst Mayr's "Essentialism Story": A case study of German idealistic morphology. *Theory in Biosciences* 124:281–307. DOI 10.1016/j.thbio.2005.11.003.

Lewens, T. 2009a. What is wrong with typological thinking? *Philosophy of Science* 76:355–71. DOI 10.1086/649810.

Lewens, T. 2009b. Evo-devo and "typological thinking": An exculpation. *Journal of Experimental Zoology* 312B:789–96. DOI 10.1002/jez.b.21292.

Lewontin, R. C. 1964. Models, mathematics and metaphors. In *Form and strategy in science. Studies dedicated to Joseph Henry Woodger on the occasion of his seventieth birthday*, ed. J. R. Gregg, and F. T. C. Harris, 274–96. Dordrecht: D. Reidel. ISBN 978-9401036030.

Lindberg, D. C. 2007. *The beginnings of Western science. The European scientific tradition in philosophical, religious, and institutional context, prehistory to a.d. 1450*. Chicago: University of Chicago Press. ISBN 978-0226482057.

Lindley, J. 1835. *A key to structural, physiological, and systematics botany*. London: Longman.

Lindley, J. 1836. *The vegetable kingdom: or The structure, classification, and uses of plants, illustrated upon the natural system*. London: Bradbury & Evans. ISBN 978-1371123994 (2016 reprint).

Lindroth, S. 1983. The two faces of Linnaeus. In *Linnaeus. The man and his work*, ed. T. Fraingsmyr, 1–62. Berkeley (CA): University of California Press. ISBN 9780520045682.

Lines, J. L., and T. R. Mertens. 1970. *Principles of biosystematics*. Chicago: Educational Methods. ISBN 978-1682862650.

Linnaeus, C. 1735. *Systema naturae, sive regna tria naturae systematice proposita per classes, ordines, genera, & species*. Leiden: Haak.

Linnaeus, C. 1736. *Fundamenta botanica: quae majorum operum prodromi instar theoriam scientiae botanice*. Amsterdam: Salomonem Schouten. ISBN 978-9060460641 (2010 reprint).

Linnaeus, C. 1751. *Philosophia botanica in qua explicantur fundamenta botanica cum definitionibus partium, exemplis terminorum, observationibus rariorum, adiectis figuris aeneis*. Stockholm: Godofr. Kiesewetter. ISBN 978-1104629731 (2009 reprint).

Linnaeus, C. 1766. *Systema naturae: per regna tria natura, secundum classes, ordines, genera, species, cum characteribus, differentiis, synonymis, locis*, Vol. 1, Part I. Ed. duodecima reformata. Holmiae: Impensis direct. Laurentii Salvii. ISBN 9780243506941 (2016 reprint).

Linnaeus, C. 1767. *Systema naturae*, Vol. 1, Part II. Ed. duodecima reformata. Holmiae: Impensis direct. Laurentii Salvii. ISBN 9780243506941 (2016 reprint).

Look, B. 2009. Leibniz and Locke on real and nominal essences. In *Branching off: The early moderns in quest for the unity of knowledge*, ed. V. Alexandrescu, 380–409. Bucharest: Zeta Books. ISBN 978-9731997421.

Lopez, A., S. Atran, J. Coley, et al. 1997. The tree of life: Universal and cultural features of folkbiological taxonomies and inductions. *Cognitive Psychology* 32:251–95. DOI 10.1006/cogp.1997.0651.

Loreau, M. 2010. Linking biodiversity and ecosystems: Towards a unifying ecological theory. *Philosophical Transactions of the Royal Society, B Biological Sciences* 365:49–60. DOI 10.1098/rstb.2009.0155.

Löther, R. 1972. *Die Beherrschung der Mannigfaltigkeit. Philosophische Grundlagen der Taxonomie*. Jena: Gustav Fischer. www.zvab.com/Beherrschung-Mannigfaltigkeit-Philosophische-Grundlagen-Taxonomie-Löther/12728736685/bd.

Lotsy, J. P. 1931. On the species of the taxonomist in its relation to evolution. *Genetica* 13:1–16. DOI 10.1007/BF01725037.

Love, A. C. 2007. Functional homology and homology of function: Biological concepts and philosophical consequences. *Biology and Philosophy* 22:691–708. DOI 10.1007/s10539-007-9093-7.

Love, A. C. 2009. Typology reconfigured: From the metaphysics of essentialism to the epistemology of representation. *Acta Biotheoretica* 57:51–75. DOI 10.1007/BF00 385837.

Lovejoy, A. O. 1936. *The great chain of being: A study of the history of an idea.* Cambridge (MA): Harvard University Press. ISBN 978-0674361539.

Løvtrup, S. 1973, Classification, convention and logics. *Zoologica Scripta* 2:119–6l. DOI 10.1111/j.1463–6409.1974.tb00739.x.

Løvtrup, S. 1975. On phylogenetic classification. *Acta Zoologica Cracowienzia* 20:499–523. DOI 10.1111/j.1463–6409.1975.tb00724.x.

Løvtrup, S. 1977. Phylogenetics: Some comments on cladistic theory and methods. In *Major patterns of vertebrate evolution*, ed. M. K. Hecht, P. C. Goody, and B. M. Hecht, 80–122. New York: Plenum Press. ISBN 978-0306356148.

Lubischew, A. A. 1969. Philosophical aspects of taxonomy. *Annual Review of Entomology* 14:19–38. DOI 10.1146/annurev.en.14.010169.000315.

Ludwig, D. 2017. Indigenous and scientific kinds. *The British Journal for the Philosophy of Science* 68:187–212. DOI 10.1093/bjps/axv031.

Lukashevich, V. K. 1991. [*Scientific method: Its structure, justification, development.*] Minsk: Navuka i Tekhnika (in Russian).

Luschei, E. C. 1962. *The logical system of Lesniewski.* Amsterdam: North Holland Publ. Co. ISBN 978-0720422160.

Lyubarsky, G. Yu. 1992. [Biostylistics and the problem of classification of the life forms.] *Zhurnal Obshchei Biologii* 53:649–61. ISSN 0044-4596 (in Russian).

Lyubarsky, G. Yu. 1996. *Archetype, style and rank in biological systematics.* Moscow: KMK Sci. Pres. ISBN 5873170053 (in Russian).

Lyubarsky, G. Yu. 2015. [*The birth of science. Analytical morphology, classification system, scientific method.*] Moscow: YaSK Publishing (in Russian).

Lyubarsky, G. Yu. 2018. [*Origins of hierarchy: The history of taxonomic rank.*] Moscow: KMK Sci. Press. ISBN 978-5950082962 (in Russian).

Lyubishchev, A. A. 1923. [On the form of the natural system of organisms.] *Communiucations of Biological Research Institute of Perm University* 2:99–110 (in Russian).

Lyubishchev, A. A. 1968. [Problems of systematics.] In [*Problems of evolution*], Vol. 1, 7–29. Novosibirsk: Nauka (in Russian).

Lyubishchev, A. A. 1972. [To the logic of systematics.] [*Problems of evolution*], Vol. 2, 45–68. Novosibirsk: Nauka (in Russian).

Lyubishchev, A. A. 1975. [On some postulates of general systematics. Theoretical applications of the methods of mathematical logic.] [*Notes of LOMI Scientific Seminar*], 49: 159–175 (in Russian).

MacCormac, E. R. 1985. *A cognitive theory of metaphor.* Cambridge (MA): The MIT Press. ISBN 9780262132121.

MacLeay, W. S. 1819. *Horae entomologicae, or Essay on the annulose animals*, Vol. 1, Pt. 1. London: S. Bagster. DOI 10.5962/bhl.title.48636.

Maddalon, M. 2003. Recognition and classification of natural kinds. In *Nature knowledge: Ethnoscience, cognition, and utility*, ed. S. Glauco, and O. Gherardo, 23–37. New York: Berghan Books Publ. ISBN 978-1571818225.

Magnol, P. 1689. *Prodromus historiae generalis plantarum in quo familiae plantarum per tabulas disponuntur.* Monspelij: ex typograpjia Gabrielis & Honorati Pech. https://gallica.bnf.fr/ark:/12148/bpt6k98412v.image.

Magnus, P. D. 2014. John Stuart Mill on taxonomy and natural kinds. *Journal of the International Society for the History of Philosophy of Science* 5:269–80. https://philarchive.org/archive/MAGJSM.

Mahner, M. 1993. What is a species? *Journal for General Philosophy of Science* 24:103–26. DOI 10.1007/BF00769517.

Mahner, M., and M. Bunge. 1997. *Foundations of biophilosophy*. Frankfurt: Springer Verlag. ISBN 978-3662033685.

Makovelsky, A. O. 2004. [*History of logic.*] Moscow: Kuchkovo Pole (in Russian).

Mallet, J. 2001. Species, concept of. In *Encyclopedia of biodiversity*, Vol. 5, ed. S. Levin, 427–40. London: Academic Press. DOI 10.1016/B978-0-12-384719-5.00131-3.

Mallet, J. 2013. Subspecies, semispecies, superspecies. In *Encyclopedia of biodiversity*, 2d Ed. Vol. 5, ed. S. Levin, 45–8. London: Academic Press. DOI 10.1016/B0-12-226865-2/00261-3.

Mamchur, E. A. 2004. [*Objectivity of science and relativism (To discussions in modern epistemology).*] Moscow: Institute of Philosophy RAN (in Russian).

Margolis, E., and S. Laurence. 2011. Concepts. In *The Stanford encyclopedia of philosophy*, ed. E. N. Zalta. https://plato.stanford.edu/entries/ concepts/.

Marradi, A. 1990. Classification, typology, taxonomy. *Quality & Quantity* 24:129–57. DOI 10.1007/BF00209548.

Martynov, A. V. 2011. [*Ontogenetic systematics, and a new model of bilaterian evolution.*] Moscow: KMK Sci. Press. ISBN 978-5873177509 (in Russian).

Martynov, A. V. 2012. Ontogenetic systematics: The synthesis of taxonomy, phylogenetics, and evolutionary developmental biology. *Paleontological Journal* 46:833–64. DOI 10.1134/S0031030112080072.

Matthen, M. 2007. Defining vision: What homology thinking contributes. *Biology and Philosophy* 22:675–89. DOI 10.1007/s10539-007-9088-4.

Mavrodiev, E. V. 2002. [Once again about "Gregg's paradox" and its decision.] *Zhurnal Obshchei Biologii* 63:236–8. ISSN 0044-4596 (in Russian).

Maxwell, S. J., B. C. Congdon, and T. L. Rymer. 2020. Essentialistic pluralism: The theory of spatio-temporal positioning of species using integrated taxonomy. *Proceedings of the Royal Society of Queensland* 124:1–17. ISSN 0080-469X.

Mayden, R. L. 1997. A hierarchy of species concepts: The denouement in the saga of the species problem. In *Species. The units of biodiversity*, ed. M. F. Claridge, A. H. Dawah, and M. R. Wilson, 381–424. London: Chapman & Hall. ISBN 978-0412631207.

Mayden, R. L., and E. O. Wiley 1993. The foundations of phylogenetic systematics. In *Systematics, historical ecology, and North American freshwater fishes*, ed. R. L. Mayden, 114–85. Stanford: Stanford Univ. Press. ISBN 978-0804721622.

Mayr, E. 1942. *Systematics and the origin of species, from the viewpoint of zoologist*. New York: Columbia University Press. ISBN 978-0674862500.

Mayr, E. 1963. *Animal species and evolution*. Cambridge (MA): Harvard University Press. ISBN 978-0674865327.

Mayr, E. 1965. Classification and phylogeny. *American Zoologist* 5:165–74. DOI 10.1093/icb/5.

Mayr, E. 1969. *Principles of systematic zoology*. New York: McGraw Hill. ISBN 978-0070411432.

Mayr, E. 1982. *The growth of biological thought: Diversity, evolution, and inheritance*. Cambridge (MA): Belknap Press. ISBN 978-0674364462.

Mayr, E. 1988. *Toward a new philosophy of biology*. New York: Cambridge University Press. ISBN 978-0674896666.

Mayr, E., and P. D. Ashlock. 1991. *Principles of systematic zoology*, 2nd ed. New York: McGraw Hill. ISBN 978-0070411449.

Mayr, E., and W. J. Bock. 2002. Classifications and other ordering systems. *Journal of Zoological Systematics and Evolutionary Research* 40:169–94. DOI 10.1046/ j.1439-0469.2002.00211.x.

McCray, A. T. 2006. Conceptualizing the world: Lessons from history. *Journal of Biomedical Informatics* 39:267–73. DOI 10.1016/j.jbi.2005.08.007.

McGhee, G. R. 1999. *Theoretical morphology: The concept and its application.* New York: Columbia University Press. ISBN 978-0231106160.

McKenna, M. C., and S. K. Bell. 1997. *Classification of mammals above the species level.* New York: Columbia University Press. ISBN 978-0231110136.

McNeill, J. 1982. Phylogenetic reconstruction and phenetic taxonomy. *Zoological Journal of the Linnean Society* 74:337–44. DOI 10.1111/j.1096–3642.1982.tb01156.x.

McOuat, G. 2001. From cutting nature at its joints to measuring it: New kinds and new kinds of people in biology. *Studies in History and Philosophy of Biological and Biomedical Sciences* 32:613–45. DOI 0.1016/S0039-3681(01)00027-9.

McOuat, G. R. 1996. Species, rules and meaning: The politics of language and the ends of definitions in 19th century natural history. *Studies in the History and Philosophy of Science* 21:413–519. DOI 10.1016/0039-3681(95)00060-7.

Meacham, C. A., and T. Duncan. 1987. The necessity of convex groups in biological classification. *Systematic Botany* 12:78–90. DOI 10.2307/2419216.

Mechkovskaya, B. N. 2007. [*Semiotics: Language. Nature. Culture.*] Moscow: Akademia. ISBN 978-5-7695-4467-5 (in Russian).

Medin, D. L., and S. Atran. 2004. The native mind: Biological categorization and reasoning in development and across cultures. *Psychological Review* 111:960–83. DOI 10.1037/ 0033-295X.111.4.960.

Mednikov, B. M. 1980. [Application of the methods of genosystematics for elaboration of the systema of Chordata.] In [*Molecular bases of genosystematics*], ed. A. S. Antonov, 203–15. Moscow: Moscow University Publ. (in Russian).

Meier, R., and S. Richter. 1992. Suggestion for a more precise usage of proper names of taxa. Ambiguities related to the stem lineage concept. *Zeitschrift fur Zoologische Systematik und Evolutionsforschung* 30:81–8. DOI 10.1111/j.1439-0469.1992.tb00159.x.

Mensch, J. 2013. *Kant's organicism. Epigenesis and the development of critical philosophy.* Chicago: University of Chicago Press. ISBN 978-0226021980.

Merkulov, I. P. (ed.). 1996. [*Evolutionary epistemology: Problems and perspectives.*] Moscow: ROSSPEN. ISBN 520101884X (in Russian).

Meyen, S. V. 1973. Plant morphology in its nomothetical aspects. *Botanical Review* 39:205–60. DOI 10.1007/BF02860118.

Meyen, S. V. 1977. [Taxonomy and meronomy.] In [*Methodological issues in geological sciences*], 25–33. Kïev: Naukova Dumka (in Russian).

Meyen, S. V. 1978. [Basic aspects of the organismal typology.] *Zhurnal Obshchei Biologii* 39:495–508. ISSN 0044-4596 (in Russian).

Meyen, S. V. 1984. [Principles of historical reconstructions in biology.] In [*Systemity and evolution*], 7–32. Moscow: Nauka (in Russian).

Meyen, S. V. 1988. Principles and methods of paleontological systematics. In *Contemporary paleontology*, Vol. 1, ed. V. V. Menner and V. P. Makridin, 447–66. Moscow: Nedra. ISBN: 5-247-00152-4 (in Russian).

Meyen, S. V., and Yu. A. Shreyder. 1976. [Methodological issues of the theory of classification.] *Voprosy Filosofii*, 12:67–79. ISSN: 0042-8744 (in Russian).

Meyer, A. 1935. Über typologische und phylogenetische Systematik. In *Proceedings of the VI International Botanical Congress, Amsterdam*, Vol. 2:58–60.

Meyer-Abich, A. 1934. Ideen und Ideale der biologischen Erkenntnis. Beiträge zur Theorie und Geschichte der biologischen Ideologien. *Bios* 1:1–202. ASIN B000NJI49M.

Meyer-Abich, A. (ed.). 1949. *Biologie der Goethezeit. Klassische Abhandlungen über die Grundlagen und Hauptprobleme der Biologie von Goethe und den grossen Naturforschern seiner Zeit.* Stuttgart: Hippokrates-Verlag Marquardt. www.booklooker. de/Bücher/Adolf-Hrsg-Beiträge-von-und-über-Johann-Wolfgang-von-Goethe-Meyer-Abich+Biologie-der-Goethezeit/id/A00xiirB01ZZy.

Miall, L. C. 1912. *The early naturalists, their lives and work (1530–1789).* London: Macmillan. ISBN 978-0028492209.

Michener, C. D. 1957. Some bases for higher categories in classification. *Systematic Zoology* 6:160–73. DOI 10.2307/sysbio/6.4.160.

Michener, C. D. 1962. Some future developments in taxonomy. *Systematic Zoology* 12:151–72. DOI 10.2307/2411757

Michener, C. D., and R. R. Sokal. 1957. A quantitative approach to a problem in classification. *Evolution* 11:130–62. DOI 10.1111/j.1558–5646.1957.tb02884.x.

Mickevich, M. F. 1982. Transformation series analysis. *Systematic Zoology* 31:461–78. DOI 10.1093/sysbio/31.4.461.

Mikhailov, A. I. (ed.). 1983. [*Research on non-classical logic and formal systems.*] Moscow: Nauka (in Russian).

Mikhailov, K. E. 2003. [Typological comprehension of "biological species", and the way of stabilization of near-species taxonomy in birds.] *Ornitologia* 30:9–24 (in Russian).

Mikulinsky, S. R., and M. G. Yaroshevsky (eds.). 1977. [*Schools in science.*] Moscow: Nauka (in Russian).

Mill, J. S. 1882. *A system of logic, ratiocinative and inductive, being a connected view of the principles of evidence, and the methods of scientific investigation,* 8th ed. New York: Harper & Brothers. www.gutenberg.org/files/27942/27942-pdf.pdf.

Miller, E. H. (ed.). 1985. *Museum collections: Their roles and future in biological research.* Victoria: British Columbia Provincial Museum. ISBN 978-0771884641.

Miller, E. H. 1993. Biodiversity research in museums: A return to basics. In *Our living legacy: Proceedings of a Symposium on Biological Diversity,* ed. M. A. Fenger, and E. H. Miller, 141–73. Victoria: Royal British Columbia Museum. ISBN 978-0771893551.

Miller, G. A. 1996. Contextuality. In *Mental models in cognitive science: Essays in honour of Phil Johnson-Laird,* ed. A. Garnham, and J. Oakhill, 1–18. Erlbaum (UK): Psychology Press. ISBN 978-0863774485.

Milne-Edwards, H. 1844. Considerations sur quelques principes relatifs a la classification naturelle des animaux. *Annales des Sciences Naturelles, Ser.* 3 1:65–99. www. biodiversitylibrary.org/item/47984#page/7/mode/1up.

Minelli, A. 1994. *Biological systematics. The state of the art.* London: Chapman & Hall. ISBN 978-0-412-36440-2.

Minelli, A. 1998. Molecules, developmental modules and phenotypes: A combinatorial approach to homology. *Molecular Phylogenetics and Evolution* 9:340–7. DOI 10.1006/mpev.1997.0490.

Minelli, A. 2003. *The development of animal form: Ontogeny, morphology, and evolution.* Cambridge (MA): Cambridge University Press. ISBN 978-0511072413.

Minelli, A. 2015. Biological systematics in the evo-devo era. *European Journal of Taxonomy* 125:1–23. DOI 10.5852/ejt.2015.125.

Minelli, A. 2016. Tracing homologies in an ever-changing world. *Rivista di Estetica* 62:40–55. DOI 10.4000/estetica.1174.

Minelli, A., and G. Fusco. 2013. Homology. In *The philosophy of biology: A companion for educators,* ed. K. Kampourakis, 289–322. Dordrecht: Springer Science + Business Media. DOI 10.1007/978-94-007-6537-5_15.

Minelli, A., and T. Pradeu (eds.). 2014. *Towards a theory of development.* London: Oxford University Press. ISBN 978-0199671434.

Minelli, A., G. Fusco, and S. Sartori. 1991. Self-similarity in biological classifications. *BioSystems* 26:89–97. DOI 10.1016/0303-2647(91)90040-R.

Mirabdullaev, I. M. 1997. [Biological systematics: Phylogenetic and ecomorphological approaches.] *Vestnik Zoologii* 31:11–5. http://mail.izan.Kïev.ua/vz-pdf/1997/N4/VZ_T31_N4_1997-02-Mirabdullayev.pdf (in Russian).

Mirkin, B. M. 1985. [*Theoretical foundations of modern phytocenology.*] Moscow: Nauka (in Russian).

Miroshnikov, Yu. I. 2010. [Ontological, logical-ognostic, and axiological bases of scientific classification.] In [*New ideas in scientific classification*], ed. Yu. I. Miroshnikov, and M. P. Pokrovskiy, 78–119. Yekaterinburg: Institute of Philosophy and Law RAS. ISBN 978-5769121203 (in Russian).

Mishler, B. D. 1990. Phylogenetic analogies in the conceptual development of science. *Proceedings of the Biennial Meeting of the Philosophy of Science Association (1990)*, Vol. 2: 225–35. Chicago: University of Chicago Press. https://doi.org/10.1086/psaproc bienmeetp.1990.2.193070.

Mishler, B. D. 2009. Three centuries of paradigm changes in biological classification: Is the end in sight? *Taxon* 58:61–7. DOI 10.1002/tax.581009.

Mithen, S. 2006. Ethnobiology and the evolution of the human mind. *Journal of the Royal Anthropological Institute of Great Britain and Ireland* (N.S.) 12, Suppl. 1:S45–61. DOI 10.1111/j.1467-9655.2006.00272.x.

Mivart, G. J. 1870. On the use of the term "homology". *Annals and Magazine of Natural History, Ser.* 4, 6:113–27. www.biodiversitylibrary.org/item/93156#page/1/mode/1up.

Morgun, D. V. 2006. [Ontological sources of non-classical biology.] *Vestnik Moskovskogo Universiteta, Ser. 7 (Philosophy)* 1:42–58. ISSN 0579-9368 (in Russian).

Morison, R. 1672. *Plantarum umbelliferarum distributio nova, per tabulas cognationes et affinitatis ex libro naturae.* Oxford: Theatro Sheldoniano. ISBN 978-1240831456 (2011 reprint).

Morreau, M. 2010. It simply does not add up: Trouble with overall similarity. *Journal of Philosophy* 107:469–90. DOI 10.5840/jphil2010107931.

Morris, B. 1984. The pragmatics of folk classification. *Journal of Ethnobiology* 4:45–60. https://ethnobiology.org/sites/default/files/pdfs/JoE/4-1/Morris1984.pdf.

Morrison, D. A., M. J. Morgan, and S. A. Kelchner. 2015. Molecular homology and multiple-sequence alignment: An analysis of concepts and practice. *Australian Systematic Botany* 28:46–62. DOI 10.1071/SB15001.

Morrone, J. 2018 The spectre of biogeographical regionalization. *Journal of Biogeography* 45:1–7. DOI 10.1111/jbi.13135.

Moskvitin, A. Yu. 2013. [*Typology in social cognition: Philosophical and methodological analysis.*] Saint Petersburg: Saint Petersburg State University of Services and Economy Publ. ISBN: 978-5228006614 (in Russian).

Mishler, B. D. 1990. Phylogenetic analogies in the conceptual development of science. *Proceedings of the Biennial Meeting of the Philosophy of Science Association (1990)*, Vol. 2: 225–35. Chicago: University of Chicago Press. https://doi.org/10.1086/psaproc bienmeetp.1990.2.193070.

Moss, W. W., and J. A. Hendrickson. 1973. Numerical taxonomy. *Annual Review of Entomology* 18:227–58. ISSN 0066-4170.

Müller, G. B. 2003. Homology: The evolution of morphological organization. In *Origination of organismal form: Beyond the gene in developmental and evolutionary biology*, ed. G. B. Muller, and S. A. Newman, 52–69. Cambridge (MA): The MIT Press. ISBN 978-0262134194.

Murphy, G. L. 2002. *The big book of concepts.* Cambridge (MA): The MIT Press. ISBN 978-0262632997.

Myers, G. S. 1952. The nature of systematic biology and of a species description. *Systematic Zoology* 1:106–11. DOI 10.1093/sysbio/1.3.106.

Naef, A. 1913. Studien zur generellen Morphologie der Mollusken, Vol. 1. Über Torsion und Asymmetrie der Gastropoden. *Ergebnisse Fortschritte der Zoologie* 3:73–164. ISSN 0071-7991.

Naef, A. 1917. *Die individuelle Entwicklung organischer Formen als Urkunde ihrer Stammesgeschichte (kritische Bemerkungen uber das sogenannte "biogenetische Grundgesetz")*. Jena: Gustav Fischer. ASIN B00IV88ZEG.

Naef, A. 1919. Idealistische Morphologie und Phylogenetik (zur Methodik der systematischen Morphologie). *Zeitschrift für Vererbungslehre* 22:279–82. DOI 10.1007/BF01882544.

Naef, A. 1931. Allgemeine Morphologie. I. Die Gestalt als Begriff und Idee. In *Handbuch der vergleichenden Anatomie der Wirbeltiere*, Vol. 1, ed. L. Bolk, E. Göppert, and E. Kallius, 77–118. Berlin: Urban und Schwarzenberg. ISBN 978-3743449770.

Nalimov, V. V. 1979. [*A probabilistic model of language. On relation between natural and artificial languages.*] Moscow: Nauka (in Russian).

Nathan, M. J. 2017. Pluralism is the answer! What is the question? *Philosophy, Theory, and Practice in Biology* 11:1–14. www.du.edu/ahss/philosophy/media/documents/mnathan/nathanpluralismisthequestion.pdf.

Needham, J. G. 1911. The law that inheres in nomenclature. *Science, New Series*, 33:813–16. DOI 10.1126/science.33.856.813.

Neff, N. A. 1986. A rational basis for a apriori character weighting. *Systematic Zoology* 35:110–23. DOI 10.1093/sysbio/35.1.110.

Nei, M., and S. Kumar. 2000. *Molecular evolution and phylogenetics*. Oxford: Oxford University Press. ISBN 978-0195135855.

Nelson, G. 1978. Ontogeny, phylogeny, paleontology, and the biogenetic law. *Systematic Zoology* 27:324–45. DOI 10.2307/2412883.

Nelson, G. 1979. Cladistic analysis and synthesis: principles and definitions, with a historical note on Adanson's Familles des Plantes (1763–1764). *Systematic Zoology* 28:1–21. DOI 10.2307/2412995.

Nelson, G., and N. Platnick. 1981. *Systematics and biogeography: Cladistics and vicariance*. New York: Columbia University Press. ISBN 978-0231045742.

Nelson, G. J., and N. Platnick. 1991. Three-taxon statements: A more precise use of parsimony? *Cladistics* 7:351–66. DOI 10.1111/j.1096-0031.1991.tb00044.x.

Newmaster, S. G., R. Subramanyam, and R. F. Ivanoff. 2006. Mechanisms of ethnobiological classifications. *Ethnobotany* 18:4–26. DOI 10.1515/9781400862597.

Ng, M., and M. F. Yanofsky. 2001. Function and evolution of the plant MADS-box gene family. *Nature Reviews Genetics* 2:186–96. DOI 10.1038/35056041.

Nickelsen, K. 2006. *Draughtsmen, botanists and nature: The constructing eighteenth-century botanical illustrations*. Dordrecht: Springer. ISBN 978-1402048197.

Niklas, K. J. 1997. *The evolutionary biology of plants*. Chicago: University of Chicago Press. ISBN 978-0226580821.

Nixon, K. C., and J. M. Carpenter. 1993. On outgroup. *Cladistics* 9:413–26. DOI 10.1111/j.1096-0031.1993.tb00234.x.

Nixon, K. C., and J. M. Carpenter. 2000. On the other "phylogenetic systematics". *Cladistics* 16:298–318. DOI 10.1111/j.1096-0031.2000.tb00285.x.

Nowak, M. A. 1992. What is a quasispecies? *Trends in Ecology and Evolution* 7:118–21. DOI 10.1007/3-540-26397-7_1.

Oderberg, D. S. 2009. *Real essentialism*. Abingdon: Routledge. ISBN 978-0415872126.

Odum, E. P. 1953. *Fundamentals of ecology*. Philadelphia: Saunders. ISBN 978-0534420666.

Ogilvie, B. W. 2006. *The science of describing: Natural history in Renaissance Europe*. Chicago: University of Chicago Press. ISBN 978-0226620886.

O'Hara, R. J. 1988. Homage to Clio, or toward an historical philosophy of evolutionary biology. *Systematic Zoology* 37:142–55. DOI 10.2307/2992272.

O'Hara, R. J. 1991. Representations of the natural system in the nineteenth century. *Biology and Philosophy* 6:255–74. DOI 10.1007/BF02426840.

O'Hara, R. J. 1992. Telling the tree: Narrative representation and the study of evolutionary history. *Biology and Philosophy* 7:135–60. DOI 10.1007/BF00129880.

O'Hara, R. J. 1996. Trees of history in systematics and philology. In *Systematics biology as an historical science*, ed. M. C. C. de Pinna, and M. Ghiselin, 81–8. Milan: Museo Civico di Storia Naturale di Milano. ISBN 978-0520260856.

O'Hara, R. J. 1997. Population thinking and tree thinking in systematics. *Zoologica Scripta* 26:323–9. DOI 10.1111/j.1463–6409.1997.tb00422.x.

Oken, L. 1847. *Elements of physiophilosophy*. London: Royal Society. ISBN 978-1150213281 (2012 reprint).

Oliver, F. W. 1913. *Makers of British botany: A collection of biographies by living botanists*. London: Cambridge University Press. ISBN 978-1376729696.

Orlov, Yu. K. 1976. [Generalized Zipf–Mandelbrot law and frequency structures of information units of different levels.] In [*Computational linguistics*], ed. E. K. Guseva, 179–202. Moscow: Nauka (in Russian).

Orton, G. L. 1955. The role of ontogeny in systematics and evolution. *Evolution* 9:75–83. DOI 10.1111/j.1558–5646.1955.tb01515.x.

Osborn, H. F. 1902. Homoplasy as a law of latent or potential homology. *American Naturalist* 36:259–71. DOI 10.1086/278118.

Oskolski, A. 2011. The taxon as an ontological problem. *Biosemiotics* 4:201–22. DOI 10.1007/BF00128837.

Owen, R. 1843. *Lectures on the comparative anatomy and physiology of the invertebrate animals*. London: Longman. ISBN 978-1290925495 (2016 reprint).

Owen, R. 1848. *On the archetype and homologies of the vertebrate skeleton*. London: J. van Voorst. ISBN 978-1230235370 (2013 reprint).

Padial, J. M., S. Castroviejo-Fisher, J. Kohler, et al. 2009. Deciphering the products of evolution at the species level: The need for an integrative taxonomy. *Zoologica Scripta* 38:431–47.

Padial, J. M., A. Miralles, I. Riva, et al. 2010. The integrative future of taxonomy. *Frontiers in Zoology* 7:16. DOI 10.1186/1742-9994-7-16.

Page, R. D. M. 2005. Phyloinformatics: towards a phylogenetic database. In *Data mining in bioinformatics*, ed. J. T. L. Wang, M. J. Zaki, H. Toivonen, et al., 219–41. Stuttgart: Springer Verlag. ISBN 978-0081001073.

Pallas, P. S. 1766. *Elenchus zoophytorum: Sistens generum adumbrationes generaliores et specierum cognitarum succinatas descriptiones, cum selectis auctorum synonymis*. The Hague: Petrum van Cleef. ISBN 978-0282325183 (2018 reprint).

Panchen, A. L. 1992. *Classification, evolution, and the nature of biology*. New York: Cambridge University Press. ISBN 978-0521315784.

Panchen, A. L. 1994. Richard Owen and the concept of homology. In *Homology: The hierarchical basis of comparative morphology*, ed. B. K. Hall, 21–62. London: Academic Press. ISBN 978-0123195838.

Panova, N. S., and Yu. A. Shreyder. 1975. [The principle of duality in classification theory.] In [*Scientific and technical information*], Ser. 2: 3–10. Moscow: VINITI (in Russian).

Patterson, C. 1982. Morphological characters and homology. In *Problems of phylogenetic reconstruction*, ed. K. A. Joysey, and A. E. Friday, 21–74. London: Academic Press. ISBN 978-0123912503.

Patterson, C. 1983. How does ontogeny differ from phylogeny? In *Development and evolution*, ed. B. C. Goodwin, N. Holder, and C. C. Wylie, 1–31. Cambridge (UK): Cambridge University Press. ISBN 978-0521249492.

Patterson, C. 1988. The impact of evolutionary theories on systematics. In *Prospects in systematics*, ed. D. L. Hawksworth, 59–91. Oxford: Clarendon Press. ISBN 978-0198577072.

Pavlinov, I. Ya. 1989. [*Methods of cladistics.*] Moscow: Moscow University Publ. (in Russian).

Pavlinov, I. Ya. 1990. *Cladistic analysis (methodological problems)*. Moscow: Moscow University Publ. ISBN 5211009185 (in Russian).

Pavlinov, I. Ya. 1992. [If there is the biological species, or what is the "harm" of taxonomy.] *Zhurnal Obshchei Biologii* 53:757–67. ISSN 0044-4596 (in Russian).

Pavlinov, I. Ya. 1995. [Classification as a hypothesis: Entering the problem.] *Zhurnal Obshchei Biologii* 56:411–24. ISSN 0044-4596 (in Russian).

Pavlinov, I. Ya. 1998. [To the problem of axiomatic justification of evolutionary cladistics.] *Zhurnal Obshchei Biologii* 59:586–605. ISSN 0044-4596 (in Russian).

Pavlinov, I. Y. 2003. The new phylogenetics: An essay. *Wulfenia* 10:1–14. ISSN: 1561-882X.

Pavlinov, I. Ya. 2005. *Introduction to contemporary phylogenetics*. Moscow: KMK Sci. Press. ISBN 978-5041154813 (in Russian).

Pavlinov, I. Ya. 2006. [Classical and non-classical systematics: Where does the boundary pass?] *Zhurnal Obshchei Biologii* 67:83–108. ISSN 0044-4596 (in Russian).

Pavlinov, I. Ya. 2007a. [Phylogenetic thinking in contemporary biology.] *Zhurnal Obshchei Biologii* 68:19–34. ISSN 0044-4596 (in Russian).

Pavlinov, I. Ya. 2007b. [Etudes on metaphysics of contemporary systematics.] *Archives of the Zoological Museum of Moscow State University* 48: 123–82. ISSN 0134-8647 (in Russian).

Pavlinov, I. Ya. 2007c. [On the structure of phylogenesis and phylogenetic hypothesis.] In [*Theoretical and practical problems of studying invertebrate communities*], ed. A. I. Kafanov, 81–129. Moscow: KMK Sci. Press. ISBN: 978-5873173761 (in Russian).

Pavlinov, I. Ya. 2007d. On the structure of biodiversity: Some metaphysical essays. In *Focus on biodiversity research*, ed. J. Schwartz, 101–14. New York: Nova Science Publ. ISBN 978-1600213724.

Pavlinov, I. Ya. 2009a. [Mastering of evolutionary idea by systematics in the 19th century.] *Archives of the Zoological Museum of Moscow State University*, 50: 67–116. ISSN 0134-8647 (in Russian).

Pavlinov, I. Ya. 2009b. The species problem: Another look. In *Species and speciation: An analysis of new views and trends*, ed. A. F. Alimov, and S. D. Stepanyanz, 250–71. Saint Petersburg: Zoological Institute. ISBN 978-5873175895 (in Russian).

Pavlinov, I. Ya. 2010a. [Comments on biomorphism (ecomorphological systematics).] *Zhurnal Obshchei Biologii* 71:187–92. ISSN 0044-4596 (in Russian).

Pavlinov, I. Ya. 2010b. [Meaningful contexts of biological systematics.] In [*New ideas in scientific classification*], ed. Yu. I. Miroshnikov, and M. P. Pokrovskiy, 240–61. Yekaterinburg: Institute of Philosophy and Law RAS. http://zmmu.msu.ru/personal/pavlinov/doc/cont_syst.pdf (in Russian).

Pavlinov, I. Ya. 2011a. [How it is possible to build a taxonomic theory?] *Zoologicheskie Issledovania* 10:45–100. ISSN 1025-532X (in Russian).

Pavlinov, I. Ya. 2011b. Concepts of rational taxonomy in biology. *Biology Bulletin Reviews* 1:60–78. DOI 10.1134/S2079086411030078.

Pavlinov, I. Ya. 2012. The contemporary concepts of homology in biology: A theoretical review. *Biology Bulletin Reviews* 2:36–54. DOI 10.1134/S2079086412010057.

Pavlinov, I. Ya. 2013a. The species problem: Why again? In *The species problem: Ongoing issues,* ed. I. Ya. Pavlinov, 3–37. Rijeka: InTech Open Access Publ. DOI 10.5772/51960.

Pavlinov, I. Ya. 2013b. Taxonomic theory for nonclassical systematics. *Biology Bulletin Reviews* 3:17–26. DOI 10.1134/S2079086413010088.

Pavlinov, I. Ya. 2013c. A critical analysis of A.V. Martynov's version of ontogenetic systematics. *Thalassas* 29:23–33. DOI 10.1134/S0031030112080072.

Pavlinov, I. Ya. (ed.). 2013d. *The species problem: Ongoing issues.* Rijeka: InTech Open Access Publ. ISBN 978-9535109570.

Pavlinov, I. Ya. 2015. [*Nomenclature in systematics: History. Theory. Practice.*] Moscow. KMK Sci. Press. ISBN 978-5990715745 (in Russian).

Pavlinov, I. Ya. 2016. [Biodiversity and biocollections—problem of correspondence.] In *Aspects of biodiversity,* ed. I. Ya. Pavlinov, M. V. Kalyakin, and A. V. Sysoev, 733–86. Moscow: KMK Sci. Press. ISBN 978-5990841666 (in Russian).

Pavlinov, I. Ya. 2018. [*Foundations of biological systematics: Theory and history.*] Moscow: KMK Sci. Press. ISBN 978-5604074992 (in Russian).

Pavlinov, I. Ya. 2019. *Biological systematics: In search of the Natural System.* Moscow: KMK Scientific Press. ISBN 978-5907099951 (in Russian).

Pavlinov, I. Ya. 2020. Multiplicity of research programs in the biological systematics: A case for scientific pluralism. *Philosophies* 5:7. DOI 10.3390/philosophies5020007.

Pavlinov, I. Ya., and G. Yu. Lyubarsky. 2011. [*Biological systematics: Evolution of ideas.*] Moscow: KMK Sci. Press. ISBN: 978-5873176854 (in Russian).

Pedroso, M. 2012. Essentialism, history, and biological taxa. *Studies in History and Philosophy of Science Part C, Studies in History and Philosophy of Biological and Biomedical Sciences* 43:182–90. DOI 10.1016/j.shpsc.2011.10.019.

Pellegrin, P. 1982. *Aristotle's classification of animals: Biology and the conceptual unity of the Aristotelian corpus.* Berkeley (CA): University of California Press. ISBN 978-0520055025.

Pellegrin, P. 1987. Logical difference and biological difference: The unity of Aristotle's thought. In *Philosophical issues in Aristotle's biology,* ed. A. Gotthelf, and J. Lennox, 313–38. New York: Cambridge University Press. ISBN 978-0511552564.

Perfetti, S. 2000. *Arisostles' zoology and its Renaissance commentators (1521–1601).* Leuven: Leuven University Press. ISBN 978-9058670502.

Perminov, V. Ya. 2001. [*Philosophy and foundations of mathematics.*] Moscow: Progress-Traditsia. ISBN 5898260986 (in Russian).

Pesenko, Yu. A. 1989. [Methodological analysis of systematics. I. Formulation of the problem, and principal taxonomic schools.] *Trudy Zoologicheskogo Institute AN USSR,* 206:8–119. ISSN 0206-0477 (in Russian).

Pesenko, Yu. A. 1991. [Methodological analysis of systematics. I. II. Phylogenetic reconstructions as scientific hypotheses.] *Trudy Zoologicheskogo Institute AN USSR,* 234:61–155. ISSN 0206-0477 (in Russian).

Petry, M. J. 2001. Introduction. In *Linne C. Nemesis Divina,* 3–79. Dordrecht: Springer Science + Business Media. ISBN 978-0761823940.

Platnick, N. I. 1979. Philosophy and the transformation of cladistics. *Systematic Zoology* 28:537–46. DOI 10.2307/sysbio/28.4.537.

Platnick, N. I., and E. S. Gaffney. 1977. Systematics: A Popperian perspective. *Systematic Zoology* 26:360–5. DOI 10.1093/sysbio/26.3.360.

Plotnikov, V. I. 2010. [Typological approach.] In [*New ideas in scientific classification*], ed. Yu. I. Miroshnikov, and M. P. Pokrovskiy, 119–27. Yekaterinburg: Institute of Philosophy and Law RAS. ISBN 978-5769121203 (in Russian).

Podani, J. 2010. Monophyly and paraphyly: A discourse without end? *Taxon* 59:1011–15. DOI 10.2307/20773972.

Podani, J. 2013. Tree thinking, time, and topology: Comments on the interpretation of tree diagrams in evolutionary/phylogenetic systematics. *Cladistics* 29:315–27. DOI 0.1111/j.1096-0031.2012.00423.x.

Pogue, M. G., and Mickevich M.F. 1990. Character definitions and character state delineation: The bête noire of phylogenetic inference. *Cladistics* 6:319–61. DOI 10.1111/j.1096-0031.1990.tb00549.x.

Pokrovsky, M. P. 2014. *Introduction to the classiology*. Yekaterinburg: Institute of Geology & Geochemistry. ISBN 978-5-9433-2108-5 (in Russian).

Popov, I. Yu. 2002. "Periodical systems" in biology (a historical issue). In *Die Entstehung biologischer Disziplinen*, Vol. II, ed. U. Hossfeld, and T. Junker, 55–69. Berlin: VWB. ISBN 9783861353881.

Popov, I. Yu. 2008. [*Periodical systems and a periodical law in biology.*] Moscow: KMK Sci. Press. ISBN 978-5873175055 (in Russian).

Popper, K. 1959. *The logic of scientific discovery*. London: Hutchinson. ISBN 978-0415278447.

Popper, K. R. 1975. *Objective knowledge: An evolutionary approach*. Oxford: Oxford University Press. ISBN 978-0198750246.

Pozdnyakov, A. A. 2015. [*Philosophical foundations of classical biology: Mechanicism in evolutionism and systematics.*] Moscow: LENAND. ISBN. 978-5971019534 (in Russian).

Pratt, V. 1982. Aristotle and the essence of natural history. *History and Philosophy of the Life Sciences* 4:203–23. ISSN 0391-9714.

Pratt, V. 1984. The essence of Aristotle's zoology. *Phronesis* 29:267–78. ISSN 0031-8868.

Pratt, V. 1985. System-building in the eighteenth century. In *The light on Nature: Essays in the history and philosophy of science presented to A.C. Crombie*, ed. J. D. North, and J. J. Roche, 421–31. Dordrecht: Nijhoff. ISBN 978-9400951198.

Prigogine, I., and I. Stengers. 1984. *Order out of chaos. Man's new dialogue with nature*. London: Heinemann. ISBN 978-0553340822.

Prytkov, V. P. 2013. [Structure of scientific problem.] In [*Theory and practice of social development*], 1:44–47. http://cyberleninka. ru/article/n/struktura-nauchnoy-problemy (in Russian).

Putnam, H. 1981. *Reason, truth and history*. Cambridge (UK): Cambridge University Press. ISBN 978-0511625398.

Putnam, H. 1991. *Representation and reality*. Cambridge (MA): The MIT Press. 978-0262161084.

Quicke, D. L. J. 1993. *Principles and techniques of contemporary taxonomy*. London: Chapman & Hall. ISBN 978-9401049450.

Quine, W. V. 1996. *Ontological relativity and other essays*. New York: Columbia University Press. ISBN 978-0231083577.

Quinn, A. 2017. When is a cladist not a cladist? *Biology and Philosophy* 32:581–98. DOI 10.1007/s11229-013-0317-x.

Radnitzky, G., and W. W. Bartley (eds.). 1993. *Evolutionary epistemology, rationality, and the sociology of knowledge*. La Salle (IL): Open Court Publ. ISBN 978-0812690392.

Raikov, B. E. 1969. [*German biologist-evolutionists before Darwin.*] Leningrad: Nauka (in Russian).

Rangel, T. F., R. K. Colwell, G. R. Graves, et al. 2015. Phylogenetic uncertainty revisited: Implications for ecological analyses. *Evolution* 69:1301–12. DOI 10.1111/evo.12644.

Raposo, M. A., G. M. Kirwan, A. Carolina, et al. 2020. On the notions of taxonomic 'impediment', 'gap', 'inflation' and 'anarchy', and their effects on the field of conservation. *Systematics and Biodiversity*. DOI 10.1080/14772000.2020.1829157.

Rasnitsyn, A. P. 1983. [Phylogeny and systematics.] In [*Theoretical problems of modern biology*], 41–9. Pushchino: Acad. Sci. Publ. (in Russian).

Rasnitsyn A. P. 1996. Conceptual issues in phylogeny, taxonomy, and nomenclature. *Contributions to Zoology* 66:3–41. DOI 10.1163/26660644-06601001.

Rasnitsyn, A. P. 2002. [Evolution process and methodology of systematics.] *Trudy Russkogo Entomologicheskogo Obshchestva*, 73. 5–108. ISSN 1605–7678 (in Russian).

Rautian, A. S. 2001. [Comparative method apology: On the nature of typological knowledge.] In [*Homologies in botany: Experience and reflexion*], ed. A. K. Timonin, 65–72. Saint Petersburg: SPb Soyuz Uchenykh. ISBN 5901536010 (in Russian).

Raven, P., B. Berlin, and D. Breedlove. 1971. The origin of taxonomy. *Science* 174:1210–13. DOI 10.1126/science.174.4015.1210.

Ray, J. 1686. *Historia plantarum: Species hactenus editas aliasque insuper multas noviter inventas & descriptas complectens. In qua agitur primò de plantis in genere, earúmque partibus, accidentibus & differentiis; deinde genera omnia tum summa tum subalterna ad species usque infirmas, notis suis certis & characteristicis definita, methodo naturæ vestigiis insistente disponuntur.* London: Mariae Clark. ISBN 978-1373926371 (2016 reprint).

Ray, J. 1696. *De variis plantarum methodus dissertation brevis. In qua agitur I. De methodi origine & progrelfu. II. De notis generum characteristicis. Ill. De methodo sua in specie. IV. De notis quas reprobat & rejiciendas censet D. Tournefort. V. De methodo turneforiana.* London: S. Smith & B. Walford.

Ray, J. 1714. *The wisdom of God manifested in the works of the creation*, 6th ed. London: William Innys. ISBN 978-1372871337 (2016 reprint).

Ray, J. 1733. *Methodus plantarum emendate et aucta: In qua notae maxime characteristicae exhibentur, quibus stirpium genera tum summa, tum infima congniscuntur & se mutuo digniscuntur.* London: Christianum Andream Myntsing. ISBN 978-1272980320 (2012 reprint).

Read, D. W. 1974. Some comments on typologies in archaeology and an outline of a methodology. *American Antiquity* 39:216–42. DOI 10.2307/279584.

Reif, W.-E. 2003. Problematic issues of cladistics: 1. Ancestor recognition and phylogenetic classification. *Neues Jahrbuch fur Geologie und Palaontologie—Abhandlungen* 230:97–143. ISSN 0077-7749.

Reif, W.-E. 2005a. Problematic issues of cladistics: 12. Phylogenetic relationship. *Neues Jahrbuch fur Geologie und Palaontologie—Abhandlungen* 237:161–84. ISSN 0077-7749.

Reif, W.-E. 2005b. Problematic issues of cladistics: 17. Monophyletic taxa can be paraphyletic clades. *Neues Jahrbuch fur Geologie und Palaontologie—Abhandlungen* 238:313–54. ISSN 0077-7749.

Reif, W.-E. 2006. Problematic issues of cladistics: 21. Was Darwin a cladist? *Neues Jahrbuch fur Geologie und Palaontologie—Abhandlungen* 242:43–82. ISSN 0077-7749.

Reif, W.-E. 2007. Problematic issues of cladistics: 23. Darwin's concept of phylogenetic relationship. *Neues Jahrbuch fur Geologie und Palaontologie—Abhandlungen* 244:227–46. ISSN 0077-7749.

Reif, W.-E. 2009. Problematic issues of cladistics: 25. Fundamental theorems of phylosystematics. *Neues Jahrbuch fur Geologie und Palaontologie—Abhandlungen* 252:145–66. ISSN 0077-7749.

Remane, A. 1943. Die Bedeutung der Lebensformtypen für die Orologi. *Biologia Generalis* 17:164–82. ISSN 0366-0427.

Remane, A. 1956. *Grundlagen des Natürlichen Systems, der Vergleichenden Anatomie und der Phylogenetik. Theoretische Morphologie und Systematik.* 2. Auf l. Leipzig: Akad. Verlag. ISBN 978-3874290296.

Reydon, T. A. C. 2005. On the nature of the species problem and the four meanings of "species". *Studies in History and Philosophy of Biological and Biomedical Sciences* 36:135–58. DOI e nature of the species problem and the four meanings of 'species' I Present-day … DOI 10.1016/j.shpsc.2004.12.004.

Richards, R. 1992. The structure of narrative explanation in history and biology. In *History and evolution*, ed. H. Nitecki, and D. Nitecki, 19–54. New York: State University of New York Press. ISBN 978-0791412121.

Richards, R. A. 2010. *The species problem: A philosophical analysis*. New York: Cambridge University Press. ISBN 978-0521196833.

Richards, R. A. 2016. *Biological classification: A philosophical introduction*. Cambridge (UK): Cambridge University Press. ISBN 10:1107687845.

Richardson, M. K., A. Minelli, M. Coates, et al. 1998. Phylotypic stage theory. *Trends in Ecology and Evolution* 13:158. DOI 10.1016/s0169-5347(98)01340-8.

Riddle, B. R., and D. J. Hafner. 1999. Species as units of analysis in ecology and biogeography: Time to take the blinders off. *Global Ecology and Biogeography* 8:433–41. DOI 10.1046/j.1365-2699.1999.00170.x.

Riedel, A., K. Sagata, Y. R. Suhardjono, et al. 2013. Integrative taxonomy on the fast track— towards more sustainability in biodiversity research. *Frontiers in Zoology* 10:15. www.frontiersinzoology.com/content/10/1/15.

Riedley, M. 1986. *Evolution and classification: Reformation of cladism*. New York: Longman. ISBN 978-0582444973.

Riegner, M. F. 2013. Ancestor of the new archetypal biology: Goethe's dynamic typology as a model for contemporary evolutionary developmental biology. *Studies in History and Philosophy of Biological and Biomedical Sciences* 44:735–44. DOI 10.1016/j.shpsc.2013.05.019.

Riel, R., and R. Gulick. 2014. Scientific reduction. In *The Stanford encyclopedia of philosophy*, ed. E. N. Zalta. https://plato.stanford.edu/entries/scientific-reduction/#RepOntRed.

Rieppel, L. 2016. Museums and botanical gardens. In *A companion to the history of science*, ed. B. Lightman, 238–51. Oxford: Wiley Blackwell. ISBN 978-1118620779.

Rieppel, O. 1985. Ontogeny and hierarchy of types. *Cladistics* 1(3):234–246.

Rieppel, O. 1988a. Louis Agassiz (1807–1873) and the reality of natural groups. *Biology and Philosophy* 3:29–47. DOI 10.1007/BF00127627.

Rieppel, O. 1988b. *Fundamentals of comparative biology*. Basel: Birkhauser Verlag. ISBN 978-0817619565.

Rieppel, O. 1989. Ontogeny, phylogeny, and classification. In *Phylogeny and the classification of fossil and recent organisms*, ed. N. Schmidt-Kittler, and R. Willmann, 63–82. Hamburg: Verlag Paul Prey. ISBN 978-3490144966.

Rieppel, O. 1990. Ontogeny—A way forward for systematics, a way backward for phylogeny. *Biological Journal of the Linnean Society* 39:177–91. DOI 10.1111/j.1095–8312.1990.tb00510.x.

Rieppel, O. 1992. Homology and logical fallacy. *Journal of Evolutionary Biology* 5:701–15. DOI 10.1046/j.1420-9101.1992.5040701.x.

Rieppel, O. 2003. Popper and systematics. *Systematic Biology* 52:271–80. DOI 10.1080/10635150390192.

Rieppel, O. 2004. The language of systematics, and the philosophy of "total evidence". *Systematics and Biodiversity* 2:9–19. DOI 10.1017/S147720000400132X.

Rieppel, O. 2005a. The philosophy of total evidence and its relevance for phylogenetic inference. *Papeis Avulsos de Zoologia* 45:77–89. DOI 10.1590/S0031-10492005000800001.

Rieppel, O. 2005b. Monophyly, paraphyly, and natural kinds. *Biology and Philosophy* 20:465–87. DOI 10.1007/s10539-004-0679-z.

Rieppel, O. 2006. On concept formation in systematics. *Cladistics* 22:474–92. DOI 10.1111/j.1096-0031.2006.00114.x.

Rieppel, O. 2007. The nature of parsimony and instrumentalism in systematics. *Journal of Zoological Systematics and Evolutionary Research* 45:177–83. DOI 0.1111/j.1439-0469.2007.00426.x.

Rieppel, O. 2008. Hypothetico-deductivism in systematics: fact or fiction? *Papeis Avulsos de Zoologia* 48:253–63. DOI 10.1590/S0031-10492008002300001.

Rieppel, O. 2009. Total evidence in phylogenetic systematics. *Biology and Philosophy* 24:607–22. DOI 10.1111/j.1096-0031.1991.tb00045.x.

Rieppel, O. 2010a. The series, the network, and the tree: Changing metaphors of order in nature. *Biology and Philosophy* 25:475–96. DOI 10.1186/1471-2148-5-33.

Rieppel, O. 2010b. New essentialism in biology. *Philosophy of Science* 77:662–73. DOI 10.1086/656539.

Rieppel, O. 2016. *Phylogenetic systematics. Haeckel to Hennig*. Boca Raton (FL): Taylor & Francis. ISBN 978-0367876456.

Rieppel, O., and M. Kearney. 2002. Similarity. *Biological Journal of the Linnean Society* 75:59–82. DOI 10.1046/j.1095-8312.2002.00006.x.

Rieppel O., M. Rieppel, and L. Rieppel. 2006. Logic in systematics. *Journal of Zoological Systematics and Evolutionary Research* 44:186–92. DOI 0.1111/j.1439-0469.2006.00370.x.

Ritterbush, P. C. 1969. Art and science as influences on the early development of natural history collections. *Proceedings of the Biological Society of Washington* 82, 561–78. DOI 10.1126/science.168.3932.726.

Robson, G. C. 1928. *The species problem: An introduction to the study of evolutionary divergence in natural populations*. Edinburgh: Oliver & Boyd. ASIN: B0006EUWFI.

Rodoman, B. B. 1999. [*Territorial areas and networks. Essays on theoretical geography.*] Smolensk: Oekumena. ISBN: 5-935200015 (in Russian).

Rogers, D. J. 1963. Taximetrics—New name, old concept. *Brittonia* 15:285–90. DOI 10.1007/BF02909606.

Rogers, D. P. 1958. The philosophy of taxonomy. *Mycologia* 50:326–32. DOI 10.1080/00275514.1958.12024730.

Rohlf, F. J., and R. R. Sokal. 1981. Comparing numerical taxonomic studies. *Systematic Zoology* 30:459–90. DOI 10.1093/sysbio/30.4.459.

Rollins, R. C. 1965. On the bases of biological classification. *Taxon* 14:1–6. DOI 10.2307/1216700.

Rosch, E., C. B. Mervis, W. D. Gray, et al. 1976. Basic objects in natural categories. *Cognitive Psychology* 8:382–439. DOI 10.1016/0010-0285(76)90013-X.

Rosenberg, A. 1985. *The structure of biological science*. New York: Cambridge University Press. ISBN 978-0521275613.

Rosenberg, A. 2020. *Reduction and mechanism*. Cambridge (UK): Cambridge University Press. ISBN 978-1108742313.

Rossolimo, O. L. 1979. [Geographical variability, environmental gradient, and adaptive organization of mammals.] *Archives of the Zoological Museum of Moscow State University* 18:44–75. ISSN 0134-8647 (in Russian).

Roth, V. L. 1982. On homology. *Biological Journal of the Linnean Society* 22:13–29. DOI 10.1111/j.1095-8312.1984.tb00796.x.

Roth, V. L. 1988. The biological basis of homology. In *Ontogeny and systematics*, ed. C. J. Humphries, 1–26. New York: Columbia University Press. ISBN 9780231063708.

Roth, V. L. 1991. Homology and hierarchies: Problems solved and unresolved. *Journal of Evolutionary Biology* 4: 167–94. DOI 10.1046/j.1420-9101.1991.4020167.x.

Rozanova, M. A. 1946. [*Experimental foundations of plant systematics.*] Moscow: Acad. Sci. Publ. (in Russian).

Rozov, M. A. 1995. [Classification and theory as knowledge systems.] In [*On the way to the theory of classification*], ed. S. S. Rozova, 81–127. Novosibirsk: Novosibirsk University Publ. ISBN 5887420049 (in Russian).

Rozov, M. A. 2002. [*Philosophy and theory of history. Prolegomen.*] Moscow: Logos. ISBN 978-5971064305 (in Russian).

Rozova, S. S. 1980. [Philosophical reflection on the classification problem.] *Voprosy Filosofii* 8:49–56. ISSN: 0042-8744 (in Russian).

Rozova, S. S. 1986. [*Classification problem in modern science.*] Moscow: Nauka (in Russian).

Rozova, S. S. 2014. [*Theory of social relay in epistemology and philosophy of science.*] Saarbrucken: Lambert Academic Publ. 978-365918226 (in Russian).

Rupke, N. A. 1993. Richard Owen's vertebrate archetype. *Isis* 84:231–51. DOI 10.1086/356461.

Ruse, M. 1973. *The philosophy of biology.* London: Hutchinson University Library. ISBN 978-0091152215.

Ruse, M. 1979. Falsifiability, consilience, and systematics. *Systematic Zoology* 28:530–6. DOI 10.2307/sysbio/28.4.530.

Russell, G. 2019. Logical pluralism. In *The Stanford encyclopedia of philosophy*, ed. E. N. Zalta. https://plato.stanford.edu/archives/sum2019/entries/logical-pluralism/.

Ruzavin, G. I. 2005. [Abduction and methodology of scientific search.] *Epistemology and Philosophy of Science* 6:148–56. ISSN: 1811-833X (in Russian).

Ruzhentsev, V. E. 1960. [Principles of systematics, system and phylogeny of Paleozoic ammonoids.] *Trudy Paleontologicheskogo Institute AN USSR*, 83:5–331. ISSN 0376-1444 (in Russian).

Rybnikov, K. L. 1994. [*History of mathematics.*] Moscow: Moscow University Publ. ISBN 5211020685 (in Russian).

Sachs, J. 1906. *History of botany, 1530–1860.* Oxford: Clarendon Press. ISBN 978-1408603901 (2007 reprint).

Saether, O. A. 1979. Underlying synapomorphies and anagenetic analysis. *Zoologica Scripta* 8:305–12. DOI 10.1111/j.1463–6409.1979.tb00644.x.

Saether, O. A. 1982. The canalized evolutionary potential: Inconsistencies in phylogenetic reasoning. *Systematic Zoology* 32:343–59. DOI 10.1093/sysbio/32.4.343.

Saether, O. A. 1986. The myth of objectivity—post-Hennigian deviations. *Cladistics* 2:1–13. DOI 10.1111/j.1096-0031.1986.tb00438.x.

Salthe, S. N. 1985. *Evolving hierarchical systems: Their structure and representation.* New York: Columbia University Press. ISBN 978-0231060165.

Salthe, S. N. 1993. *Development and evolution: Complexity and change in biology.* Cambridge (MA): The MIT Press. ISBN 978-0262193351.

Salthe, S. N. 2012. Hierarchical structures. *Axiomathes* 22:355–83. DOI 10.1007/s10516-012-9185-0.

Sander, K. 1983. The evolution of patterning mechanisms: Gleanings from insect embryogenesis and spermatogenesis. In *Development and evolution*, ed. B. C. Goodwin, N. Holder, and C. C. Wylie, 187–91. Cambridge (UK): Cambridge University Press. ISBN 978-0521249492.

Sanderson, M. J., and L. Hufford (eds.). 1996. *Homoplasy. The recurrence of similarity in evolution.* San Diego: Academic Press. ISBN 978-0123907523.

Santos, C. M. D., and D. S. Amorim. 2007. Why biogeographical hypotheses need a well supported phylogenetic framework: A conceptual evaluation. *Papéis Avulsos de Zoologia* 47:63–73. DOI 10.1590/S0031-10492007000400001.

Sarkar, S., and C. Margules. 2001. Operationalizing biodiversity for conservation planning. *Journal of Biosciences* 27 Suppl. 2:299–308. DOI 10.1007/BF02704961.

Sattler, R. 1994. Homology, homeosis, and process morphology in plants. In *Homology: The hierarchical basis of comparative morphology*, ed. B. K. Hall, 423–75. New York: Academic Press. ISBN 978-0123195838.

Sattler, R. 1996. Classical morphology and continuum morphology: Opposition and continuum. *Annals of Botany* 78:577–81. https://pdfs.semanticscholar.org/953a/7422054acd22c976 7717ece42faf5edd91e3.pdf.

Savva, G., J. Dicks, and I. N. Roberts. 2003. Current approaches to whole genome phylogenetic analysis. *Briefings in Bioinformatics* 4:63–74. DOI 10.1093/bib/4.1.63.

Scherer, R. 2012. *Multiple fuzzy classification systems*. Berlin: Springer-Verlag. ISBN 978-3642306044.

Schierwater, B., and K. Kuhn. 1998. Homology of Hox genes and the zootype concept in early metazoan evolution. *Molecular Phylogenetics and Evolution* 9:375–81. DOI 10.1006/mpev.1998.0489.

Schindewolf, O. H. 1969. Über der "Typus" in morphologischer und phylogenetischer Biologie. *Abhandlungen der Matematisch-Naturwissenschaftlichen Klasse* 4:58–131. DOI 10.1111/j.1439-0469.1971.tb00898.x.

Schlick-Steiner, B. C., F. M. Steiner, B. Seifert, et al. 2010. Integrative taxonomy: A multisource approach to exploring biodiversity. *Annual Review of Entomology* 55:421–38. DOI 10.1146/annurev-ento-112408-085432.

Schmidt-Kittler, N., and K. Vogel (eds.). 1991. *Constructional morphology and evolution*. Heidelberg: Springer. ISBN 978-3642761560.

Schoch, R. M. 1986. *Phylogeny reconstruction in paleontology*. New York: Van Nostrand Reinhold. ISBN 978-0442279677.

Scholtz, G. 2005. Homology and ontogeny: Pattern and process in comparative developmental biology. *Theory in Biosciences* 124:121–43. DOI 10.1007/BF02814480.

Scholtz, G. 2010. Deconstructing morphology. *Acta Zoologica* (Stockholm) 91:44–63. DOI 10.1111/j.1463-6395.2009.00424.x.

Schoute, J. C. 1949. *Biomorphology in general*. Amsterdam: North-Holland Pub. Co. ISBN 978-0472086849.

Schram, F. R. 2004. The truly new systematics—megascience in the information age. *Hydrobiologie* 519:1–7. DOI 10.1023/B:HYDR.0000026601.89333.aa.

Schuh, R. T. 2000. *Biological systematics. Principles and applications*. Ithaca: Cornell University Press. ISBN 978-0801447990.

Schulze, E.-D., and H. A. Mooney. 1994. Ecosystem function of biodiversity: A summary. In *Ecosystem function of biodiversity*, ed. E.-D. Schulze, and H. A. Mooney, 497–510. New York: Springer. ISBN 978-3642580017.

Schwartz, J. H., and B. Maresca. 2007. Do molecular clocks run at all? A critique of molecular systematics. *Biological Theory* 1:357–71. DOI 10.1162/biot.2006.1.4.357.

Schwarz, S. S. 1980. [*Ecological patterns of evolution.*] Moscow: Nauka (in Russian).

Scotland, R. W. 2010. Deep homology: A view from systematics. *Bioessays* 32:438–49. DOI 10.1002/bies.200900175.

Scott, W. B. 1896. Paleontology as a morphological discipline. In *Biological lectures delivered at the Marine Biological Laboratory of Wood's Holl in the summer session of 1894*, 43–61. Boston: Ginn. DOI 10.1126/science.4.85.177.

Scott-Ram, N. R. 1990. *Transformed cladistics: Taxonomy and evolution*. New York: Cambridge University Press. ISBN 978-0521055130.

Semple, C., and M. Steel. 2003. *Phylogenetics*. Oxford: Oxford University Press. ISBN 978-0198509424.

Sepkoski, J. J. 1996. Patterns of Phanerozoic extinction: A perspective from global data bases. In *Global events and event stratigraphy in the Phanerozoic*, ed. O. H. Walliser, 35–51. Berlin: Springer Verlag. ISBN 978-3642796340.

Serebryakov, I. G. 1962. [*Ecological morphology of plants. Life forms of angiosperms and coniferals.*] Moscow: Vysshaia Shkola (in Russian).

Sereno, P. C. 2005. The logical basis of phylogenetic taxonomy. *Systematic Biology* 54:595–619. DOI 10.1080/106351591007453.

Shapiro, S. 1997. *Philosophy of mathematics: Structure and ontology*. New York: Oxford University Press. ISBN 978-0195139303.

Shatalkin, A. I. 1983. [Methodological aspects of application of mathematical methods in systematics.] In [*Theory and methodology of biological classifications*], 46–55. Moscow: MOIP (in Russian).

Shatalkin, A. I. 1988. *Biological systematics*. Moscow: Moscow University Press. ISBN 978-5211001459 (in Russian).

Shatalkin, A. I. 1990a. [Similarity and classification.] *Zhurnal Obshchei Biologii* 51:610–18. ISSN 0044-4596 (in Russian).

Shatalkin, A. I. И. 1990b. [Similarity and homology.] *Zhurnal Obshchei Biologii* 51:841–9. ISSN 0044-4596 (in Russian).

Shatalkin, A. I. 1991. [Claudistics and evolutionary systematics: Points of divergence.] *Zhurnal Obshchei Biologii* 52:584–97. ISSN 0044-4596 (in Russian).

Shatalkin, A. I. 1993. [Aristotle and systematics. On the foundations of typology.] *Zhurnal Obshchei Biologii* 54:243–52. ISSN 0044-4596 (in Russian).

Shatalkin, A. I. 1994. [The typological approach in systematics.] *Zhurnal Obshchei Biologii* 55:661–72. ISSN 0044-4596 (in Russian).

Shatalkin, A. I. 1995. [Hierarchies in systematics: A set theory model.] *Zhurnal Obshchei Biologii* 56:277–90. ISSN 0044-4596 (in Russian).

Shatalkin, A. I. 1996. Essentialism and typology. In *Contemporary systematics: Methodological aspects*, ed. I. Ya. Pavlinov, 23–154. Moscow: Moscow University Press. ISBN 978-5873176175 (in Russian).

Shatalkin, A. I. 2002. [The problem of archetype and modern biology.] *Zhurnal Obshchei Biologii* 63:275–91. ISSN 0044-4596 (in Russian).

Shatalkin, A. I. 2003. [Regulatory genes in development and morphotype problem in insect systematics.] In [*Readings on the memory of N.A. Kholodkovsky*], 56:5–109. ISSN 1606–8858 (in Russian).

Shatalkin, A. I. 2012. [*Taxonomy. Grounds, principles, and rules.*] Moscow: KMK Sci. Press. ISBN: 978-5873178476 (in Russian).

Shchedrovitsky, G. P. 2004. [*The problem of logic of scientific research and analysis of the structure of science.*] Moscow: Put. ISBN 5875900873 (in Russian).

Shelah, S. 1990. *Classification theory*, 2nd ed. Amsterdam: North Holland. ISBN 978-0080880242.

Shreyder, Yu. A. 1983. [Systematics, typology, classification.] In [*Theory and methodology of biological classifications*], ed. S. A. Shreider, and B. S. Shornikov, 90–100. Moscow: Nauka (in Russian).

Shreyder, Yu. A., and A. A. Sharov. 1982. [*Systems and models.*] Moscow: Radio i Svyaz (in Russian).

Shuman, A. N. 2001. [*Philosophical logic: Origins and evolution.*] Minsk: EkonomPress. ISBN 985-6479266 (in Russian).

Sigwart, J. D. 2018. *What species mean: A user's guide to the units of biodiversity.* Boca Raton: CRC Press. ISBN 978-1498799379.

Simpson, G. G. 1970. Uniformitarianism. An inquiry into the principle, theory, and method in geohistory and biohistory. In *Essays in evolution and genetics in honor of Theodosius Dobzhansky,* ed. M. K. Hecht, and W. C. Steere, 43–95. Amsterdam: North-Holland Publ. ISBN 978-14615-9587-8.

Simpson, G. G. 1961. *Principles of animal taxonomy.* New York: Columbia University Press. ISBN 978-0231024273.

Simpson, G. G., and A. Roe. 1939. *Quantitative zoology.* New York: McGraw-Hill. ISBN 978-0486432755 (1970 reprint).

Sklar, A. 1964. On category overlapping in taxonomy. In *Form and strategy in science. Studies dedicated to Joseph Henry Woodger on the occasion of his seventieth birthday,* ed. J. R. Gregg, and F. T. C. Harris, 395–401. Dordrecht: D. Reidel. ISBN 978-9401036030.

Skvortsov, A. K. 1967. The main stages in the development of the conception of species. *Bulleten Moskovskogo obshestva ispytatelej prirody, Biol.* 72:11–27 (in Russian).

Slack, J. M. W., P. W. H. Holland, and C. F. Graham. 1993. The zootype and the phylotypic stage. *Nature* 361:490–92. DOI 10.1038/361490a0.

Slater M. 2013. *Are species real? An essay on the metaphysics of species.* New York: Palgrave MacMillan. ISBN 978-0230393233.

Slaughter, M. 1982. *Universal languages and scientific taxonomy in the seventeenth century.* Cambridge (UK): Cambridge University Press. ISBN 978-0521135443.

Sloan, P. 1972. John Locke, John Ray, and the problem of the natural system. *Journal of the History of Biology* 5:1–53. DOI 10.1007/BF02113485.

Sloan, P. R. 1979. Buffon, German biology, and the historical interpretation of biological species. *British Journal of History of Science* 12:109–53. DOI 10.1017/S0007087400017027.

Sloan, P. R. 1987. From logical universals to historical individuals: Buffon's idea of biological species. In *Histoire du concept d'espèce dans les sciences de la vie,* 101–40. Paris: Fondation Singer-Polignac. ISBN 978-2900927199.

Sluys, R. 1991. Species concepts, process analysis, and the hierarchy of nature. *Experientia* 47:1162–70. DOI 10.1007/BF01918380.

Sluys, R. 1996. The notion of homology in current comparative biology. *Journal of Zoological Systematics and Evolutionary Research* 34:145–52. DOI 10.1111/j.1439-0469.1996.tb00820.x.

Smaling, A. 1992. Varieties of methodological intersubjectivity: The relations with qualitative and quantitative research and with objectivity. *Quality and Quantity* 26:169–80. DOI 10.1007/BF02273552.

Smirnov, E. S. 1923. [On the structure of systematic categories.] *Russian Zoological Journal* 3:358–89. ISSN 0044-5134 (in Russian).

Smirnov, E. S. 1924. Problem der exakten Systematik und Wege zu ihrer Loesing. *Zoologische Anzeiger* 61:1–14. ISSN 0044-5231.

Smirnov, E. S. 1925. The theory of type and the natural system. *Zeitschrift für Induktive Abstammungs- und Vererbungslehre* 37:28–66. ISSN: 1617–4615.

Smirnov, E. S. 1938. [Construction of species from a taxonomic point of view.] *Zoological Journal,* 47:387–418. ISSN 0044-5134 (in Russian).

Smirnov, E. S. 1959. [Homology and taxonomy.] In [*Issues in morphology and phylogeny of vertebrates*], 68–78. Moscow: Acad. Sci. Publ. (in Russian).

Smirnov, E. S. 1969. [*Taxonomic analysis.*] Moscow: Moscow University Publ. (in Russian).

Smith, A. B. 1994. *Systematics and the fossil record: Documenting evolutionary patterns.* Oxford (UK): Blackwell Scientific Publ. ISBN 978-0632036424.

Smith, V. 2013. Cybertaxonomy. *The Future of Scholarly Communication* 1:63–74. DOI 10.29085/9781856049610.007.

Sneath, P. H. A. 1958. Some aspects of Adansonian classification and of the taxonomic theory of correlated features. *Annals of Microbiology and Enzymology* 8:261–8. ISSN 1869–2044.

Sneath, P. H. A. 1961. Recent developments in theoretical and quantitative taxonomy. *Systematic Zoology* 10:118–37. DOI 10.2307/2411596.

Sneath, P. H. A. 1963. Mathematics and classification from Adanson to the present. In *Adanson: The bicentennial of Michael Adanson's "Familles des Plantes"*, part. 2, ed. G. H. M. Lawrence, 471–98. Pittsburgh: Hunt Bot. Library. www.huntbotanical.org/admin/uploads/14-hibd-adanson-pt2-pp471-498.pdf.

Sneath, P. H. A. 1995. Thirty years of numerical taxonomy. *Systematic Biology* 44:281–98. DOI 10.1093/sysbio/44.3.281.

Sneath, P. H. A., and R. R. Sokal. 1973. *Numercial taxonomy. The principles and methods of numerical classification.* San Francisco: W.H. Freeman. ISBN 978-0716706977.

Sneed, J. D. 1979. *The logical structure of mathematical physics*, 2nd ed. Reidel: Springer. ISBN: 978-90-277-1059-8.

Sober, E. 1984. Sets, species, and evolution. *Philosophy of Science* 51:334–41. DOI 10.1016/j.sjbs.2017.04.013.

Sober, E. 2000. *Philosophy of biology*, 2nd ed. Boulder: Westview Press. ISBN 978-0813340654.

Söderström, A. 1925. *Homologie, Homogenie und Homoplasie. Eine Kritik, ein Protest und ein Turanschlag.* Leipzig: F. B. Kolher.

Sokal, R. R. 1962. Typology and empiricism in taxonomy. *Journal of Theoretical Biology* 3:230–67. DOI 10.1016/S0022-5193(62)80016-2.

Sokal, R. R. 1966. Numerical taxonomy. *Scientific American* 215:106–16. ISSN 0036-8733.

Sokal, R. R., and J. H. Camin. 1965. The two taxonomies: Areas of agreement and conflict. *Systematic Zoology* 14:176–95. DOI 10.2307/2411548.

Sokal, R. R., and R. H. A. Sneath. 1963. *Principles of numerical taxonomy.* San Francisco: W.H. Freeman. ASIN B0006AYNO8.

Sokolov, V. V. 2001. [*Medieval philosophy.*] Moscow: Editorial URSS. ISBN: 978-5382012131 (in Russian).

Solbrig, O. T. 1970. *Principles and methods of plant biosystematics.* New York: Macmillan. ISBN 978-0024137005.

Sonnhammer, E. L. L., and E. V. Koonin. 2002. Orthology, paralogy and proposed classification for paralog subtypes. *Trends in Genetics* 18:619–20. DOI 10.1016/s0168-9525(02)02793-2.

Sorabji, R. (ed.). 1990. *Aristotle transformed: The ancient commentators and their influence.* London: Cornell University Press. ISBN 978-0801424328.

Spemann, H. 1915. Zur Geschichte und Kritik des Begriffs der Homologie. In *Allgemeine Biologie*, ed. C. Chun, and W. Johannsen, 63–85. Berlin: B. G. Teubner. ISBN 978-1167729775 (2010 reprint).

Spencer, Q. 2016. Genuine kinds and scientific reality. In *Natural kinds and classification in scientific practice*, ed. C. Kendig, 184–99. Abingdon: Routledge. ISBN 978-1138344839.

Spirova, E. M. 2006. [Hermeneutic circle.] In [*Knowledge. Understanding. Skill*], 2: 198–204 (in Russian).

Sprengel, C. 1808. *Curtii Sprengel historia rei herbariae.* [Auctor] Amstelodami: sumtibus tabernae librariae et artium. ISBN 978-1167710209 (2010 reprint).

Stace, C. A. 1989. *Plant taxonomy and biosystematics*, 2nd ed. London: Cambridge University Press. ISBN 978-0521427852.

Stafleu, F. A. 1963. Adanson and his "Familles des plantes". In *Adanson: The bicentennial of Michel Adanson's "Familles des plantes"*, part 1, ed. G. H. M. Lawrence, 123–264. Lehre: J. Cramer. www.huntbotanical.org/admin/uploads/05-hibd-adanson-pt1-pp123-264.pdf.

Stafleu, F. A. 1969. A historical review of systematic biology. In *Systematic biology: Proceedings of the International Conference*, 16–44. Washington (DC): National Academy of Sciences. ISBN 978-0309016926.

Stafleu, F. A. 1971. *Linnaeus and Linnaeans*. Utrecht: A. Oosthoek. ISBN 978-9060460641.

Stamos, D. N. 1996. Was Darwin really a species nominalist? *Journal of the History of Biology* 29:127–44. DOI 10.1007/BF00129698.

Stamos, D. N. 2003. *The species problem. Biological species, ontology, and the metaphysics of biology*. Oxford: Lexington Books. ISBN 978-0739107782.

Stamos, D. N. 2005. Pre-Darwinian taxonomy and essentialism—A reply to Mary Winsor. *Biology and Philosophy* 20:79–96. DOI 10.1007/s10539-005-0401-9.

Stamos, D. N. 2007a. *Darwin and the nature of species*. New York: State University of New York Press. ISBN 978-0791469385.

Stamos, D. N. 2007b. Popper, laws, and the exclusion of biology from genuine science. *Acta Biotheoretica* 55:357–75. DOI 0.1007/s10441-007-9025-6.

Starobogatov, Ya. I. 1989. [Natural system, artificial systems and some principles of phylogenetic and systematic research.] *Trudy Zoologicheskogo Institute AN USSR*, 206:191–222. ISSN 0206-0477 (in Russian).

Starostin, B. A. 1970. [*Plant phylogenetics and its development*.] Moscow: Nauka (in Russian).

Stearn, W. T. 1961. *Botanical gardens and botanical literature in the eighteenth century*. Pittsburgh: Hunt Foundation. ASIN B002JMJL4I.

Stebbins, G. L. 1970. Biosystematics: An avenue towards understanding evolution. *Taxon* 19:205–14. DOI 10.1111/aen.12158.

Steigerwald, J. 2002. Goethe's morphology: Urphanomene and aesthetic appraisal. *Journal of the History of Biology* 35:291–328. DOI 10.1023/A:1016028812658.

Stekolnikov, A. A. 2003. [The problem of truth in biological systematics.] *Zhurnal Obshchei Biologii* 64:357–68. ISSN 0044-4596 (in Russian).

Stepin, V. S. 2005. *Theoretical knowledge*. Heidelberg: Springer. ISBN 978-1402030468.

Sterner, B. 2014. Well-structured biology: Numerical taxonomy's epistemic vision for systematics. In *The evolution of phylogenetic systematics*, ed. A. Hamilton, 213–44. Berkeley (CA): University of California Press. ISBN 978-0756502461.

Stevens, P. F. 1984a. Hauy and A.-P. Candolle: Crystallography, botanical systematics, and comparative morphology, 1780–1840. *Journal of the History of Biology* 17:49–82. DOI 10.1007/BF00397502.

Stevens, P. 1984b. Metaphors and typology in the development of botanical systematics 1690–1960, or the art of putting new wine in old bottles. *Taxon* 33:169–211. DOI 10.2307/1221161.

Stevens, P. F. 1994. *The development of biological systematics. Antoine-Laurent de Jussieu, Nature, and the Natural System*. New York: Columbia University Press. ISBN 978-0231064408.

Stevens, P. F. 1997. How to interpret botanical classifications: Suggestions from history. *Journal of BioScience* 47:243–50. DOI 10.2307/1313078.

Steward, J. H. 1955. *Theory of culture change: The methodology of multilinear evolution*. Urbana (IL): University of Illinois Press. ISBN 978-2252002957.

Strickland, H. E. 1841. On the true method of discovering the natural system in zoology and botany. *Annals and Magazine of Natural History*, Ser. 5, 6:184–94. DOI 10.1080/03745484009443283.

Striedter, G. F., and R. G. Northcutt. 1991. Biological hierarchies and the concept of homology. *Brain, Behavior and Evolution* 38:177–89. DOI 10.1159/000114387.

Stuessy, T. F. 1997. Classification: More than just branching patterns of evolution. *Aliso: A Journal of Systematic and Evolutionary Botany* 15:113–24. DOI 10.5642/aliso.19961502.06.

Stuessy, T. F. 2009. *Plant taxonomy. The systematic evaluation of comparative data*, 2nd ed. New York: Columbia University Press. ISBN 978-0231147125.

Stuessy, T. F., and E. Hörandl. 2014. Evolutionary systematics and paraphyly: Introduction. *Annals of the Missouri Botanical Garden* 100:2–5. DOI 10.3417/2012083.

Stuessy, T. F., and C. Konig. 2008. Patrocladistic classification. *Taxon* 57:594–601. www.mobot.org/plantscience/resbot/EvSy/PDF/Stuessy-König-Patrocl2008.pdf.

Styles, B. T. (ed.). 1987. *Infraspecific classification of wild and cultivated plants*. New York: Oxford University Press. ISBN 978-0198577010.

Suárez, M. 2010. Scientific representation. *Philosophy Compass* 5:91–101. DOI 10.1111/j.1747-9991.2009.00261.x.

Subbotin, A. L. 2001. [*Classification.*] Moscow: Institute of Philosophy RAN. ISBN 5201020461 (in Russian).

Surov, I. A. 2002. Quantum cognitive triad. Semantic geometry of context representation. *Neurons and Cognition* (q-bio.NC): arXiv:2002.11195. https://arxiv.org/abs/2002.11195.

Suzuki, D. G., and S. Tanaka. 2017. A phenomenological and dynamic view of homology: Homologs as persistently reproducible modules. *Biological Theory* 12:169–80. DOI 10.1007/s13752-017-0265-7.

Swainson, W. 1835. *The cabinet cyclopedia: Natural history, a treatise on the geography and classification of animals*. London: Longman. ISBN 978-1164131328 (2010 reprint).

Swofford, D., G. J. Olsen, P. J. Waddell, et al. 1996. Phylogenetic inference. In *Molecular systematics*, 2nd ed., ed. D. M. Hillis, C. Moritz, and B. K. Mable, 407–514. Sunderland: Sinauer Assoc. ISBN 978-0878932825.

Swoyer, C. 2006. Conceptualism. In *Universals, concepts, and qualities: New essays on the meaning of predicates*, ed. E. S. Trawson, and A. Chakrabarti, 127–54. Routledge: CRC Press. ISBN 978-0754650324.

Sylvester-Bradley, P. C. 1952. *The classification and coordination of infraspecific categories*. London: Syst. Assoc.

Szalay, F., and W. Bock. 1991. Evolutionary theory and systematics: Relationships between process and patterns. *Zeitschrift für Zoologische Systematik und Evolutions-forschung* 29:1–39. DOI 10.1111/j.1439-0469.1991.tb00442.x

Szucsich, N. U., and C. S. Wirkner. 2007. Homology: A synthetic concept of evolutionary robustness of patterns. *Zoologica Scripta* 36:281–9. DOI 10.1111/j.1463-6409.2007.00275.x.

Takhtajan, A. L. 1970. [Biosystematics: Past, present and future.] *Botanical Journal* 55:331–45. ISSN 006-8136 (in Russian).

Talmy, L. 2000. *Toward a cognitive semantics. Vol. 1. Concept structuring systems*. Cambridge (MA): MIT Press. ISBN 978-0262700962.

Tatarinov, L. P. 1977. [Classification and phylogeny.] *Zhurnal Obshchei Biologii* 38:676–89. ISSN 0044-4596 (in Russian).

Taylor, B. 2006. *Models, truth, and realism*. Oxford (UK): Oxford University Press. ISBN 978-0199286690.

Taylor, P. M. 1990. *The folk biology of the Tobelo people. A study in folk classification.* Washington (DC): Smithsonian Institute Press. ISBN 978-0835743259.

Tchaikovsky, Yu. V. 1990. [*Elements of evolutionary diatropics.*] Moscow: Nauka. ISBN 5020080861 (in Russian).

Terentiev, P. V. 1957. [On the applicability of the subspecies concept in the study of geographic variability.] *Vestnik Leningrad University, Ser. 2* 21:75–81. ISSN 1025–8604 (in Russian).

Tetenyi, P. 2013. Homology of biosynthetic routes: The base in chemotaxonomy. In *Chemistry in botanical classification: Medicine and natural sciences*, ed. G. Bendz, and J. Santesson, 67–80. New York: Academic Press. ISBN 978-0123942494.

Theophrastus. 1916. *Enquiry into plants*, Vol. 1: Books 1–5. Cambridge (MA): Harvard University Press. ISBN 978-0674990883.

Thompson, W. R. 1952. The philosophical foundations of systematics. *Canadian Entomologist* 84:1–16. DOI 10.4039/Ent841-1.

Thorp, W. H. 1940. Ecology and the future of systematics. In *The new systematics*, ed. J. Huxley, 341–64. London: Oxford University Press. ISBN 978-0403017867.

Timiryazev, K. A. 1904. [An essay on Darwin's theory.] In [*Theory of development*], ed. V. A. Fausek, 49–143. Saint Petersburg: Brokhaus & Efron (in Russian).

Timonin, K. A. (ed.). 2001. [*Homologies in botany: Experience and reflexion.*] Saint Petersburg: Saint Petersburg Union of Scientists. ISBN 5901536010 (in Russian).

Tipton, J. A. 2014. *Philosophical biology in Aristotle's Parts of Animals.* Dordrecht: Springer Internat. Publ. ISBN 978-3319014210.

Todd, C. W. 2006. The current status of baraminology. *Creation Research Society Quarterly* 43:149–58. https://citeseerx.ist.psu.edu/viewdoc/download;jsessionid=86545BD20410 43A14D70C0C35DFE0699?doi=10.1.1.174.2435&rep=rep1&type=pdf.

Tournefort, Pitton de, J. 1694. *Élémens de botanique, ou Methode pour connoître les plantes*, Vol. 1. Paris: l'Imprimerie Royale. ISBN 978-1272619824 (2012 reprint).

Troll, W. 1951. Biomorphologie und Biosystematik als typologische Wissenschaften. *Studium Generale* 4:376–89. ISSN 0039-4149

Tschulok, S. 1910. *Das System der Biologie in Forschung und Lehre.* Jena: Gustav Fischer. ISBN 978-1167931680 (2010 reprint).

Tuomikoski, R. 1967. Notes on some principles of phylogenetic systematics. *Annales Entomologici Fennici* 33:137–47. ISSN 0003-4428.

Turesson, G. 1922. The species and the variety as ecological units. *Hereditas* 3:100–13. DOI 10.1111/j.1601–5223.1922.tb02727.x.

Turrill, W. B. 1925. Species. *Journal of Botany British and Foreign* 63: 359–66. http://archive.bsbi.org.uk/Journal_of_Botany_1925.pdf.

Turrill, W. B. 1938. The expansion of taxonomy with special reference to spermatophyta. *Biological Reviews* 13:342–73. DOI 10.1111/j.1469-185X.1938.tb00522.x.

Turrill, W. B. 1940. Experimental and synthetic plant taxonomy. In *The new systematics*, ed. J. Huxley, 47–71. London: Oxford University Press. ISBN 978-0403017867.

Tversky, A. 1977. Features of similarity. *Psychological Review* 84:327–52. DOI 10.1037/0033-295X.84.4.327.

Tversky, B. 1989. Parts, partonomies, and taxonomies. *Developmental Psychology* 25:983–95. DOI 0.1037/0012-1649.25.6.983.

Uemov, A. I. 1974. [*Analogy in the practice of scientific research.*] Moscow: Nauka (in Russian).

Uexküll, J., von. 2010. The theory of meaning. In *Essential readings in biosemiotics. Anthology and commentary*, ed. D. Favareau, 81–114. Dordrecht: Springer Science + Business Media. ISBN 978-1402096495.

Uranov, A. A. 1979. [*Methodological foundations of plant systematics (in their historical development).*] Moscow: MGPI (in Russian).

Urmantsev, Yu. A. 1988. [*General systems theory: Its state, application, and development perspectives.*] Moscow: Mysl: 38–124 (in Russian).

Valentine, D. H., and A. Löve. 1958. Taxonomic and biosystematic categories. *Brittonia* 10:153–66. DOI 10.2307/2804945.

Van Valen, L. M. 1971. Adaptive zones and the orders of mammals. *Evolution* 25:420–8. DOI 10.1111/j.1558-5646.1971.tb01898.x.

Vandamme, P., B. Pot, M. Gillis, et al. 1996. Polyphasic taxonomy, a consensus approach to bacterial systematics. *Microbiological Reviews* 60:407–38. DOI 10.1007/s12088-007-0022-x.

Vasilchenko, I. T. 1960. [Experimental taxonomy and the main directions of its development.] *Botanical Journal* 45:1585–99 (in Russian).

Vasiliev, N. A. 1989. [*Imaginary logic.*] Moscow: Nauka. ISBN 5020079464 (in Russian).

Vasilieva, L. N. 1989. [Typological school of taxonomy.] In [*Methodological problems of biology and ecology*], 26–43. Vladivostok: Far-East State University Publ. (in Russian).

Vasilieva, L. N. 1992. [*Platonism in systematics.*] Vladivostok: Far-East Branch Acad. Sci. Publ. (in Russian).

Vasilieva, L. N. 1998. [Hierarchical model of evolution.] *Zhurnal Obshchei Biologii* 59:5–23. ISSN 0044-4596 (in Russian).

Vasilieva, L. N. 2003. [Essentialism and typological thinking in biological systematics.] *Zhurnal Obshchei Biologii* 64: 99–111. ISSN 0044-4596 (in Russian).

Vasilieva, L. N. 2007. [Linnaean hierarchy and "extensional thinking."] *Archives of the Zoological Museum of Moscow State University* 48: 183–212. ISSN 0134-8647 (in Russian).

Vasyukov, V. L. 2005. [*Quantum logic.*] Moscow: PerSe. ISBN 5929201420 (in Russian).

Vavilov, N. I. 1931. [The Linnean species as a system.] *Trudy of Applied botany, Genetics, and Selection*, 26:109–34. ISSN 2227-8834 (in Russian).

Velichkovsky, B. M. 2006. [*Cognitive science. Foundations of psychology of knowing.*] Moscow: Academia Publ. ISBN 5769529830 (in Russian).

Vences, M, J. M. Guayasamin, A. Miralles, et al. 2013. To name or not to name: Criteria to promote economy of change in Linnaean classification schemes. *Zootaxa* 3636:201–44. DOI 10.11646/zootaxa.3636.2.1.

Vernon, K. 2001. A truly taxonomic revolution? Numerical taxonomy 1957–1970. *Studies in History and Philosophy of Biological and Biomedical Sciences* 32:315–41. DOI 10.1016/S1369-8486(01)00007-3.

Vicq d'Azyr, F. 1786. *Traité d'anatomie et de physiologie, avec des planches colorées*, Vol. 1. Paris: Imprim. de Franç. Amb. Didot. 123 p. https://gallica.bnf.fr/ark:/12148/bpt6k1513765b.

Vinogradov, V. A. 1982. [Functional typological criteria and genealogical classification of languages.] In [*Theoretical foundations of the classification of languages of the world: Problems of kinship*], ed. B. A. Serebrennikov, 258–312. Moscow: Nauka (in Russian).

Vityaev, E. E., and B. C. Kostin. 1992. [Natural classification as a law of nature.] In [*Intellectual support for activities in the complex subject areas*], 4:107–15. Novosibirsk: Siberian Branch Acad. Sci. Publ. (in Russian).

Vogel, E. C., S. A. Bordignon, R. Trevisan, et al. 2017. Implications of poor taxonomy in conservation. *Journal for Nature Conservation* 36:10–13. DOI 10.1016/j.jnc.2017.01.003.

Voigt, W. 1973. *Homologie und Typus in der Biologie. Weltanschaulich-philosophische und erkenntnistheoretisch-methodologische Probleme*. Jena: Gustav Fischer.

Vollmer, G. 1975. *Evolutionäre Erkenntnistheorie. Angeborene Erkenntnisstrukturen im Kontext von Biologie, Psychologie, Linguistik, Philosophie und Wissenschaftstheorie.* Stuttgart: S. Hirzel. ISBN 3-7776-0275-2.

Voronin, Yu. A. 1985. [*Classification theory and its applications.*] Novosibirsk: Nauka (in Russian).

Voss, E. G. 1952. The history of keys and phylogenetic trees in systematic biology. *Journal of the Scientific Laboratories, Denison University* 43:1–25. ISSN 0096-3755.

Voyshvillo, E. K. 1989. [*Concept as a form of thinking: Logical and epistemological analysis.*] Moscow: Moscow State University Publ. (in Russian).

Waddington, C. H. 1962. *New pattern in genetics and development.* New York: Columbia University Press. ISBN 978-1125278345.

Wägele, J.-W. 2005. *Foundations of phylogenetic systematics.* Munich: Dr. Friedrich Pfeil Verlag. ISBN 978-3899370560.

Wagner, G. P. 1989. The biological homology concept. *Annual Review of Ecology, Evolution and Systematics* 20:51–69. DOI 10.1146/annurev.es.20.110189.000411.

Wagner, G. P. 1994. Homology and the mechanisms of development. In *Homology, the hierarchical basis of comparative biology*, ed. B. K. Hall, 273–99. San Diego: Academic Press. ISBN 978-0123195838.

Wagner, G. P. 2001. Characters, units, and natural kinds: An introduction. In *The character concept in evolutionary biology*, ed. G. P. Wagner, 1–10. San Diego (CA): Academic Press. ISBN 978-0127300559.

Wagner, G. P. 2007. The developmental genetics of homology. *Nature Reviews Genetics* 8:473–9. DOI 10.1038/nrg2099.

Wagner, G. P. 2014. *Homology, genes, and evolutionary innovation.* Princeton (NJ): Princeton University Press. ISBN 978-0691156460.

Wagner, G. P., and P. F. Stadler. 2003. Quasi-independence, homology and the unity of type: A topological theory of characters. *Journal of Theoretical Biology* 220:505–27. DOI 10.1006/jtbi.2003.3150.

Wake, D. B. 1992. Homology and homoplasy. In *Keywords and concepts in evolutionary developmental biology*, ed. B. K. Hall, and W. M. Olson, 191–201. Cambridge (MA): Harvard University Press. ISBN 978-0674022409.

Walsh, D. 2006. Evolutionary essentialism. *British Journal for the Philosophy of Science* 57:425–48. DOI 10.1093/bjps/axl001.

Wang, X. 2002. Taxonomy, truth-value gaps and incommensurability: A reconstruction of Kuhn's taxonomic interpretation of incommensurability. *Studies in the History and Philosophy of Sciences* 33:465–85. DOI 10.1016/S0039-3681(01)00039-5.

Warming, E. 1908. Om planterigest lifsformer. Copenhagen: Festskr. udg. University Kjobenhavn.

Warnow, T. J. 2017. *Computational phylogenetics. An introduction to designing methods for phylogeny estimation.* www.semanticscholar.org/paper/Computational-Phylogenetics%3A-An-Introduction-to-for-Warnow/2d26f4f0ab1d14ecd9ecca1f8751115be7ee7541.

Wartofsky, M. W. 1979. *Models: Representation and scientific understanding.* Boston: Springer Science + Business Media. ISBN 978-9027707369.

Watrous, J. E., and Q. D. Wheeler. 1981. The out-group comparison method of character analysis. *Systematic Zoology* 30:1–11. DOI 10.1093/sysbio/30.1.1.

Wayne, F. 2000. Baraminology: Classification of created organisms. *Creation Research Society Quarterly* 37:82–91. DOI 10.1111/j.1420-9101.2010.02039.x.

Weber, M. 1904. *On the methodology of social sciences.* Glencoe (IL): The Free Press.

Webster, G. 1993. Causes, kinds and forms. *Acta Biotheoretica* 41:275–87.

Webster, G., and B. Goodwin. 1996. *Form and transformation: Generative and relational principles in biology.* Cambridge (UK): Cambridge University Press.

Weinberg, S. 1992. *Dreams of a final theory: The scientist's search for the ultimate laws of nature.* New York: Pantheon.

Wheeler, Q. D. 2001. Systematics, overview. In *Encyclopedia of biodiversity*, Vol. 5, ed. S. Levin, 569–88. www.sciencedirect.com/ science/article/pii/B0122268652002650.

Wheeler, Q. D. (ed.). 2008. *The new taxonomy.* Boca Raton (FL): CRC Press. ISBN 978-0849390883.

Wheeler, Q. D. 2009. Revolutionary thoughts on taxonomy: Declarations of independence and interdependence. *Zoologia* 26:1–4. www.scielo.br/scielo.php?script=sci_arttext&pid=S1984-46702009000100001.

Wheeler, Q. D., and A. Hamilton. 2014. The New Systematics, the New Taxonomy, and the future of biodiversity studies. In *The evolution of phylogenetic systematics*, ed. A. Hamilton, 287–301. Berkeley (CA): University of California Press. ISBN 978-0520276581.

Wheeler, Q. D., and R. Meier (ed.). 2000. *Species concepts and phylogenetic theory: A debate.* New York: Columbia University Press. ISBN 978-0231101431.

Wheeler, Q. D., S. Knapp, D. W. Stevenson, et al. 2012. Mapping the biosphere: Exploring species to understand the origin, organization and sustainability of biodiversity. *Systematics and Biodiversity* 10:1–20. DOI 10.1080/14772000.2012.665095.

Wheeler, Q. D., P. H. Raven, and E. O. Wilson. 2004. Taxonomy: Impediment or expedient? *Science* 303:285. DOI 10.1126/science.303.5656.285.

Wheeler, W. C. 2001. Homology and DNA sequence data. In *The character concept in evolutionary biology*, ed. G. P. Wagner, 303–17. San Diego (CA): Academic Press. ISBN 978-0127300559.

Wheeler, W. C. 2016. Computational aspects of the phylogenetic analysis of comparative sequence data. In *Aspects of biodiversity*, ed. I. Ya. Pavlinov, M. V. Kalyakin, and A. V. Sysoev, 99–115. Moscow: KMK Sci. Press. ISBN 978-5990841666.

Whewell, W. 1847. *The philosophy of the inductive sciences: Founded upon their history.* London: John W. Parker. ISBN 978-1230389066 (2013 reprint).

Whitehead, A. N. 1925. *Science and the modern world.* New York: Macmillan. ISBN 978-0684836393

Wiley, E. O. 1979. An annotated Linnean hierarchy, with comments on natural taxa and competing systems. *Systematic Zoology* 28:308–37. DOI 10.1093/sysbio/28.3.308.

Wiley, E. O. 1981. *Phylogenetics: The theory and practice of phylogenetic systematics.* New York: John Wiley. ISBN 978-0471059752 (1967 reprint).

Wiley, E. O. 2009. Patrocladistics, nothing new. *Taxon*, 58:2–6. DOI 10.1002/tax.581002.

Wiley, E. O., and R. L. Mayden. 1981. The evolutionary species concept. In *Species concepts and phylogenetic theory: A debate*, ed. Q. D. Wheeler, and R. Meier, 70–89. New York: Columbia University Press. ISBN 978-0231101431.

Wilkins, J. S. 1998. The evolutionary structure of scientific theories. *Biology and Philosophy* 13:479–504. DOI 10.1023/A:1006507411225.

Wilkins, J. S. 2006. The concept and causes of microbial species. *History and Philosophy of Life Science* 28:389–408. http://philsci-archive.pitt.edu/id/eprint/3426.

Wilkins, J. S. 2007. The dimensions, modes and definitions of species and speciation. Biology and Philosophy 22:247–66. DOI 10.1007/s10539-006-9043-9.

Wilkins, J. S. 2009. *Species: A history of the idea.* Berkeley (CA): University of California Press. ISBN 978-0520260856.

Wilkins, J. S., and Ebach M. C. 2014. *The nature of classification. Relationships and kinds in the natural sciences.* Houndmills: Palgrave Macmillan. ISBN 978-0230347922.

Williams, D. M. 1993. A note on molecular homology: Multiple patterns from single datasets. *Cladistics* 9:233–45. DOI 10.1111/j.1096-0031.1993.tb00221.x.

Williams, D. M., and M. C. Ebach. 2008. *Foundations of systematics and biogeography.* New York: Springer Science. ISBN 978-1441944450.

Williams, D. M., and S. Knapp (eds.). 2010. *Beyond cladistics: The branching of a paradigm.* Berkeley (CA): University of California Press. ISBN 978-0520267725.

Williams, W. T., and M. B. Dale. 1965. Fundamental problems in numerical taxonomy. *Advances in Botanical Research* 2:35–68. DOI 10.1016/S0065-2296(08)60249-9.

Willner, W., K. Hubler, and M. A. Fischer. 2014. Return of the grades: To objectivity in evolutionary classification. *Preslia* 86:233–43. www.preslia.cz/P143Willner.pdf.

Wilson, E. O. (ed.). 1988. *Biodiversity.* Washington (DC): National Academy Press. ISBN 978-0309037396.

Wilson, E. O. 2005. Systematics and the future of biology. *Proceedings of the National Academy of Sciences* 102 (suppl 1):6520–1. DOI 10.1073/pnas.0501936102.

Wilson, E. O., and W. L. Brown. 1953. The subspecies concept and its taxonomic application. *Systematic Zoology* 2:97–111. DOI 10.2307/2411818.

Wilson, R. A. (ed.). 1999. *Species: New interdisciplinary essays.* Cambridge: The MIT Press. ISBN 978-0262731232.

Winkler, H. 1988. An examination of concepts and methods in ecomorphology. In *Acta XIX Congressus Internationalis Ornithologici*, ed. H. Ouellet, 2246–53. Ottawa: Nat. Mus. Nat. Sci. ISBN 978-0776601960.

Winsor, M. P. 1979. Louis Agassiz and the species question. *Studies in History of Biology* 3:89–111. DOI 10.1080/13825577.2012.703821.

Winsor, M. P. 2001. Cain on Linnaeus: The scientist-historian as unanalysed entity. *Studies in the History and Philosophy of Biological and Biomedical Science* 32:239–254. DOI 10.1016/S1369-8486(01)00010-3.

Winsor, M. P. 2003. Non-essentialist methods in pre-Darwinian taxonomy. *Biology and Philosophy* 18:387–400. DOI 10.1023/A:1024139523966.

Winsor, M. P. 2004. Setting up milestones: Sneath on Adanson and Mayr on Darwin. In *Milestones in systematics*, ed. D. M. Williams, and P. L. Forey, 1–18. Boca Raton (FL): CRC Press. ISBN 978-0415280327.

Winsor, M. P. 2006a. Linnaeus' biology was not essentialist. *Annals of the Missouri Botanical Garden* 93:2–7. DOI 10.3417/0026-6493(2006)93[2:LBWNE]2.0.

Winsor, M. P. 2006b. Creation of the essentialism story: An exercise in metahistory. *History and Philosophy of the Life Sciences* 28:149–74. http://citeseerx.ist.psu.edu/viewdoc/download?doi=10.1.1.1067.3753&rep=rep1&type=pdf.

Winsor, M. P. 2009. Taxonomy was the foundation of Darwin's evolution. *Taxon* 58:1–7. DOI 10.1002/tax.581007.

Winston, M. E., R. Chaffin, and D. Herrmann. 1987. A taxonomy of part–whole relations. *Cognitive Science* 11:417–44. DOI 10.1207/s15516709cog1104_2.

Winther, R. G. 2020. The structure of scientific theories. In *The Stanford encyclopedia of philosophy*, ed. E. N. Zalta. https://plato.stanford.edu/archives/win2016/entries/structure-scientific-theories/.

Witteveen, J. 2018. Typological thinking: Then and now. *Journal of Experimental Zoology (Molecular and Developmental Evolution)* 330:123–31. DOI 10.1002/jez.b.22796.

Woodger, J. H. 1937. *The axiomatic method in biology.* Cambridge (UK): Cambridge University Press. ASIN B00085P2C8.

Woodger, J. H. 1952. From biology to mathematics. *British Journal for the Philosophy of Science* 3:1–21. DOI 10.1093/bjps/III.9.1.

Wray, G. A. 1999. Evolutionary dissociations between homologous genes and homologous structures. In *Homology*, ed. G. Bock, and G. Cardew, 189–203. Chichester: John Wiley. ISBN 0471984930.

Wray, G. A., and E. Abouheif. 1998. When is homology not homology? *Current Opinion in Genetics & Development* 8:675–80. DOI 10.1016/s0959-437x(98)80036-1.

Yudakin, A. P. 2007. [*Essays on evolutionary typology.*] Moscow: Humanitary. ISBN 978-5913670243 (in Russian).

Yudin, E. G. 1997. [*Methodology of science. Systemity. Activity.*] Moscow: Editorial URSS. ISBN 5901006070 (in Russian).

Yuichi, A. 2017. The general concept of species. *Journal of Philosophical Ideas*, Spec. Iss.: 89–120.

Zabrodin, V. Yu. 1989. [On the problem of naturalness of classifications: Classification and law.] In [*Classification in modern science*], ed. A. I. Kochergin, and S. S. Mitrofanova, 59–73. Novosibirsk: Nauka (in Russian).

Zabulionite, A.-K. I. 2011. [Type-image and universality of the typological method in the natural philosophy of J.V. Goethe.] In [*Ethnic society and interethnic culture*] 4:18–33. ISSN: 2072–3091 (in Russian).

Zachos, F. E. 2011. Linnean ranks, temporal banding, and time-clipping: Why not slaughter the sacred cow? *Biological Journal of the Linnean Society* 103:732–4. DOI 10.1111/j.1095-8312.2011.01711.x.

Zachos, F. E. 2016. *Species concepts in biology. Historical development, theoretical foundations and practical relevance.* Basel: Springer. ISBN 978-3319449661.

Zadeh, L. A. 1992. Knowledge representation in fuzzy logic. In *An introduction to fuzzy logic applications in intelligent systems*, ed. R. R. Ygaer, and L. A. Zadeh, 1–25. New York: Springer Science + Business Media. ISBN 978-1461536406.

Zakharov, B. P. 2005. [*Transformational typological systematics.*] Moscow: KMK Sci. Press. ISBN 978-5041154622 (in Russian).

Zaluziansky, A. 1592. *Methodi herbariae libri tres.* Prague: Georgij Daczennni. ASIN B01JJABJD8 (1940 reprint).

Zander, R. H. 2013. *A framework for post-phylogenetic systematics.* St. Louis: Zetetic Publ. ISBN 978-1492220404.

Zarenkov, N. A. 1976. [*Lectures on the theory of systematics.*] Moscow: Moscow University Publ. (in Russian).

Zarenkov, N. A. 1988. [*Theoretical biology.*] Moscow: Moscow University Publ. 233 c [In Russian].

Zarenkov, N. A. 2009. [*Biosymmetric.*] Moscow: Editorial URSS. ISBN: 978-5397000819 (in Russian).

Zavadsky, K. M. 1968. [*Species and speciation.*] Leningrad: Nauka (in Russian).

Zavarzin, G. A. 1995. [A paradigm shift in biology.] *Vestnik RAS*, 65:8–23. ISSN 0869-5873 (in Russian).

Zimmermann, W. 1931. Arbeitsweise der botanischen Phylogenetik und anderer Gruppierungswissenschaften. In *Handbuch der biologischen Arbeitsmethoden*, Vol. 3, 1 (9), ed. E. Abderhalden, 941–1053. Berlin: Urban & Schwarzenberg.

Zimmermann, W. 1934. Research on phylogeny of species and of single characters (Sippenphylogenetik und Merkmalsphylogenetik). *American Naturalist* 68:381–4. DOI 10.1086/280558.

Zirkle, C. 1959. Species before Darwin. *Proceedings of the American Philosophical Society* 103:636–44. ISSN 0003-049X.

Zuev, V. V. 2002. [*The problem of reality in biological taxonomy.*] Novosibirsk: Novosibirsk State University Publ. ISBN. 594356084X (in Russian).

Zuev, V. V. 2015. [*Introduction to the theory of biological taxonomy.*] Moscow: Infra-M. ISBN. 978-5160106281 (in Russian).

Index

Printed in the United States
by Baker & Taylor Publisher Services